Waste Management: Concerns and Challenges

Waste Management: Concerns and Challenges

Edited by **Adele Cullen**

New York

Published by NY Research Press,
23 West, 55th Street, Suite 816,
New York, NY 10019, USA
www.nyresearchpress.com

Waste Management: Concerns and Challenges
Edited by Adele Cullen

International Standard Book Number: 978-1-63238-505-5 (Hardback)

Printed in the United States of America.

Contents

Preface

Over the recent decade, advancements and applications have progressed exponentially. This has led to the increased interest in this field and projects are being conducted to enhance knowledge. The main objective of this book is to present some of the critical challenges and provide insights into possible solutions. This book will answer the varied questions that arise in the field and also provide an increased scope for furthering studies.

Waste management refers to the practice of managing waste from its source till its disposal. It focuses on regulating and monitoring the transportation, collection, treatment and disposal of waste in an efficient way. The main objective of this system is to minimize the adverse effects caused by waste on environment and living beings and also to control its rapid growth. This book will discuss the issues related to waste management and will propose some of the theories to control it. It will provide the readers with in-depth knowledge about the growing problems related to waste management. The topics included in this text are very crucial for students, researchers, environmentalists, ecologists and all those who are interested in this field.

I hope that this book, with its visionary approach, will be a valuable addition and will promote interest among readers. Each of the authors has provided their extraordinary competence in their specific fields by providing different perspectives as they come from diverse nations and regions. I thank them for their contributions.

Editor

Transforming Municipal Waste into a Valuable Soil Conditioner through Knowledge-Based Resource-Recovery Management

MH Golabi[1*], **Kirk Johnson**[1], **Takeshi Fujiwara**[2] and **Eri Ito**[2]

[1]*College of Natural and Applied Sciences, University of Guam, Mangilao, Guam 96923, USA*

[2]*Solid Waste Management Research Center, Okayama University, 3-1-1 Tsushima Naka, Okayama, 7008530, Japan*

*Corresponding author: MH Golabi, College of Natural and Applied Sciences, University of Guam, Mangilao, Guam 96923, USA
E-mail: mgolabi@uguam.uog.edu

Abstract

Guam is a small, isolated tropical island in the western Pacific with a population of over 160,000 people. Although population growth and life style have been shown to have strong effects on the character and generation of waste, very little is known about consumption patterns and behavior of the people of Guam in this regard. Currently landfilling is the only discard method available to the island. Placement of huge volumes of organic waste material in landfills not only causes environmental problems for the island but in fact constitutes loss of valuable resources that could be composted and made available for land application as a soil amendment in forest lands, farm fields, and home gardens. Composting on the other hand reduces both the volume and the mass of the raw material while transforming it into a valuable soil conditioner. Here we present some of the results of survey questionnaires that was developed and conducted over the past two years that is anticipated to help waste operating managers and decision makers to determine societal consumption behavior and residential life style as the first step toward development of an effective waste-management strategy for the island of Guam. In this regards, we are also presenting an example of a large scale composting method developed in Isfahan, Iran, for recycling of organic wastes of municipal origin.

Keywords: Waste management; Environmental crisis; Landfill; Waste recycling; Composting; Zero-waste Management strategy

Introduction

Rapid increases in the volume and variety of solid and hazardous waste as a result of continuous economic growth, urbanization, and industrialization are a burgeoning problem for national and local governments, which must ensure effective and sustainable management of waste [1]. Between 2007 and 2011, global generation of municipal waste has been estimated to have risen by 37.3%, equivalent to roughly an 8% increase per year [1]. The EU has estimated that its 25 member states produce 700 million tons of agricultural waste annually [1].

As reported by the United Nations Environmental Program [1], developing countries face difficult challenges to proper management of their waste; most effort is devoted to reducing the final volume and to generating sufficient funds for waste management. If most of the waste could be diverted through material and resource recovery, then a substantial reduction in final volumes of waste could be achieved, and the recovered material and resources could be used to generate revenue to fund waste management. This scenario forms the premise for Integrated Solid Waste Management; a system based on the 3R (Reduce-Reuse-Recycle) principle [1].

Appropriate segregation and recycling systems have been shown to divert significant quantities of waste from landfills and to convert them into resources [2]. On Guam, over the past several decades, solid-waste generation and disposal have transitioned from a concern needing a remedy to a crisis of monumental proportions.

Although Guam is a small, isolated tropical island with a population of just over 160,000 people, the island generates more than 90,000 tons of waste material each year [3]. The need for a comprehensive solid-waste management and recycling plan is therefore urgent if Guam is to minimize cost and avoid the undesirable environmental effects of legal and illegal dump sites. A comprehensive waste management would also allow for the use of recyclable as well as green and other organic refuse that is currently discarded in landfills as sources of producing organic soil conditioner for a sustainable agricultural cropping system in Guam and the other island in the Micronesian region.

Waste reduction and recycling are fundamental to any future waste-management strategy on Guam and other islands of Micronesia. Accurate information on waste generation, especially waste characteristics, is also needed for study of the feasibility of such strategies on Guam and its neighbors. Unfortunately, presently available data on Guam [4,5] are not reliable enough for development of a comprehensive management and recycling strategies, and information is lacking on social behavior and life style that may strongly affect the character and production of waste.

Basic data from the unregulated local Ordot landfill over the past several years suggest that Guam residents produce on average more waste per capita than the rest of the United States [6]. Because any waste-management policy must find meaning and purpose within the framework of consumption patterns if it is to be effective at all. Residents must come to know and understand what, how much, and more importantly why they consume and hence must become aware of the impact that such consumption has on their island's environment and economy and on the social and cultural life of their community.

Residents' education about awareness of the types and amounts of waste generated and its handling are essential parts of any waste-management strategy that would be economically feasible, culturally acceptable, and environmentally sustainable while maintaining the integrity of the island's natural resources.

Survey questionnaire for educational purposes

To collect the necessary data, we have developed and used questionnaires as a surveying tool. The results are expected to help us understand the social behavior and the residents' life style as the first step toward the development of a sound and effective waste-management strategy for the island of Guam.

To determine not only the composition of waste by components but also citizens' consciousness and knowledge about waste reduction and recycling, we developed a citizen questionnaire designed for statistical analysis not only as a management tool but also as a way of educating the general public. The contents of the survey questionnaire included (1) educational background of head of the household, (2) public awareness of environmental problems associated with waste, (3) waste characteristics and willingness to segregate its components, and (4) participation in reduce-reuse-recycle activities.

By analyzing survey results, we hoped to develop a model of waste generation and citizens' consciousness of reduce-reuse-recycle principles for use in a comprehensive waste-management strategy. We expect that the survey results representing a true sample of the citizens of Guam will aid in the distribution and processing (collecting, compiling) of waste necessary to obtain high recovery ratios.

Determination of the waste generation and characterization

Although reports by the European Commission [7] and studies by Martinho and Silveira [8] recommend sampling of waste containers placed in public areas (e.g., apartment complexes) as an ideal sampling technique, doing so would entail higher costs than did our survey.

Our approach not only obtained up-to-date data but also contributed to education of the public about waste management while promoting "zero waste." It also educated the public, private sector, as well as government agencies about composting and recycling of large-scale organic wastes. The survey approach has revealed that up to 77% of household waste on Guam is organic (food stuffs, yard wastes, newspapers, etc., Figure 1) in nature. This humongous amounts of wastes generated in a small island could easily be recycled through large-scale composting, as it is done at the Isfahan 'composting factory' described below (Figures 2–10). The remaining 23% of the nonorganic waste material like plastic bottles, cans, durable goods, etc., could also be recycled, leading to a "zero waste" management strategy that might require no land-filling cost for the community.

We therefore introduce here the idea of zero-waste management by presenting an example adopted in the city of Isfahan, Iran. We hope that presentation of such an example will lead local government leaders and the private sectors on Guam and the other islands of Micronesia, as well as major cities around the world, to consider adopting such a strategy.

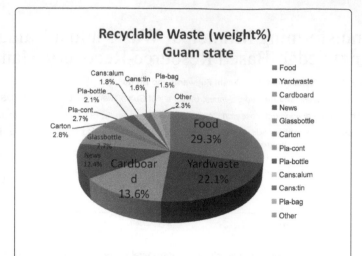

Figure 1: Recyclable waste produced by a Guam household. Pla-cont, plastic containers; pla-bottle, plastic bottles; Pla-bag, plastic bags; News, Newspapers; alum, aluminum cans.

Zero Waste Management Strategies

Different zero-waste management strategies and techniques have been developed and adopted in different countries [9], but the strategy used in the city of Isfahan, Iran, which includes large-scale mechanical composting as a major component is of particular interest to Guam and the neighboring islands in the Micronesian region. The Isfahan composting operation is not only a recycling facility but also an organic fertilizer production plant that uses the organic waste generated by the city of Isfahan as a major source for its production lines.

Isfahan compost facility and recycling techniques

The Isfahan compost facility has two production lines, each able to handle 750 tons of garbage per day. Garbage is gathered at night at the transfer stations and sent to the recycling facility in both small and large hauling trucks (Figure 2). Upon delivery, the garbage is pushed to the conveyors by front-end loaders, and the receiving station is cleared as soon as trucks leave the site (Figure 3). Conveyors then carry the trash to vibrating screens that separate plastics, glass, cloth, etc. The trash is then carried on the conveyors to hammer mills, equipped with double rotors and anti-explosives, which break up the large pieces included in the bulk of trash (Figure 4).

A second set of vibrators is used to unpack organic materials and loosen everything else. Next, two magnets separate metals from the rest (Figure 5). The metals, aluminum cans, etc., all fall into a separate compartment (Figure 6) and are transferred to the other section of the facility for further processing and/or recycling/packaging. The remaining garbage is then carried by vibrating conveyors to drum sieves for screening through different mesh sizes and for further screening (Figure 7).

Figure 2: An Isfahan compost factory waste-gathering and transportation vehicle.

Figure 3: A large front-end loader pushes the garbage to the conveyors, immediately and cleaning the drop-off site.

HAMMER MILL SET IN ISFAHAN COMPOST FACTORY - IRAN
دستگاه خرد کن در کارخانه کمپوست اصفهان

Figure 4: A hammer mill equipped with double rotors and anti-explosives, which is used for breaking up large pieces.

MAGNET OF ISFAHAN COMPOST FACTORY

Figure 5: Magnets used to separate metals form the remainder of the waste stream.

GATHERED METALS BY MAGNET IN ISFAHAN COMPOST FACTORY - IRAN
ضایعات فلزی جمع آوری شده توسط مگنت در کارخانه کمپوست اصفهان

Figure 6: The metals trapped by the magnets fall into a separate compartment for transfer to other sections of the plant for further processing.

Additional sorting

Additional drums further screen plastics, glass bottles, wood, cloth, paper, etc. At this stage, items like fabric and cloth are separated, placed in a special compartment, and hauled away to be pressed into bales, which are sold for processing into pulp and other biodegradable material [9].

Separation of organic matter

Organic materials (technically waste containing carbon, including paper, plastics, wood, food wastes, and yard wastes; and other materials derived from plant or animal sources and decomposable by microorganisms [1] pass through drums of 50-mm mesh size and are then sent on separate conveyors to fermentation sections (Figure 8), where it is placed in windrows and mixed for aeration every four days

(Figures 9 and 10) until the material become a mature compost and ready for marketing after a comprehensive laboratory testing.

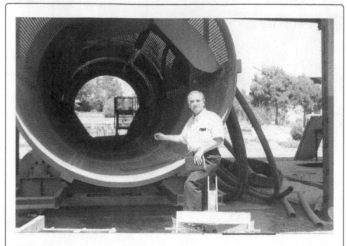

Figure 7: Drum sieves, used to screen garbage through mesh into different size classes for further screening on vibrating conveyors.

Figure 8: Organic materials being conveyed to composting windrows for the fermentation process.

Figure 9: Composting windrow during the fermentation and turning process.

Figure 10: Mixing and turning of organic waste by a compost turner, which turns each windrow every four days.

Facility production capacity

Each year, from over 270,000 tons of input, the facility produces approximately 30,000 tons of fine mature compost, of which are sold to farmers, ranchers, horticulturists, and private gardeners as soil amendment/soil conditioner. Also, the facility produces about 12,000 tons of coarser compost as mulch, which is sold to municipalities for landscaping and for maintaining green spaces in many parks in Isfahan. The remaining biodegradable materials (fabrics, discarded cloth, etc.) are also sold to other cities within Iran as well as other neighboring countries for use in the pulp production and other similar manufacturing.

It is worth mentioning that all the products are tested at the factory's state-of-the-art laboratory for quality assurance before they are shipped to vendors and general customers and users of compost and mulch.

Figure 11 provides a flow-chart summary of the steps from waste collection to final compost production at the Isfahan 'composting factory.

Relevance of the Isfahan composting technology for Guam

The survey project described above served as a part of a comprehensive approach that includes increasing public awareness of comprehensive waste-management strategies for the island of Guam as well as the other islands in the western Pacific.

In addition to waste characterization by means of the survey questionnaires, presentation of the "Isfahan Waste Management System" also provides a knowledge-based foundation that we hope will lead to adaptation of the technology for Guam and neighboring islands. Because of the amount of the waste generated on Guam and the limited space available for landfilling, the Isfahan waste management technology appears to be the most practical and feasible method that Guam and other islands in Micronesian could possibly adopt as a sustainable waste-management strategy.

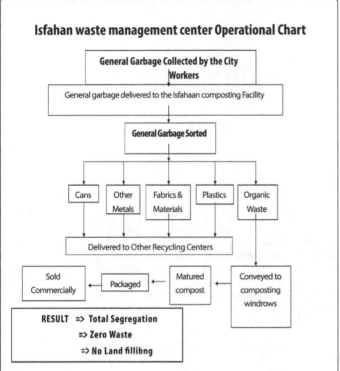

Figure 11: Chart showing the steps from waste collection to final compost production.

Acknowledgements

The authors would like to acknowledge the cooperation of the engineers as well as the managers at the Isfahan "composting factory" for providing the tour and description of the steps and stages during the operations at the facility, to the senior author few years ago. The authors would also like to acknowledge the efforts of the chief engineer and staff for providing the pictures and related information that are being presented here and at related conferences.

References

1. United Nations Environmental Program (2009) Developing Integrated Solid Waste Management (ISWM) Plan Training Manual. Division of Technology, Industry and Economics, International Environmental Technology Centre Osaka, Shiga Japan 4:1-172.

2. Bureau of Statistics and Plans (2010) Guam Census Population Counts.

3. Guam Solid Waste Authority (2014) Guam Solid Waste Authority Services.

4. Guam Solid Waste Receivership Information Center (2009).

5. Guam Integrated Solid Waste Management Plan (2006).

6. Johnson K (2009) Waste generation rate for Guam residents personal observation.

7. European Commission documentation (2004) Methodology for the Analysis of Solid Waste (SWA-Tool) User version. Project: SWA-Tool, Development of a Methodological Tool to Enhance the Precision & Comparability of Solid Waste Analysis Data. 57.

8. Martinho MGM, Silveira AI, Branco FDEM (2008) Report: New guidelines for characterization of municipal Solid Waste: the Portuguese case. Waste Management and Research 265: 486-490.

9. Isfahan Composting Factory. Steps taken during the zero waste management strategy as described by the general manger.

2

Adsorption Mechanisms and Transport Behavior between Selenate and Selenite on Different Sorbents

Michelle MV Snyder[1] and Wooyong Um[1,2*]

[1]*Energy and Environment Directorate, Pacific Northwest National Laboratory, Richland, WA 99354, USA*

[2]*Division of Advanced Nuclear Engineering, Pohang University of Science and Technology (POSTECH), Pohang, South Korea (ROK)*

*Corresponding author: Wooyong Um, Energy and Environment Directorate, Pacific Northwest National Laboratory, Richland, WA 99354, USA
E-mail: wooyong.um@pnnl.gov

Abstract

Adsorption of different oxidation species of selenium (Se), selenate (SeO_4^{2-}) and selenite (SeO_3^{2-}), with varying pHs (2-10) and ionic strengths (I=0.01 M, 0.1 M and 1.0 M $NaNO_3$) was measured on quartz, aluminum oxide, and synthetic iron oxide (ferrihydrite) using batch reactors to obtain a more detailed understanding of the adsorption mechanisms (e.g., inner- and outer-sphere complex). In addition to the batch experiments with single minerals contained in native Hanford Site sediment, additional batch adsorption studies were conducted with native Hanford Site sediment and groundwater as a function of 1) total Se concentration (from 0.01 to 10 mg L^{-1}) and 2) soil to solution ratios (1:20 and 1:2 grams per mL). Results from these batch studies were compared to a set of saturated column experiments that were conducted with natural Hanford sediment and groundwater spiked with either selenite or selenate to observe the transport behavior of these species. Both batch and column results indicated that selenite adsorption was consistently higher than that of selenate in all experimental conditions used. These different adsorption mechanisms between selenite and selenate result in the varying mobility of Se in the subsurface environment and explain the dependence on the oxidation species.

Keywords: Selenite; Selenate; Inner-sphere complex; Outer-sphere complex; Adsorption

Introduction

Selenium (Se) is required for adequate nutrition, but at high concentrations Se can be toxic to humans and animals. In humans, exposure to high concentrations of Se can result in loss of hair and fingernails, cause numbness in the fingers or toes, and weaken the body's nervous and circulatory systems [1]. Similar to humans, animals also have a narrow range between deficient and toxic concentrations of Se. Contamination of Se, caused by irrigation drainage, was discovered in 1983 at Kesterson Reservoir in California and was linked to deformities in waterfowl [2]. Because of the environmental and human health risks associated with Se, the U.S. Environmental Protection Agency (EPA) has determined the Maximum Contaminant Level (MCL) for Se in drinking water to be 0.05 parts per million (ppm) [1]. The MCL for Se is sometimes exceeded in areas that have a large amount of mining, industrial activity, and seleniferous soils [1,3]. Furthermore, the planned disposal of immobilized low-activity waste (ILAW) glass containing [79]Se ($t_{1/2}=2.9x10^5$ years) [4], located in the 200 East area of the Hanford Site, southeastern Washington State, may also pose a significant risk for radioactive Se release [5]. Dissolution of the vitrified radioactive waste matrix containing [79]Se, a fission product of [235]U, could result in Se release into the environment. Depending on the rate of glass dissolution, Se that is released from the host matrix can be transported through the vadose zone and can contaminate the underlying sediment and groundwater aquifer. To prevent spreading of radioactive Se contamination, a more in depth understanding of the migration of Se in the environment is required [5-7].

The ability to predict the fate and transport of toxic and radioactive Se in subsurface environments requires an understanding of the sorption processes that occur at the mineral-water interface. The effect of sorption on contaminant mobility is determined by molecular level processes involving the sorption complex, which can be divided into three distinct mechanisms: 1) adsorption (inner-sphere vs. outer-sphere complex), 2) absorption, and 3) surface precipitation [8,9]. Although each individual process is extremely important, adsorption is expected to play the largest role in the mobility of most contaminants, including Se.

The process of adsorption occurs when adsorbate accumulates on the surface of a solid (adsorbent) at the solid-water interface, where the ions, liquid, or gas molecules are held onto the surface without the development of a three-dimensional molecular arrangement [9]. Adsorption not only influences the distribution of contaminants between the aqueous phase and particulate matter, but it also affects the electrostatic properties of particles and the reactivity of mineral surfaces [10].

Oxyanions selenite (SeO_3^{2-}) and selenate (SeO_4^{2-}) are dominant in both mild and strong oxidizing conditions and are the most mobile aqueous species of Se [11]. Numerous studies have been conducted to evaluate the adsorption behavior of these oxidation species, SeO_3^{2-} and SeO_4^2 [12-15]. Many of these studies focused on the effects of various environmental factors on SeO_3^{2-} and SeO_4^{2-} adsorption. For example, Balistrieri and Chao studied the effects of pH, temperature, total Se concentrations, and competing anions on the adsorption of SeO_3^{2-} on goethite (FeOOH) [12]. Results demonstrated that SeO_3^{2-} adsorption increased with decreasing pH and an increase in the solid to solution ratio. However, adsorption of SeO_3^{2-} decreased with an increase in the total Se concentration and in the presence of competing anions Ffisuch as phosphate, silicate or citrate which bind strongly to the goethite surface. Zhang and Sparks conducted studies to determine the mechanisms of SeO_3^{2-} and SeO_4^{2-} at the goethite/water interface [16]. Their results suggested that SeO_3^{2-} forms a stronger inner-sphere surface complex whereas SeO_4^{2-} forms a weaker bonded outer-sphere surface complex. Gonzalez et al. [17] studied an iron/manganese oxide nanomaterial as a potential material for the treatment technology and removal of SeO_3^{2-} and SeO_4^{2-}. Results found sorption on the nanomaterial to be pH independent (between pH 2 to 6) and sorption of both oxyanions to be effected by the presence of phosphate with a larger decrease in SeO_4^{2-} sorption than SeO_3^{2-}. The addition of sulfate only effected the sorption of SeO_4^{2-} and the addition of nitrate and chloride had no effect on the sorption of either SeO_3^{2-} or SeO_4^{2-}. Additional studies conducted to determine sorption onto simple mineral systems include Zonkhoeva [18] that studied the sorption of SeO_3^{2-} on natural zeolites and Yoon, et al. [19] which studied sorption of SeO_4^{2-} onto zero-valent iron. In general, studies that compared adsorption of SeO_3^{2-} and SeO_4^{2-} consistently found that SeO_3^{2-} is more rapidly adsorbed than SeO_4^{2-} and lower pH conditions favor stronger Se adsorption [20].

Although the information discussed above represents a majority of the research on the adsorption of SeO_3^{2-} and SeO_4^{2-}, few studies are available on the transport of these oxyanions under real environmental conditions. Therefore, the purpose of this study was to evaluate how changes in the Se redox state (SeO_3^{2-} or SeO_4^{2-}) will affect adsorption and transport of Se in native Hanford Site soil and groundwater conditions. Hanford Site groundwater contains several of the anions expected to compete with Se for available adsorption sites. To evaluate the adsorption and transport behavior of SeO_3^{2-} or SeO_4^{2-}, a set of batch adsorption and saturated column experiments were conducted with Hanford sediment and groundwater spiked with either SeO_3^{2-} or SeO_4^{2-}. In addition to these experiments, a set of batch adsorption experiments using pure single minerals [e.g., quartz (SiO_2), aluminum oxide (Al_2O_3), and iron oxide (2-line ferrihydrite, ($Fe(OH)_3$)] contained in native Hanford sediments were also conducted as a function of ionic strength and pH. The two principal objectives of these experiments were 1) to determine the mineral phase that has the largest effect on Se adsorption and 2) to use macroscopic methods to determine whether or not SeO_3^{2-} and SeO_4^{2-} form an inner- or outer-sphere complex on the mineral surface. These and other results obtained from this study will improve our understanding of the release and transport of Se from the radioactive waste glass that will be buried in the Integrated Disposal Facility (IDF) on the Hanford Site.

Materials and Methods

Hanford sediment characterization

Hanford sediment was utilized as an adsorbent to determine the adsorption of SeO_3^{2-} and SeO_4^{2-}. The sediment was collected from borehole C3177 (299-E24-21) located in the center of the 200 East Area, northeast corner of the ILAW disposal site, at the Hanford Site Nuclear Reservation located in southeastern Washington state [21]. The sediment, C3177-110, was obtained from borehole C3177 by homogenizing four two-foot long (71.0 cm) cores to provide sufficient volume of representative Hanford sediment.

The Hanford formation[1] can be subdivided into three main layers within the ILAW disposal facility and consists of pebble to boulder-sized gravel and fine to coarse grained sand with interbedded, thin silt and/or clay beds [22]. Composite C3177-110 is representative of Hanford formation layer II and is dominated by sand. Particle size distribution was determined by wet sieve and hydrometer methods, and revealed that the sediment is comprised of gravel (3.85 wt%), sand (88.6 wt%), silt (6.19 wt%), and clay (1.36 wt%) [21]. X-ray diffraction (XRD) analysis revealed that the Hanford sediment is dominated by quartz and feldspars with lesser amounts of chlorite, mica and amphibole, which is typical of Hanford Site sediments (Table 1). Additional sediment characterization included moisture content, carbon content, BET surface area, bulk sediment composition determined by x-ray fluorescence (XRF), and mineralogical content of clay determined by X-ray diffraction (XRD). These results are summarized in Tables 1 and 2.

Mineral phase on bulk sample (wt%)[1]						Moisture content (wt%)	Surface area (m^2/g)		Carbon content (wt %)[2]	
Q	A	P	K	M	C	bulk	clay	total	inorganic	
45	1	25	19	9	2	2.76	5.11	41.8	0.20	0.19

Table 1: Characterization of the composite sediment including moisture content, carbon content and mineralogy of bulk sample by XRD (modified from [21]). 1) Mineral phase indicates Q (quartz), A (Amphibole), P (Plagioclase), K (K-feldspar), M (Mica), and C (Chlorite). 2) Organic carbon content is determined by difference between total and inorganic carbon contents.

Mineral phase of clay (wt %)				Bulk composition of major element oxides (wt %)									
Smectite	Illite	Chlorite	Kaolinite	SiO_2	Al_2O_3	TiO_2	Fe_2O_3	MnO	CaO	MgO	K_2O	Na_2O	P_2O_5
26	51	16	7	70.4	13.2	0.672	4.91	0.077	3.72	1.87	2.54	2.87	0.153

Table 2: Characterization of the composite sample including clay mineralogy by XRD and major elemental composition by XRF (modified from [21]).

Preparation and characterization of pure single minerals

White quartz sand (SiO_2), 50-70 mesh size (Aldrich Chemical Company) was acid treated (to remove any impurities) with analytical grade concentrated sulfuric acid (H_2SO_4) for two days, rinsed several times with double deionized (DDI) water, and oven dried at 85°C for 48 hours prior to use. The cleaning of quartz was based on a procedure used by [23].

---

[1]   The term "Hanford formation" is used informally to describe Pleistocene cataclysmic flood deposits within the Pasco Basin. It is not a formalized stratigraphic unit.

Synthetic aluminum oxide (Al$_2$O$_3$), or transition Alumina (C-33), is a porous and high-surface area synthetic aluminum oxide with a specific surface area of 110 m^{2}g^{-1} [24]. The alumina (C-33) used in this study was created by the partial dehydration of aluminum hydroxides and oxyhydroxides, where the final product is a completely dehydrated oxide (corundum) [25]. Additional characterization of this aluminum oxide can be found in reference [25].

The synthetic iron oxide [ferrihydrite, Fe(OH)$_3$] preparation was based on Schwertmann and Cornell [26]. The ferrihydrite was prepared by neutralizing 0.2 M ferric nitrate [Fe(NO$_3$)$_3$·9H$_2$O], from Aldrich Chemical Company; (analytical grade) with 1.0 M sodium hydroxide (NaOH), from J.T. Baker; (analytical grade). The resulting solid was washed by successive centrifuging and decanting with double deionized water (DDI). The solid was filtered from solution by vacuum using a 0.45 μm filter unit. Powder X-ray diffraction (XRD) patterns were obtained using a Scintag, Pad-V X-Ray Diffractometer (XRD) with a Cu source. The ferrihydrite was analyzed on the XRD using a 2-Theta range from 2° to 65°, a step size of 0.01°, and a 2 second count time at each step. The XRD pattern was identified as ferrihydrite (Figure 1) using the JADE software (MDI, Livermore, California).

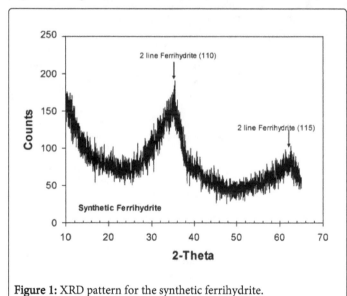

Figure 1: XRD pattern for the synthetic ferrihydrite.

Batch adsorption experiments

Groundwater used for the adsorption and column experiments with Hanford sediment was collected from the well 699-49-100C located in the 600 area of the Hanford Site. Well 699-49-100C is an uncontaminated background monitoring well. Analyses of cations and anions in this groundwater were conducted using an inductively coupled plasma-optical emission spectroscopy (ICP-OES) and an Ion Chromatograph (IC), respectively. The pH was measured with a solid-state pH electrode and a pH meter calibrated with buffers 7 and 10. The electrical conductivity (EC) was measured using a Pharmacia Biotech Conductivity Monitor. Approximately 2 to 3 milliliters of filtered sample were measured in the conductivity meter and compared to potassium chloride standards with a range of 0.001 to 1.0 M. Chemical compositions as well as the pH and EC values for this groundwater are shown in Table 3.

Elements (cations)	Concentration (mg/L)	Elements (anions)	Concentration (mg/L)
Ca	59.5	HCO$_3$	169
Si	28	SO$_4$	70.6
S	26.1	Cl	18.9
Na	24.4	NO$_3$	9.44
Mg	22	F	0.45
K	7.5		
Sr	0.236		
Zn	0.283		
Cr	0.002		
Mo	0.002		
pH (measured)=8.04; electrical conductivity=0.474 mS/cm			

Table 3: Chemical composition, pH, and conductivity of groundwater (699-49-100C).

Batch adsorption experiments were conducted in duplicate for adsorption of SeO$_3^{2-}$ and SeO$_4^{2-}$ on the C3177 sediment. Particles greater than 2 mm were removed prior to using the sediment and the sediment was pre-equilibrated with the Hanford groundwater. The groundwater was added to the sediments, the sediments were shaken overnight, centrifuged, and the pH measured and then the supernatant was decanted. This was repeated three times until the pH of the supernatant groundwater did not change after contact overnight with the sediment. Pre-equilibration was done to ensure that reactions that occurred during laboratory batch sorption experiments were not due to any other reactions (e.g. dissolution) except adsorption. Batch experiments were conducted in 15 mL polypropylene centrifuge tubes with solid to solution ratio of 1:20 or 1:2 (gram per mL). Initial concentrations (Co) of SeO$_3^{2-}$ and SeO$_4^{2-}$ ranged from 0.01 to 10 mg L^{-1} and were prepared by the addition of Na$_2$SeO$_3$ or Na$_2$SeO$_4$, respectively, to the groundwater.

Additional batch adsorption experiments were conducted for SeO$_3^{2-}$ and SeO$_4^{2-}$ on single minerals such as quartz, aluminum and iron oxide under varying ionic strength (I=0.01 to 1.0 M NaNO$_3$) and pH (2 to 10) conditions. Experiments were conducted at a solid to solution ratio of 1:100 (gram per mL) with the addition of 1 mg L^{-1} of Na$_2$SeO$_3$ or Na$_2$SeO$_4$, respectively, in NaNO3 solution. The pH was adjusted by the addition of 0.01 M nitric acid (HNO$_3$) or 0.01 M sodium hydroxide (NaOH).

Hanford sediment or pure single minerals were contacted with solution spiked with SeO$_3^{2-}$ or SeO$_4^{2-}$. After shaking for 7 days on a slowly moving platform shaker, the slurry sediment sample was centrifuged to separate the solids out of solution. The final pH was measured, and the supernatant was decanted and filtered through a 0.45 μm syringe filter. Solution was analyzed for Se concentration using an inductively coupled plasma mass spectrometer (ICP-MS). The difference in C$_o$ and the final concentration of SeO$_3^{2-}$ or SeO$_4^{2-}$ in solution was used for adsorption of Se onto the sediments.

In addition, two different types of experimental control samples were also prepared to determine 1) whether or not SeO$_3^{2-}$ and SeO$_4^{2-}$

adsorbed onto the walls of the test vessels and 2) to determine the contribution of natural SeO_3^{2-} and SeO_4^{2-} released from the un-reacted Hanford sediment. The final solution was also analyzed for Se concentrations, and showed negligible Se adsorption onto the test tube walls and no natural Se release from the sediment.

Calculation of partition coefficient (Kd)

The partition coefficient (K_d) is a measure of adsorption and is defined as the ratio of the amount of an adsorbate sorbed on a solid to the amount of adsorbate still in solution at equilibrium [27]. The expression is as follows:

$$K_d = \frac{Ai}{Ci}$$

where A_i=adsorbate on the solid at equilibrium ($\mu g\ g^{-1}$); C_i=adsorbate remaining in solution at equilibrium ($\mu g\ ml^{-1}$). For the batch studies conducted during this experiment, the K_d values were calculated using the following equation:

$$K_d = \frac{Vw(Co - Ci)}{MsedCi}$$

where V_w=known volume of solution (ml); C_o=initial concentration of adsorbate ($\mu g\ mL^{-1}$); M_{sed}=known mass of sediment (μg).

Column experiments and determination of retardation factor (R_f)

Two borosilicate glass columns (Kontes Chromaflex Chromatography columns, L=10 cm, diameter=2.5 cm) were uniformly packed with sediment C3177-110 after removing gravel (>2 mm size fraction). Two individual columns were utilized to evaluate the mobility of SeO_3^{2-} and SeO_4^{2-}, respectively. An AVI Micro 210A Infusion Pump was used to control the constant flow rate. The columns were initially saturated with groundwater at a pumping rate (10 mL hour^{-1}) faster than what was used during the experiment to remove any dispersible particles and to make sure that a steady flow was maintained. Prior to conducting the Se transport experiment, the flow rate was decreased to 0.5 ml hour^{-1} until a steady flow rate was maintained. This flow rate was chosen to allow the SeO_3^{2-} and SeO_4^{2-} to remain in contact with sediment in the column for approximately one day. Prior to introducing Se, a bromide (Br) was added to the groundwater and introduced as a nonreactive tracer (no adsorption to the sediment) in each column to provide a comparison of Br mobility to the SeO_3^{2-} and SeO_4^{2-} transport results as well as hydrodynamic dispersion of the flow. Once completion of Br breakthrough in the column was attained, approximately 7 mg L^{-1} of SeO_3^{2-} or SeO_4^{2-}, respectively, spiked in groundwater was introduced to each column. The Se was pumped through the columns for approximately 13 pore volumes and the amount of Se in the effluent was measured using an inductively coupled plasma optical emission spectrometer (ICP-OES). Transport parameters were determined by curve fitting (based on the advection-dispersion equation) to the measured breakthrough curves (BTCs) using the CXTFIT code [28,29]. Breakthrough curves (BTCs) were graphically represented by plotting the relative concentration, C/C_o, for SeO_3^{2-} or SeO_4^{2-} versus pore volumes eluted. Both equilibrium and non-equilibrium (two site/two region) models were applied to analyze the experimental column breakthrough data [29]. The equilibrium model described in dimensionless terms is

$$R\frac{\partial C}{\partial T} = \frac{1}{P}\frac{\partial^2 C}{\partial Z^2} - \frac{\partial C}{\partial Z} \tag{3a}$$

where

$$T = \frac{vt}{L}, \quad Z = \frac{x}{L}, \quad C = \frac{c}{co}, \quad P = \frac{vL}{D}, \quad R = 1 + \frac{\rho BKd}{\theta} \tag{3b}$$

and T=dimensionless time equal to pore volume; t=time (T); L=column length (L); v=linear pore water velocity (LT^{-1}); Z=dimensionless distance; x=distance from the input (L); C=relative concentration between initial (c_o) and effluent (c) concentrations (ML^{-3}); P=Peclet number [-]; D=hydrodynamic dispersion coefficient (L^{2}T^{-1}); R[-]=retardation factor determined by the equation containing bulk density (ρ_b) (ML^{-3}), porosity (Θ) (L^{3}L^{-3}), and distribution coefficient (K_d) (M^{-1}L^3). The non-equilibrium model is described by the following equations with dimensionless terms:

$$\beta R\frac{\partial C1}{\partial T} + (1-\beta)R\frac{\partial C2}{\partial T} = \frac{1}{P}\frac{\partial^2 C1}{\partial Z^2} - \frac{\partial C1}{\partial Z} \tag{4a}$$

$$(1-\beta)R\frac{\partial C2}{\partial T} = \omega(C1 - C2) \tag{4b}$$

$$T = \frac{vt}{L}, \quad Z = \frac{x}{L}, \quad C1 = \frac{cm}{co}, \quad C2 = \frac{cim}{co} \tag{4c}$$

$$P = \frac{vmL}{Dm} = \frac{vL}{D}, \quad vm = \frac{q}{\theta m}, \quad \omega = \frac{aL}{vm\theta m}, \quad \beta = \frac{\theta m + f\rho bKd}{\theta + \rho bKd} \tag{4d}$$

where β=dimensionless mobile fraction with mobile water fraction (f); C1=solute concentration in mobile water (ML^{-3}); C2=solute concentration in immobile water (ML^{-3}); ω=dimensionless mass transfer coefficient between mobile and immobile water regions; α=mass transfer coefficient [T^{-1}]; q=volumetric flow velocity [LT^{-1}]. The column parameters are summarized below in Table 4.

Columns	Porosity [-]	Porewater velocity (cm/day)	Pore volume (cm^3)	Bulk density (g/cm^3)	D^1 (cm^2/day)	Retardation factor (R$_f$)	Kd (mL/$_g$)
Selenite	0.29	8.17	15.3	1.59	7.09	3.10	0.38
Selenate	0.29	8.17	15.7	1.68	5.56	1.20	0.034

Table 4: Transport parameters of selenite and selenate columns. 1Hydrodynamic dispersion coefficient.

Results and Discussion

Batch selenium adsorption

Adsorption of Se on quartz under varying pHs (2-11) and ionic strengths (I=0.01 M to 1.0 M NaNO$_3$) resulted in negligible adsorption of SeO_4^{2-} regardless of pH and ionic strength conditions (Figure 2).

Even at low pH (<4) and ionic strength (I=0.01 M NaNO$_3$) conditions, no adsorption of SeO_4^{2-} was found on the quartz surfaces. Selenite (SeO_3^{2-}) showed a minor amount of adsorption (<10%) onto the quartz at pH values less than 5 and only at low ionic strengths (I=0.01 M and 0.1 M NaNO$_3$). Because the adsorption was so small (<10%) it is difficult to see a trend in the data. The only thing we can conclude from the data trend is that SeO_3^{2-} showed a slight high adsorption at certain pH, especially at low pH even at high ionic strength condition. An increase in SeO_3^{2-} adsorption at low pH is due to the low point of zero charge (PZC) of quartz (pHPZC ~2.82) [30].

At low pH conditions the surface adsorption sites for quartz are positively charged through protonation of surface charge, which is consistent with the fact that oxide and hydroxide surfaces are positively charged at lower pH and negatively charged at higher pH than that of the PZC value of the adsorbent [31]. Based on the increased SeO_3^{2-} adsorption to quartz at low ionic strengths (I=0.01 and 0.1 M) and low pH conditions, (although minor), SeO_3^{2-} is considered to have slightly higher adsorption affinity to quartz than SeO_4^{2-} which exhibits no detectable adsorption onto quartz at all conditions studied.

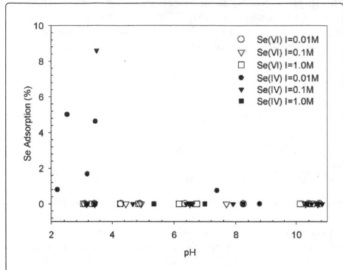

Figure 2: Adsorption of Se (%) versus pH with varying ionic strengths (I=0.01M, 0.1 and 1.0 M NaNO₃) on quartz.

Adsorption of SeO_3^{2-} on synthetic ferrihydrite at varying pH (4-8) and at fixed ionic strength (I=0.1 M NaNO₃) resulted in nearly 100% adsorption (98.4% to 99.9%) (Figure 3). However, SeO_4^{2-} displayed adsorption ranging from 6 to 82%, with decreasing adsorption as pH increased, with higher percentages of uptake at the lower pH ranges (<7). The high adsorption of SeO_4^{2-} at pH<7 was attributed to the higher PZC of iron oxide (pH$_{PZC}$=8.5 to 8.8) compared to that of quartz (pH$_{PZC}$ ~2.82) [31]. There were more discernable adsorption differences between SeO_3^{2-} and SeO_4^{2-} as the pH changed, indicating that SeO_3^{2-} has a higher adsorption affinity to Fe oxide similar to the Se adsorption results on quartz. In addition, the elevated adsorption of SeO_3^{2-} compared to SeO_4^{2-} onto the Fe oxide at all pH conditions also suggests a higher adsorption affinity of SeO_3^{2-} which agrees with the results of Se adsorption on the quartz.

Adsorption results for Se onto aluminum oxide under different ionic strengths (I=0.01 to 1.0 M NaNO₃) and varying pHs (4 - 10) are shown in Figure 4. Selenite adsorption on aluminum oxide was greater than 96% at pH<8.5 with no discernable effect from varying ionic strength (I=0.01 and 1.0 M NaNO₃). High adsorption affinity of SeO_3^{2-} on alumina oxide at low pHs was attributed to the high surface area (110 m^2 g^{-1}) and PZC of alumina oxide (pH$_{PZC}$=8.5) [25], resulting in positive surface charges due to protonation of alumina oxide surface at lower pH (pH<8.5). Decreasing adsorption of SeO_3^{2-} was found as pH increased, especially at higher pHs (>8.0) because of the negatively charged alumina surfaces at high pHs. In contrast, SeO_4^{2-} adsorption onto aluminum oxide was largely affected by changes in ionic strength from 0.01 M to 1.0 M NaNO₃. Selenate

adsorption decreased as both ionic strength and pH increased (Figure 3).

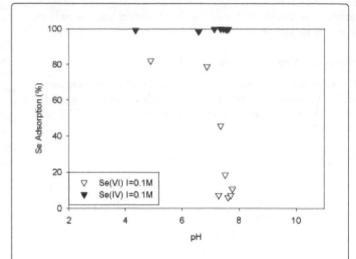

Figure 3: Adsorption of Se (%) versus pH with ionic strengths I=0.1M NaNO₃ on ferrihydrite.

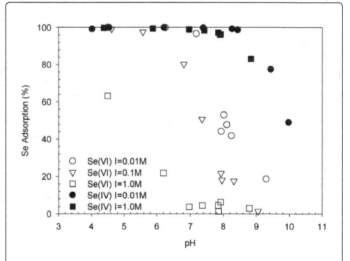

Figure 4: Adsorption of Se (%) versus pH with varying ionic strengths (I=0.01M, 0.1M and 1.0M NaNO₃) on aluminum oxide.

As ionic strength increased from I=0.01 M to 1.0 M NaNO₃, there is an increase in concentration of anions (e.g., NO_3^-) from the background electrolyte, which competes with SeO_4^{2-} for available adsorption sites. The ionic-strength dependence of SeO_42- adsorption suggests an outer-sphere surface complex, while the SeO_3^{2-} adsorption suggests formation of an inner-sphere surface complex because of the independent adsorption behavior of SeO_3^{2-} with respect to varying background ionic strengths. Due to a weaker electrostatic bond, outer-sphere complexes are less stable and considered to be relatively weakly binding. A strong dependence on ionic strength is typically found in species that form outer-sphere surface complexes [31]. Conversely, the ionic strength independence of SeO_3^{2-} adsorption suggests a stronger covalent bond indicative of an inner-sphere complex and direct binding of SeO_3^{2-} to the solid surface without the presence of co-adsorbed water molecules [31]. Based on the general comparison of

SeO_3^{2-} and SeO_4^{2-} adsorption on three different single minerals, SeO_3^{2-} showed a much higher adsorption affinity on all the adsorbents, suggesting a higher adsorption affinity due to a stronger bonded inner-sphere surface complex.

The adsorption of Se on the Hanford sediment was also measured using Hanford natural groundwater (Table 3). The results of Se adsorption (K_d values) were between 0 and 3 mL g^{-1} for SeO_4^{2-}, and slightly higher values for SeO_3^{2-} ranging from 0 to 22 mLg^{-1}, respectively (Figures 5 and 6). A decrease in the adsorption K_{ds} for both SeO_3^{2-} and SeO_4^{2-} as the initial Se concentration increased signifies the potential for nonlinear adsorption behavior for Se onto Hanford sediments. At fixed solid to solution ratio and initial Se concentration conditions, SeO_3^{2-} K_{ds} were higher than those found for SeO_4^{2-}, which agrees with the previous Se adsorption experiments conducted on pure single minerals.

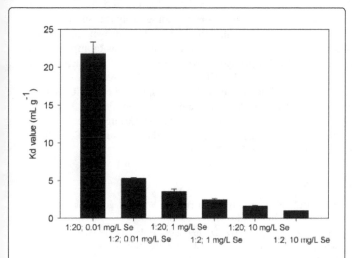

Figure 5: Selenite batch K_d results (mL g^{-1}) with varying solid to solution ratios (1:20, 1:2) and Se concentrations (0.01, 1 and 10 mg L^{-1}).

Selenium transport in column experiments

The BTCs for the nonreactive Br tracer and SeO_4^{2-} in groundwater through Hanford sediment is shown in Figure 7. The Br transport was not retarded by the Hanford sediment, resulting in a retardation value (R_f) of 1.0. The hydrodynamic dispersion coefficient for the packed columns was obtained from the CXTFIT curve fit on the Br BTC and fixed in the equilibrium model that was applied to the SeO_4^{2-} BTC. The transport of SeO_4^{2-} was slightly retarded when compared to the Br BTC and reached a relative concentration (C/C_o) of 1.0 after only 2.0 pore volumes, resulting in a R_f value of 1.2 (Table 4). The comparable retardation of Br ($R_f=1.0$) and SeO_4^{2-} ($R_f=1.2$) indicates weak adsorption affinity of SeO_4^{2-} on the Hanford sediment and was consistent with the previous batch adsorption results. Minor tailing was noticed for the SeO_4^{2-} transport at pore volume>15 (Figure 7) and showed an asymmetric BTC for SeO_4^{2-}, which suggests the presence of immobile regions resulting in physical non-equilibrium. However, the non-equilibrium model did not improve the fit results when compared to the equilibrium model, indicating that the non-equilibrium conditions were not significant for SeO_4^{2-} transport in the column.

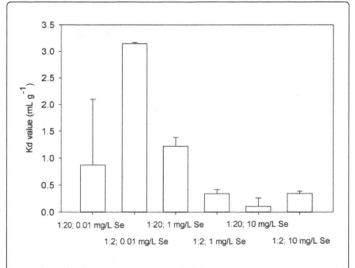

Figure 6: Selenate batch K_d results (mL g^{-1}) with varying solid to solution ratios (1:20, 1:2) and Se concentrations (0.01, 1 and 10 mg L^{-1}).

The BTCs for Br and SeO_3^{2-} with the equilibrium model fit are shown in Figure 8 and Table 4. Because the BTCs for Br and SeO_3^{2-} are fairly symmetric, an equilibrium model was used for both the Br and SeO_3^{2-} BTCs. The BTC for SeO_3^{2-} showed more retardation (with respect to Br transport) and did not reach complete breakthrough until approximately 6.5 pore volumes (Figure 8). The CXTFIT model fit to the SeO_3^{2-} column experiment resulted in an R_f value of 3.1. The SeO_3^{2-} column experiment displayed a higher adsorption resulting in more retarded transport of SeO_3^{2-} compared to the low adsorption of SeO_4^{2-} which displayed negligible retardation.

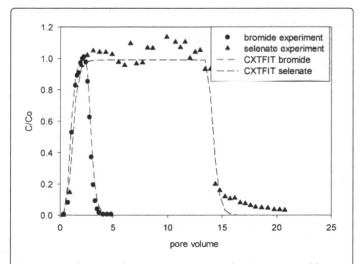

Figure 7: Selenate column experiment and CXTFIT equilibrium model results.

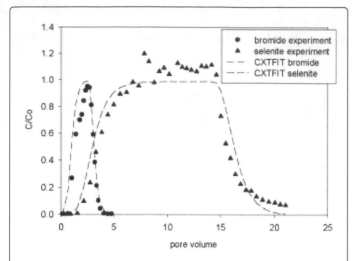

Figure 8: Selenite column experiment and CXTFIT equilibrium model results.

The calculated K_{ds} based on the retardation values obtained from the column experiments (Table 4) showed the same trend for adsorption affinity between SeO_3^{2-} and SeO_4^{2-} as observed in the batch adsorption experiments. The smaller calculated K_d values for selenitie and selenate from the column experiments compared to those obtained from the batch experiments are considered to be a result of the different solid to solution ratios and contact times between the two sets of experiments. The batch and column experiments showed the same trends for SeO_3^{2-} and SeO_4^{2-} adsorption, indicating different adsorption affinities due to different adsorption mechanisms forming inter- vs. outer-sphere complexes for SeO_3^{2-} and SeO_4^{2-}, respectively. These different adsorption mechanisms should be considered in fate and transport predictions and future performance assessment activities should utilize the different mobility attributes for the Se species, SeO_3^{2-} and SeO_4^{2-}, at the Hanford Site.

Conclusions

The batch adsorption experiments on quartz, aluminum oxide and iron oxide (ferrihydrite) resulted in a higher adsorption affinity for SeO_3^{2-} than SeO_4^{2-} with higher adsorption at the lower pHs. Particularly important are the batch adsorption results on the Fe and Al oxides, which showed a higher adsorption capacity for Se, especially at circum neutral pH conditions. The SeO_3^{2-} batch adsorption results showed an ionic strength independence, while the SeO_4^{2-} adsorption showed ionic strength dependence (i.e., adsorption increased as the ionic strength decreased). Ionic strength dependence or independence in adsorption results indicates that SeO_4^{2-} displays an outer-sphere adsorption surface complex and SeO_3^{2-} forms a stronger bonded inner-sphere adsorption surface complex. The batch adsorption experiments on the Hanford sediment with natural Hanford groundwater at pH between 7.0 to 8.5 resulted in a maximum K_d value of 21.8 and 3.14 mg L^{-1} for SeO_3^{2-} and SeO_4^{2-}, respectively. In agreement with the batch adsorption experiments, the column experiments yielded negligible adsorption of SeO_4^{2-} with an R_f of 1.2, while SeO_3^{2-} showed higher adsorption with a R_f of 3.1. All experimental results at the same background conditions showed higher adsorption for SeO_3^{2-} than SeO_4^{2-} concluding that SeO_4^{2-}, the

oxidized form of Se, will be more mobile than SeO_3^{2-} in the environment.

The pure single mineral experiments resulted in minor amounts of Se adsorption on the quartz and much higher percentages of adsorption on aluminum and iron oxides. Hanford soils are dominated by quartz with lesser amounts of aluminum- and iron-bearing minerals. The composition of Hanford sediments leads to the conclusion that there will likely not be a high percentage of Se adsorption on Hanford sediment, particularly at high pH conditions similar to natural Hanford soil (7.5-8.5).

Acknowledgments

This study was conducted in support of the ILAW project with funding from the Office of River Protection (DOE) managed by CH2M-HILL Hanford Company. Eric Clayton and Steven Baum (PNNL) are also acknowledged for providing analytical support. The authors would like to thank Jeff Serne, Chris Brown and Dawn Wellman (PNNL) for their comments and suggestions to improve this manuscript. PNNL is operated by Battelle for the U.S. Department of Energy under contracts DE-AC006-76RLO 1830. A portion of this research was carried out at the POSTECH supported by WCU (World Class University) and BK21+ programs at the Division of Advanced Nuclear Engineering (DANE) in POSTECH through the National Research Foundation of Korea funded by the Ministry of Education, Science and Technology (R31-30005).

References

1.   EPA (2012) National Primary Drinking Water Regulations-Inorganic Chemicals. US Environmental Protection Agency.

2.   Engberg RA, Westcot DW, Delamore M, Holz DD (1998) Federal and State Perspectives on Regulation and Remediation of Irrigation-Induced Selenium Problems.

3.   Mayland HF (1994) Selenium in Plant and Animal Nutrition. In: Frankenberger JWT, Sally Benson (edn) Selenium in the Environment Marcel Dekker Inc New York.

4.   Lockheed Martin (2002) Chart of the Nuclides (16edn).

5.   Mann FM, Puigh RJ, Rittmann PD, Kline NW, Voogd JA, et al. (1998) Hanford Immobilized Low-Activity Tank Waste Performance Assessment. US Department of Energy Richland WA.

6.   Kaplan DI, Serne RJ (2000) Geochemical Data Package for the Hanford Immobilized Low-Activity Tank Waste Performance Assessment (ILAW PA). Pacific Northwest National Laboratory.

7.   McGrail B, Bacon D, Serne R, Pierce E (2003) A Strategy to Assess Performance of Selected Low-Activity Waste Forms in an Integrated Disposal Facility. Pacific Northwest National Laboratory.

8.   Sposito G (1982) On the Surface Complexation Model of the Oxide-Aqueous Solution Interface. J Colloid Interface Sci 91: 329-340.

9.   Sposito G (1986) Distinguishing Adsorption from Surface Precipitation. Geochem Process Miner Surf 323: 217-228.

10. Stumm W (1992) Chemistry of the Solid-Water Interface: Processes at the Mineral-Water and Particle-Water Interface in Natural Systems. John Wiley and Sons Inc. Canada.

11. Breynaert E, Scheinost AC, Dom D, Rossberg A, Vancluysen J, et al. (2010) Reduction of Se(IV) in boom clay: XAS solid phase speciation. Environ Sci Technol 44: 6649-6655.

12. Balistrieri LS, Chao TT (1987) Selenium Adsorption by Goethite. Soil Sci Soc Am J 51: 1145-1151.

13. Boyle-Wight EJ, Katz LE, Hayes KF (2002a) Macroscopic Studies of the Effects of Selenate and Selenite on Cobalt Sorption to Al2O3. Environ Sci Technol 36: 1212-1218.

14. Boyle-Wight EJ, Katz LE, Hayes KF (2002b) Spectroscopic Studies of the Effects of Selenate and Selenite on Cobalt Sorption to Al2O3. Environ Sci Technol 36: 1219-1225.

15. Duc M, Lefèvre G, Fédoroff M (2006) Sorption of selenite ions on hematite. J Colloid Interface Sci 298: 556-563.

16. Zhang P, Sparks DL (1990) Kinetics of Selenate and Selenite Adsorption/Desorption at the Goethite/Water Interface. Environ Sci Technol 24: 1848-1856.

17. Gonzalez CM, Hernandez J, Parsons JG, Gardea-Torresdey JL (2010) A Study of the Removal of Selenite and Selenate from Aqueous Solutions Using a Magnetic Iron/Manganese Oxide Nanomaterial and ICP-MS. Microchemical Journal 96: 324-329.

18. Zonkhoeva EL, Sanzhanova SS (2011) Infrared Spectroscopy Study of the Sorption of Selenium(IV) on Natural Zeolites. Russian Journal of Physical Chemistry 85: 1233-1236.

19. Yoon IH, Kim KW, Bang S, Kim MG (2011) Reduction and Adsorption Mechanisms of Selenate by Zero-valent Iron and Related Iron Corrosion. Applied Catalysis B: Environmental 104: 185-192.

20. Sharmasarkar S, Vance GF (2002) Selenite-Selenate Sorption in Surface Coal Mine Environment Adv Environ Res 7: 87-95.

21. Horton DG, Schaef HT, Serne RJ, Brown CF, Valenta MM (2003) Geochemistry of Samples From Borehole C3177 (299-E24-21). Pacific Northwest National Laboratory.

22. Reidel SP (2002) Geologic Data Package for 2005 Integrated Disposal Facility Waste Performance Assessment. Pacific Northwest National Environment.

23. Bickmore BR, Nagy KL, Young JS, Drexler JW (2001) Nitrate-cancrinite precipitation on quartz sand in simulated Hanford tank solutions. Environ Sci Technol 35: 4481-4486.

24. Papelis C, Roberts PV, Leckie JO (1995) Modeling the rate of cadmium and selenite adsorption on micro- and mesoporous transition aluminas. Environ Sci Technol 29: 1099-1108.

25. Papelis C (1992) Cadmium and Selenite Adsorption on Porous Aluminum Oxides: Equilibrium Rate of Uptake and Spectroscopic Studies Department of Civil Engineering and the Committe on Graduate Studies Stanford University, Stanford.

26. Schwertmann U, Cornell RM (2000) Iron Oxides in the Laboratory: Preparation and Characterization. Weinheim: Wiley-VCH.

27. EPA (1999) Understanding Variation in Partition Coefficient, Kd, Values Volume I: The Kd Model Methods of Measurement and Application of Chemical Codes. Office of Radiation and Indoor Air and Office of Environmental Restoration.

28. Parker JC, Vangenuchten MT (1984) Determining transport parameters from laboratory and field tracer experiments. Virginia Agricultural Experiment Station Bulletin 1-96.

29. Toride N, Leij FJ, Genuchten MT (1999) The CXTFIT Code for Estimating Transport Parameters from Laboratory or Field Tracer Experiments. US Salinity Laboratory. US Department of Agriculture Riverside California.

30. Goyne KW, Zimmerman AR, Newalkar BL, Kormarneni S, Brantley SL, et al., (2002) Surface Charge of Variable Porosity Al2O3(s) and SiO2(s) Adsorbents. Journal of Porous Materials 9: 243-256.

31. Langmuir D (1997) Aqueous Environmental Geochemistry. Prentice-Hall Inc. Upper Saddle River New Jersey.

3

Importance of Municipal Waste Data Reliability in Decision Making Process Using LCA Model- Case Study Conducted in Timok County (Serbia)

Hristina Stevanovic Carapina[1], Jasna Stepanov[1], Dunja C Prokic[1*], Ljiljana Lj Curcic[1], Natasa V Zugic[1] and Andjelka N Mihajlov[2]

[1]*Faculty of Environmental Governance and Corporative Responsibility, Educons University, SremskaKamenica, Serbia*

[2]*Faculty of Technical Sciences, University of Novi Sad, Serbia*

*Corresponding author: Dunja C Prokic, Faculty of Environmental Governance and Corporative Responsibility, Educons University, Sremska Kamenica, Serbia
E-mail: dunjasavic@yahoo.com

Abstract

This paper presents the consequences of using various waste management data for the municipal waste management planning process in the Timok County (Serbia) using life cycle analysis. Life cycle analysis was performed using the integrated waste management model which is designed to analyze different waste management options for the implementation of integrated waste management system. The results of this model are presented through two indicators: global warming potential and costs of selected waste management system. The aim of this paper is to examine and to discuss contribution of life cycle analysis in decision making process in waste management sector and to emphasize importance of waste data in waste management planning as well as prediction of the environmental performance and economic cost of an integrated waste management system. The application of data analysis on the evaluation of life cycle solid waste management system in the Timok County (Serbia) is an innovative approach to this problem, necessary to make the right, professional decisions in the field of waste management not only in the Republic of Serbia, but also in European union candidate, potential candidate countries as well as any other countries in the world. Results indicate significant differences in global warming values, respectively the impact of waste disposal options on global warming and the costs of such system, depending on the data source that was used. Bearing in mind the implications of waste data can play in a variety of sectors and decision making process, this research refers to the need for reliable data.

Keywords: Municipal waste management; Data life cycle assessment; IWM-2 model; Global warming potential; Costs

Introduction

Municipal waste management is one of the priorities of environmental management [1] in all countries that are members of the EU as well as in the Republic of Serbia, which started the EU Accession Negotiations Process. In the Republic of Serbia, environmental protection, especially municipal waste management, is acknowledged as one of the most demanding and the most complex chapters when it comes to approximation of legislation, investments needed and adopting technologies necessary to reach the EU standards [2]. The long term environmental strategy of the Republic of Serbia, based on principles of sustainable development, is practically unachievable without considering the problems of planning and resolving inadequate waste management [3]. The Law on Waste Management (LWM) [4] and by-laws introduced the obligation of waste generators report on the quantities and composition of generated waste. Furthermore, the LWM stipulates that each municipality develop a municipal waste management plan; subsequently, municipalities must then organize themselves into regions and prepare regional waste management plans based on the local plans [5] and accurate data on the quantities and composition of waste. The importance of planning in municipal waste management is reflected in the fact that the management plans have to integrate the most appropriate option for the environment, taking into account economic, technical, social and environmental factors [6]. Analysis of a different waste management options allows decision makers to use

different instruments to consider more acceptable options and make decisions about the optimal option to satisfy their specific needs. The bases for decision-making process are waste generation and waste composition data in a given territory within a certain time.

The National Waste Management Strategy (NWMS, 2003) in Serbia [7] has identified the lack of data on composition and quantities of municipal waste generated across the country. In most municipalities in the Republic of Serbia there is a lack of information of waste qualitative and quantitative analysis, i.e. data base of quantities, characteristics, especially content and classification of waste. Most companies which have the duty of waste collection and its transport to the disposal site do not perform measuring procedure of waste quantities, nor do they have proper equipment for performing this procedure.

For this reason in the Republic of Serbia there are no quality baseline data for making professional decisions for the implementation of integrated waste management. The sources of data on the waste can be found in various national, regional and local documents and databases (Statistical Office of the Republic of Serbia, Serbian Environmental Protection Agency, regional and local waste management plans, local strategies). The diversity of data on the quantity and composition of municipal solid waste generated in local governments or regions, which are located in different documents, refer to different sizing of necessary infrastructure capacity for implementation of modern waste management system in their territories. In the final instance, this fact has a financial implication. This paper presents the consequences of using various data in the municipal waste management planning process in municipalities of the Timok County through the application of LCA (LCA-Life Cycle Assessment) study. The application of data analysis on the evaluation of life cycle solid waste management system in the municipalities Zajecar, Boljevac, Knjazevac, Majdanpek, Kladovo, Negotin and Bor (Timok County, Serbia) is an innovative approach to this problem, necessary to make the right, professional decisions in the field of waste management. The aim of this paper is to examine and to discuss contribution of LCA analysis in decision making process in waste management sector and importance of waste data in prediction of the environmental performance and economic cost of an integrated waste system.

Life cycle assessment model as a tool for planning solid waste management

Life cycle assessment is an analytical tool for evaluating environmental impacts. This tool can be used in decision-making process during the selection of appropriate options for municipal solid waste management [8]. Framework for Life Cycle Assessment (LCA) is described in ISO 14040 standard. LCA study is conducted through: the development of an inventory of relevant inputs and outputs of a given product system; evaluation of potential impacts on the environment given the inputs and outputs and interpretation of the results of inventory analysis and impact of assessment phases in relation to the objectives of the study placed [9]. LCA was initially developed for evaluating the whole life cycle of products, including extraction of resources, production, distribution, use and disposal [10].

In the waste management sector, LCA can be applied to compare the environmental performance of alternative waste treatment systems and identify focus areas for system performance improvement. All life cycle is seen through approach "from cradle to grave." The cycle begins ("the cradle") from the moment when some product is identified as a waste in households, and it is usually the dustbin. The "grave" is the final disposal of the product; often back into the earth (as emissions to water, air, soil) as landfill. Waste is inevitable "product" of society. Analysis based on the application of LCA studies on waste management claim that the waste management should be conducted so that the risk to human health is minimized and on recognition that the waste is actually a resource [8]. A prerequisite to create LCA studies for product or waste is good input data [11]. The requirements in terms of data quality are defined by the standard ISO 14044 and should include the following [9]:

- time-related coverage: age of data and the minimum length of time over which data should be collected;
- geographical coverage: geographical area from which data for unit processes should be collected to satisfy the goal of the study;
- technology coverage: specific technology or technology mix;
- precision: measure of the variability of the data values for each data expressed (e.g. variance);
- completeness: percentage of flow that is measured or estimated;
- representativeness: qualitative assessment of the degree to which the data set reflects the true population of interest (i.e. geographical coverage, time period and technology coverage);
- consistency: qualitative assessment of whether the study methodology is applied uniformly to the various components of the analysis;
- reproducibility: qualitative assessment of the extent to which information about the methodology and data values would allow

an independent practitioner to reproduce the results reported in the study;
- sources of the data;
- uncertainty of information (e.g. data, models and assumptions).

Therefore, the efficacy of LCA studies on waste management system depends on the quality of data that need to be characterized by quantitative and qualitative aspects, as well as the methods used to collect and merge these data.

Waste management system in municipalities of the Timok county

Timok County includes municipalities of: Zajecar, Boljevac, Knjazevac, Majdanpek, Kladovo, Negotin and Bor. This County belongs to parts of the mountain range Carpatho-Balkanides [12]. The specified area of Eastern Serbia occupies an area of over 3.000 km2, which corresponds to 30% of the area of the massif of the Carpatho-Balkanides [13].

Existing environmental problem of the Timok County is the fact that the total waste (municipal, industrial and hazardous), is dumped without pretreatment. The only method of waste management in the region is disposal. Waste is disposed on the municipal disposal sites, which are more or less neglected [6]. This non-systematic disposal has caused great pollution not only in municipal landfills, but also on the whole territory of the region. In addition, the city dump presents a risk to human health and the environment because they do not meet the basic measures of protection in accordance with the sanitary and technical standards. Therefore, almost all local landfills consider that unsanitary [12].

Existing environmental problem of the Timok County is the fact that the total waste (municipal, industrial and hazardous), is dumped without pretreatment. The only method of waste management in the region is disposal. Waste is disposed on the municipal disposal sites, which are more or less neglected [6]. This non-systematic disposal has caused great pollution not only in municipal landfills, but also on the whole territory of the region. In addition, the city dump presents a risk to human health and the environment because they do not meet the basic measures of protection in accordance with the sanitary and technical standards. Therefore, almost all local landfills consider that unsanitary [12].

According to the National Waste Management Strategy [2] the Waste Management Act [5], as the optimal solution for the disposal of waste is proposed to set up a regional waste management, which includes the construction of regional sanitary landfills. In accordance with these requirements, the municipality of Zajecar is expressed as a leader in the regional organization of municipalities of the Timok County. Taking into account its central position, Zajecar joined the preliminary analysis of potential locations for the regional sanitary landfill on its territory where it showed an appropriate location "Halovo" [14].

Experimental Section

As a starting point for this paper are used data on quantities and composition of waste resulting in the municipality of Zajecar, Boljevac, Knjazevac, Majdanpek, Kladovo, Negotin and Bor which are potential members of the waste management region. The total population of the County is about 260.000 inhabitants. Data are taken from the following documents:

Waste Management Strategy for the period 2010-2019 [2] (further on: the Strategy)

Regional Waste Management Plan for municipalities ZajecarBoljevac, Bor, Kladovo, Majdanapek, Negotin, and Knjazevac [13], (further on: Regional Plan)

Data from the Statistical Office of the Republic of Serbia - Census of Population and Housing in the Republic of Serbia in 2011 [14] and Statistics of waste and waste management in the Republic of Serbia [15,16] (further on: the Statistics)

Determining the composition of the waste and the amount of the assessment in order to define the strategy of secondary raw materials as part of the sustainable development of the Republic of Serbia [17]. (further on: a Project)

The data are shown in Table 1 and processed using IWM-2 model- Integrated Waste Management model. From Table 1 it is evident that the present data shows considerable variability between different data sources in all categories, i.e., in terms of population, the amount and composition of MSW-Municipal Solid Waste, and to all municipalities.

The model IWM-2 is designed for use in the development of LCA studies, specifically intended to handle issues related to waste. LCA analysis performs using the IWM-2 model and provides an understanding of the life cycle of waste and its impact on the environment. The model is intended for researchers, waste managers and policy makers as a useful tool that will help in analyzing the different options for the introduction of integrated waste management. IWM-2 model requires input data relating to the quantity and composition of waste [18]. The results of this model are presented in the form of emissions to environmental media, generated during the life cycle of waste and resource consumption for different waste management options, including collecting, sorting, biological waste treatment, thermal treatment of waste, recycling and disposal. The functional unit is the total amount of waste generated by the residents of each municipality in tonnes per year.

City / Data sources	1	2	3	4
Kladovo				
Population	22.640	23.613	20.635	23.613
Average number of person per household	2,8	2,8	2,8	2,8
Amount of generated (kg/person/year)	317	183	360	114
Composition (% by weight) :	12,6	20	15,8	14,4
paper	5,4	20	5,3	4,6
glass	2,4	5	2,7	2,8
metal	12,8	30	15	15,1
plastic	5,6	5	5,6	7,5
textiles	49,7	20	42,9	42,1
organics	11,5	/	12,7	13,5
other				
Total diesel fuel consumption for transport 38.571 litres/year				
Negotin				
Population	41.380	43.418	36.879	43.418
Average number of person per household	2,8	2,8	2,4	2,8
Amount of generated (kg/person/year)	317	208	360	114
Composition (% by weight) :	12,6	3,5	15,8	14,4
paper	5,4	6	5,3	4,6
glass	2,4	1,5	2,7	2,8
metal	12,8	5	15	15,1
plastic	5,6	6	5,6	7,5
textiles	49,7	31	42,9	42,1
organics	11,5	47	12,7	13,5
other				
Total diesel fuel consumption for transport 43.200 litres/year				
Majdanpek				
Population	21.691	23.703	18.179	23.703
Average number of person per household	2,8	2,8	2,5	2,8
Amount of generated (kg/person/year)	317	311	360	114
Composition (% by weight) :	12,6	37,5	15,8	14,4
paper	5,4	7,5	5,3	4,6
glass	2,4	6	2,7	2,8
metal	12,8	12,5	15	15,1
plastic	5,6	4	5,6	7,5
textiles	49,7	17,5	42,9	42,1
organics	11,5	15	12,7	13,5
other				
Total diesel fuel consumption for transport 30.720 litres/year				
Bor				
Population	55.817	55.817	48.155	55.817
Average number of person per household	2,9	2,9	2,9	2,9
Amount of generated (kg/person/year)	317	183	360	114
Composition (% by weight) :	12,6	9,6	15,8	14,4
paper	5,4	2,5	5,3	4,6
glass	2,4	2	2,7	2,8
metal	12,8	11,5	15	15,1
plastic	5,6	4,5	5,6	7,5
textiles	49,7	55,5	42,9	42,1
organics	11,5	14,4	12,7	13,5
other				
Total diesel fuel consumption for transport 18.023 litres/year				
Zaječar				
Population	63.398	65.969	58.547	65.969
Average number of person per household	2,9	2,9	2,7	2,9
Amount of generated (kg/person/year)	317	333	360	114

Composition (% by weight) :	12,6	9	15,8	14,4
paper	5,4	5	5,3	4,6
glass	2,4	2	2,7	2,8
metal	12,8	12	15	15,1
plastic	5,6	3	5,6	7,5
textiles	49,7	30	42,9	42,1
organics	11,5	39	12,7	13,5
other				

Total diesel fuel consumption for transport 8.135 litres/year

Boljevac

Population	14.610	15.849	12.865	15.849
Average number of person per household	3	3	2,7	3
Amount of generated (kg/person/year)	317	370	360	114
Composition (% by weight) :	12,6	1	15,8	14,4
paper	5,4	3	5,3	4,6
glass	2,4	5	2,7	2,8
metal	12,8	3	15	15,1
plastic	5,6	1	5,6	7,5
textiles	49,7	4	42,9	42,1
organics	11,5	83	12,7	13,5
other				

Total diesel fuel consumption for transport 6.933 litres/god

Knjaževac

Population	34.435	37.172	30.902	37.172
Average number of person per household	2,8	2,8	2,6	2,8
Amount of generated (kg/person/year)	317	598	360	105
Composition (% by weight) :	12,6	10	15,8	9,5
paper	5,4	5	5,3	2,9
glass	2,4	6	2,7	1,9
metal	12,8	15	15	12,3
plastic	5,6	5	5,6	4,9
textiles	49,7	30	42,9	52,5
organics	11,5	29	12,7	16
other				

Total diesel fuel consumption for transport 36.000 litres/year

Table 1: Presentation of data and data sources that are used when creating LCA studies for the Timok County.

For the purposes of this research, and in order to discuss the role of data reliability in the process of decision-making on integrated waste management, two indicators are shown in studies, using the IWM-2 model that was used for its analysis:

GWP-Global Warming Potential based on the calculation of the amount of air emissions (emissions that contribute to global warming, CO_2, CH_4 and N_2O) and

- costs of selected waste management system (collection, transport and disposal) that a resident of the municipality paid annually
- The indicators are calculated on the basis of data from these different data sources. Comparative analysis was performed for three waste management options, including:
- disposal of mixed materials/waste to local unsanitary landfill (landfill without landfill gas collection or leachate collection)
- disposal of mixed waste to local sanitary landfill (landfill with landfill gas collection and energy recovery and leachate collection and treatment)
- disposal of mixed waste from all municipalities to the regional sanitary landfill (landfill with landfill gas collection and energy recovery and leachate collection and treatment).

Data on the amount of diesel fuel needed for waste transport (to a local landfill, as well as to the future regional sanitary landfills) were obtained from the records of the municipal utilities, responsible for the collection and transportation of waste.

Distances regional landfill "Halovo" from local landfills were calculated according to regional waste management plans for the Timok County.

Results

Based on the basic data for the municipality, using the IWM-2 model, analyzed indicators were specified as:

- GWP calculated according to the model of the IPCC-Intergovernmental Panel on Climate Change, and
- information about the level of costs per capita, required for the implementation of a waste management system that includes disposal of mixed materials/waste to local unsanitary landfill, local sanitary landfill and the regional sanitary landfill.

As global warming is a priority criterion in policy, this indicator should always be included in any decision-supporting assessment of the environmental impacts of waste management options [19]. GWP values for waste management in each municipality in Timok County for two different scenarios: disposal of mixed waste to local unsanitary landfill (landfill without landfill gas collection or leachate collection) and disposal of mixed waste to local sanitary landfill (landfill with landfill gas collection and energy recovery and leachate collection and treatment) were shown in Figure 1. It can be seen that the GWP varies depending on the data source and deviations were also up to 96%.

Table 2 presents the differences (%) in the GWP values between a minimum and maximum value of the municipalities, depending on the source of the initial data for two options: sanitary and unsanitary waste disposal. The data presented in Table 2 for GWP, calculated for waste disposal at the regional sanitary landfill "Halovo", showed that GWP varies up to 65% according to different sources.

For the purpose of this research, waste disposal fees were calculated using the IWM-2. Waste disposal fees present fees which inhabitants pay for the waste disposal on the local unsanitary and sanitary landfill. The results are shown in Figure 2.

Table 3 presents differences in waste management fees per person per year (%) between the minimum and maximum amount of a fee, according to data from various sources and in the case of waste disposal to the local unsanitary and sanitary landfill.

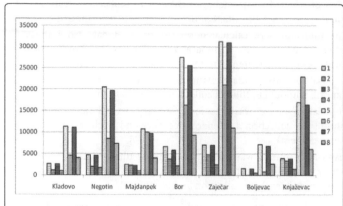

Figure 1: GWP values (tonnes) for Timok municipalities in case of waste disposal to sanitary (1-4) and unsanitary local landfill (5-8) (Data processed according to the results of IWM-2 model).

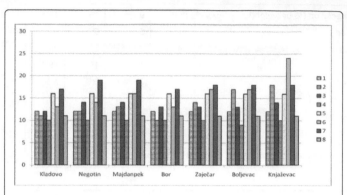

Figure 2: Waste management fees per capita in municipalities of the Timok County in the case of waste landfilling on sanitary (1-4) and unsanitary landfill (5-8) (eur/capita/year) (Data processed according to the results of IWM-2 model).

City	Kladovo	Negotin	Majdanpek	Bor	Zaječar	Boljevac	Knjaževac
Unsanitary landfill							
GWP	4/11,26	7,3/20,4	4/10,7	0,9/27	11,1/31,7	0,8/7,2	6,6/22,9
%	65	64	63	96	65	89	71
Sanitary landfill							
GWP	1,1/2,6	1,8/4,8	0,9/2,5	2,1/6,6	2,5/7,5	0.2/1,6	1,4/5,3
%	58	62	64	68	67	87	74
Regional sanitary landfill							
GWP	10,1-29						
%	65						

Table 2: Differences in GWP values (103 tons) per municipalities of the Timok County in the case of waste disposal on sanitary and non-sanitary landfill (%).

City	Kladovo	Negotin	Majdanpek	Bor	Zaječar	Boljevac	Knjaževac
Unsanitary landfill							
Fees	17-Nov	19-Nov	19-Nov	17-Nov	18-Nov	18-Nov	24-Nov
Increase %	35	42	42	35	39	39	54
Sanitary landfill							
Fees	12-Oct	14-Oct	14-Oct	13-Oct	14-Oct	17-Sep	18-Oct
Increase %	17	29	29	23	29	47	44
Regional sanitary landfill							
Fees	14-Oct						
Increase %	29						

Table 3: Shows differences in the amount of waste management fees (minimum/maximum value, eur) per person, per year in municipalities of the Timok County for waste disposal to the local unsanitary and sanitary landfill (%).

According to data presented in the Table 3, it can be concluded that differences in the amount of waste management fees per capita are up to 54% in some municipalities of the Timok County, taking into account various data sources.

This can be represented as follows: According to data from one source, an inhabitant of the municipality of Knjazevac would pay for waste landfilling on unsanitary landfill 54% more than for the used data from other sources for the calculation.

GWP values, calculated for waste landfilling on regional sanitary landfill "Halovo" for all municipalities of the Timok County, using different data sources, are shown in Figure 3.

Fees paid by inhabitants of the Timok County for collection, transport and waste disposal on sanitary landfill "Halovo", calculated according to different data sources are presented in Figure 4.

Discussion

From the results of GWP (Figure 1 and Table 2), it can be seen that the GWP value count, based on data from various sources, differ drastically, even up to 96% (e.g. the Municipality of Bor) for both waste disposal options (waste disposal on local sanitary and unsanitary landfills) and up to 65% for waste disposal on regional sanitary landfill "Halovo." Several studies also shown that the sources used for the inventory analysis varied [21-23]. These differences affect the reliability and validity of used data and direct potential data for customers' evaluation [21]. This indicates the need for updating and uniforming data about waste amount in Serbia and other developing countries. Figure 1 present that sanitary waste disposal method, with technical measures that it includes, is a better solution from the point of *greenhouse gas* (GHG) emission (e.g. the Municipality of Zajacar

77%). It is also evident that GWP values may vary depending on methods of waste treatment as well [23-25].

The municipal cost for waste collection and treatment depends on collection schemes and reprocessing operation [26]. IWM-2 software provides an opportunity to assess waste management fees (using transportation and waste disposal costs typical for EU countries), so Figure 2 and Table 3 show fees value for the collection and waste disposal on local unsanitary and sanitary landfill for inhabitants of the Timok County, according to the data from different data sources. On the basis of the present data, it should be noted that there are significant differences in waste management fees per capita, depending on the waste disposal method and used data sources. According to various data sources, differences in waste management fees per capita can be up to 54% as shown in the Table 3. It is important to conclude that sanitary waste disposal requires lower waste management fees per capita, taking into account all technical requirements for sanitary waste landfilling (using landfill gas etc.). The economics of energy recovery from landfill gas associated with CO_2 reductions are shown to be significantly better than other alternative energy forms [27]. All of this is very important in decision-making process, because waste management fee is one of the most critical parameters for decision on the establishment of a waste management system in the municipality. Using waste management data from various sources indicates that there is an inconsistency, which is reflected in the results for GWP and waste management fees per capita and have to be taken into account when decisions are made on the implementation of adequate waste management system.

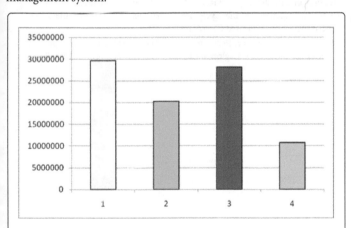

Figure 3: GWP values for the entire Timok County in the case of waste disposal on the regional sanitary landfill "Halovo" (tons) (Data processed according to the results of IWM-2 model).

Conclusions

From the above analysis, it is clear that prerequisite for the implementation of a waste management system is reliability data on a national and local level. The analysis was conducted using the IWM-2 model, and it showed significant differences in GWP values, respectively the impact of waste disposal options on global warming and the costs of such system per capita, depending on the data source that was used. The establishment of valid data on waste is a prerequisite for making decisions as follows:

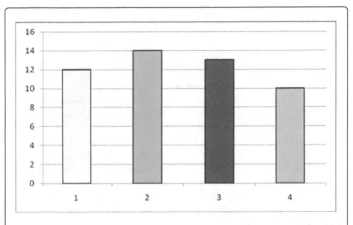

Figure 4: Waste management fees for waste disposal on regional sanitary landfill "Halovo" (eur /capita/year) (Data processed according to the results of IWM-2 model)

- designer to design the optimal and efficient facility,
- decision-makers to make timely and proper decision,
- state authorities to report on the state of the environment and waste management practices in their territories, and
- inhabitants to establish waste management fees.

At this point, the Republic of Serbia has to make significant efforts to establish a valid database on waste at national, regional and local level.

The aim of the study is not to assess the reliability of data from various sources, but to draw attention to the need for reliable data, bearing in mind the implications of these data can play in a variety of sectors.

Acknowledgement

The results presented in the paper are an output from research Ministry of Education, Science and Technological Development of the Republic of Serbia Project, Project number 176019.

References

1. Government of the Republic of Serbia (2014) Intergovernmental Conference on the accession of the Republic of Serbia to the EU " the opening statement of the Republic of Serbia, Brussels,.

2. Ministry for the Protection of Natural Resources and Environment (2009) Waste Management Strategy 2010 " 2009, Belgrade, Republic of Serbia.

3. StevanovicCarapina H, Stepanov J, Prokic D, Mihajlov A (2011) Emission of toxic components as a factor of the best practice options for waste management: Application of LCA (Life Cycle Assessment). Hem Ind. 65: 205-209.

4. Prokic D (2009) European environmental values for the benefit of the citizens of Serbia, with special reference to the practice of waste management, [In:] J. Milic (Eds.), Proceedings of papers: The European standards in Serbia, Centar for democracy, Belgrade, p. 64-77.

5. Ministry for the Protection of Natural Resources and Environment (2009) Waste Management Act, Belgrade, Republic of Serbia.

6. Prokic D, Mihajlov A (2012) Contamined sites. Practice of solid waste management in a developing country (Serbia). Environment Protection Engineering 38: 81-90.

7. Savic D (2008) Contribution to the Regional Planning of Solid Waste Management for The Region of Sombor, Apatin, Kula Odzaci and Bac, MSci Thesis, University of Novi Sad, Faculty of Technical Sciences, Novi Sad, Serbia.

8. Stepanov J (2012) Methodology for life cycle assessment of municipal solid waste â€" Sombor city MSci Thesis, University of Novi Sad, Faculty of Technical Sciences, Novi Sad, Serbia.

9. SRPS, ISO 14040:2008, Serbian standard, Environmental management-Life cycle assessment, Principles and Framework, Institute for standardization of Serbia, Serbia, 2008.

10. Bjarnadottir HJ, Fridriksson GB, Johnsen T, Sletsen H (2002) Guidelines for the use of LCA in the waste management sector, Nordtest, Finland.

11. Vigonl BW, Jensen AA (1995) Life cycle assessment: data quality and databases practitioner surve. Journal of Cleaner Production 3: 135-177.

12. RARIS "Regional Agency for Eastern Serbia Development, Regional development strategy for the Timok County 2011 " 2015, Zajecar, Republic of Serbia, 2011.

13. Ministry of Health and Environment, Report on the State of the Environment for 2000., and priority tasks in 2001 + for Serbia, Department of Environmental Protection, Belgrade, Republic of Serbia, 2002.

14. Regional Waste Management Plan for the Municipality: Zajecar, Boljevac, Bor, Kladovo, Majdanapek, Negotin, iKnjazevac, Department of environmental, University of Novi Sad, Faculty of Technical Sciences, Novi Sad, 2009.

15. Census of Population and Housing of the Republic of Serbia 2011, First results, Statistical Office of the Republic of Serbia, Belgrade, 2011.

16. Waste statistics and waste management in the Republic of Serbia, Statistical Office of the Republic of Serbia, Belgrade. 2012.

17. Determining the composition of the waste and the assessment of amount in order to define the strategy of secondary raw materials as part of the sustainable development of the Republic of Serbia, Department of environmental, University of Novi Sad, Faculty of Technical Sciences, Novi Sad, 2009.

18. StevanovicCarapina H, Jovovic A, Stepanov J (2011) Life Cycle Assessment as a tool in strategic planning of waste management. Educons University, Faculty of Environmental Protection, SremskaKamenica.

19. Merrild H (2009) Indicators for waste management: How representative is global warming as an indicator for environmental performance of waste management, PhD Thesis, DTU Environment Department of Environmental Engineering Technical University of Denmark, Denmark.

20. Stypka T (2005) Adopting the Integrated Waste Management Model (IWM-1) into the decision process, Proceedings of the International Polish-Swedish Seminar on Sustainable Municipal Solid Waste Management, Cracow, Poland.

21. Carapina HS, Stepanov J, Prokic D, Curcic Lj, Zugic N, et al. (2013) The importance of reliability of data on waste generation in decision making processes determining the optimal waste management options in the municipality of Bor. Recycling and Sustainable Development 6: 1-7.

22. Dahlbo H, Koskela S, Laukka J, Myllymaa T, Jouttijarvi T, et al. (2005) Life cycle inventory analyses for five waste management options for discarded newspapser. Waste Manag Res 23: 291-303.

23. Merrild H, Damgaard A, Christensen TH (2008) Life cycle assessment of waste paper management: The importance of technology data and system boundaries in assessing recycling and incineration, Resources. Conservation and Recycling 52: 1391–1398.

24. Larsen AW, Merrild H, Møller J, Christensen TH (2010) Waste collection systems for recyclables: An environmental and economic assessment for the municipality of Aarhus (Denmark). Waste Manag 30: 744–754.

25. Merrild H, Damgaard, A, Christensen TH (2009) Recycling of paper: accounting of greenhouse gases and global warming contributions. Waste Mang Res 27: 746-753

26. McDougall FR, White PR, Franke M, Hindle P (2008) Integrated solid waste management: a life cycle inventory, second edition, Blackwell Publishing, Oxford, United Kingdom.

27. Gardner N, Manley BJW, Pearson JM (1993) Gas emissions from landfills and their contributions to global warming. Applied Energy 44: 165-174.

4

How Economic Situation Affect the Decisions on the Building Energy Efficiency (BEE) Development in Malaysia[1]?

Queena K Qian[1]*, Abd Ghani Bin Khalid[2] and Edwin HW Chan[3]

[1] Endeavour Australia Cheung Kong Fellow, Center for Sustainable Design and Behaviour (sd+b), University of South Australia, Australia

[2] Professor, Faculty of Built Environment, University Technology Malaysia, Malaysia

[3] Professor, Building and Real Estate Department, The Hong Kong Polytechnic University, Hong Kong S.A.R., China

*Corresponding author: Queena K Qian, Endeavour Australia Cheung Kong Fellow, Center for Sustainable Design and Behaviour (sd+b), University of South Australia, Australia
E-mail: kun.qian@fulbrightmail.org

Abstract

This paper analyzes how the economic uncertainty affects the stakeholders' decision-makings on Building Energy Efficiency (BEE) investment in Malaysia. It studies the underlying barriers that hinder the BEE market penetration. Interviews with 30 architects who have BEE working experience with developers in Malaysia were conducted to understand the uncertainties affecting the current situation and future prospects for BEE adoption from the aspects of economic uncertainty. The result shows that the government incentives are trustworthy and welcomed by the stakeholders. The good opportunity to improve the BEE market is during the economic transition stage. Malaysian has high confidence that government incentives will not be changed easily. The study also suggests possible policy solutions for overcoming the uncertainties to attain the large-scale energy-efficient building investment.

Keywords: Building Energy Efficiency (BEE); Economic uncertainty; Market barriers; Malaysia

Introduction

Building production and usage can be significant determinants of sustainable development. Environmentally, the building sector is responsible for high-energy consumption, solid waste generation, global greenhouse gas emissions, external and internal pollution, environmental damage, and resource depletion [1-4]. In Malaysia, the energy consumed in buildings is 90% in the form of electricity [5]. This is quite alarming as Malaysia has one of the fastest growing building industries in the world [6]. The Malaysian government has put in place the plans, strategies, initiatives and incentives to diversify the sources of energy for sustainable development [7,8]. However, there exist some challenges that hamper the development of sustainable energy and green buildings. There should be more concerted efforts in formulating the policies and strategies to further accelerate the deployment of energy efficiency sources and its related technologies. Energy Star (2013) recognizes that energy is the first step to green. Building energy efficiency (BEE) is thus one major component of Green building. The benefits from BEE are only vaguely understood by the general public and have not been widely pursued. The stakeholders still seem to hesitate about voluntarily entering the BEE market. Technological innovations contribute to BEE. However, it is estimated that with the current sophistication of technology, a better-designed policy package to promote BEE could increase effectiveness and efficiency by 40% [9]. A critical analysis of the current market situation and its future perspective will help understand the BEE policy development and rationales to better promote BEE.

This research aims to examine the role of uncertainty during economic transition that affects the stakeholders' BEE investment from TCs perspective, using the case of Malaysia. The study tries to understand the impacts of uncertainties on the decision-making of stakeholders in real practice of BEE market through interviews with the practitioners in Malaysia.

Literature Review

BEE Embraced By Green Building

As the environmental impact of building activities becomes more apparent, a movement called green building is gaining momentum. Green, or sustainable building, is the practice of creating and using healthier and more resource-efficient models of construction, renovation, operation, maintenance, and demolition (US EPA Green Building, 2008). As energy consumption in the building sector is one of the main components of total energy consumption in most countries [9-11], BEE becomes an important theme of green building, which brings together a vast array of practices and techniques to reduce the impact of buildings on energy consumption, the environment, and human health.

The Barriers to BEE Causing Uncertainties

With socio-economic progress, more building market stakeholders are getting involved and each of them looks after their own business interests which may have conflict with each other. Real estate developers generally do no more than just meeting the basic requirements of the law and policies, in order to minimize the costs of the extra work entailed by energy efficiency regulations. Contractors

---

[1] An earlier version of this paper has been presented in the SB13 Dubai conference, 8-10, Dec, 2013, Dubai. The authors appreciate the feedbacks and comments collected from the discussion for improving the paper quality.

also want to avoid these extra tasks, which require special expertise and specialized equipment that they do not typically possess. Manufacturers of BEE products want regulations to be even stricter to create greater demand in the market. Financially, building-design professionals and their institutes will not be adversely influenced by the new policies but are apt to succumb to the demands of developers due to the nature of their relationship. These conflicting interests are the main sources of the uncertainties concerns and the barriers to BEE development. The unique barriers to the BEE market, including the longevity of life causing slow innovation adaptation; Low level standardization requiring trustworthy performance; extended supply chain; bounded rationality; risk aversion; negative externalities; split incentives; public goods; cash flow constraints; information uncertainty; unaware of the hidden costs, can be established from extant studies [9,12-15].

Economic Uncertainty

Uncertainty Staley [16] plays a vital role in the stakeholders' decision-makings of their BEE investment [15]. Uncertainty due to information asymmetry is the fundamental aspect of transaction costs theory, which will be verified from interview data. On the supply side, the developer has to search detailed information about strategies of his opponents before he decides to invest in BEE. The searching process may incur high uncertainty that even prevent him from entering the market. The primary reason is that the degree of compliance of BEE code cannot be perfectly observed from the public, and some developers may exaggerate the energy efficiency performance. The extreme case is to sell the conventional building product at the price of BEE, which would fill the BEE market with a lot of fake and low-quality non-BEE products. As practical evidences show, the inability to distinguish the BEE from the non-BEEs and the constant doubt from the public further undermines the attractiveness of BEE to stakeholders and eventually leads to a "Lemon market". Moreover, the external factors, such as the stability of economic and policy environment, will also cause the concerns of the stakeholders in their decision-making process on BEE.

In Malaysia, Samari observed that the current financial incentives are not able to cover the high upfront cost of green buildings [8]. Only Architects association Malaysia has developed the green building index for Malaysia and other than that very little initiative has been made. Green activist group Malaysia has also made some initiatives to promote green cities but not much has been done. A zoom-in focus of the building industry in Malaysia gives an insightful perspective of BEE market development in developing Asian countries. Through interviews with the professionals we may have a better understanding the impacts of economic uncertainty to BEE market.

Methodology

Interview with the Architects in Malaysia

In-depth interviews to solicit views and issues regarding BEE investment were conducted with the architects, who have BEE working experience with developers in Malaysia. The research team travelled to Malaysia in 2012 and met the potential interviewee, through a CPD event and guest lecture on BEE to the Malaysian Institute of Architects (Pertubuhan Akitek Malaysia-PAM)–which was contacted through University Technology of Malaysia (UTM), and Construction Industry Development Board CIDB of Malaysia. After screening to ensure the interviewees with BEE working experience, the team interviewed 30 architects, who have actively worked on BEE projects for major real estate development companies in Malaysia. The purpose of the data collection was to get the first hand opinions of architects with BEE project experience about the role of economic uncertainty in BEE investment, especially during the economic transition period. The data analysis also provides a better picture of BEE market development relating to a specific institution in the Malaysia case, and provides reference for designing rational policy. The responses from these architects are important to understand the market/business expectation from a more objective perspective.

Interview Questions Design

The hypotheses and the interview questions were developed based on the literature review as presented above and through pilot discussions among the research team members and with 2 experts in industry and 2 academics. The relations between the three aspects of economic uncertainties, two hypotheses (H), and three interview questions (Q) are listed in Table 1 below. The purpose of these interviews is to understand how the uncertainty affects the BEE investment decisions by a case study of Kuala Lumpur of Malaysia with the local architects' viewpoint, which ascertains the impact of economic transition to BEE development in practice. What is the impact of economic transition on the BEE development (to the developer–H1; to the government–H2)? Is it a challenge or an opportunity? How do the developers' concerns change in an economic downturn or upturn? What should government be alert to during such periods and how can it develop the most effective policies to promote BEE accordingly? These are the main issues that are addressed in interview questions Q1- Q3.

Empirical Analysis and Interview Results

Table 1 shows the major opinions (extraction of the top ranked points from the collection of answers) of the interviews, which have been summarized and grouped under a few dominating points in the "Summary of the Key Responses" in Table 1. It was an interview exercise where the respondents could give several options or views to one question. The rate of respondents with the views close to the summarizing key point is shown in the right hand column of the table. The % rate shown for the answers of each question shows the weighted similar opinions to each interview question among the interviewees, which cannot be taken as comparison with another question in absolute value or importance. The interviewees are free to have multiple answers to each interview question, as long as they do not conflict with each other. Therefore the percentage of the different views to each question does not necessarily add up to 100%. Those Key Responses highlighted in bold letters are the significant issues to be discussed in more details in the discussion section.

H1; The economic context (upturn or downturn economic transition) affects the concerns of the real estate developers about BEE investment.		
Questions	Summary of the Key Responses	

Q1 At "economic transition" period/stage (upturn or downturn), what are the new challenges or opportunities to developers in investing GB/BEE? How do shifts (upturn or downturn) in the economy change the developers' major concerns (neutral, positive, or negative) and in which aspects?	It depends on the planning, priority and value judgment of the corporation and individual decision-maker.	
	When products become harder to be sold in the market (less demand, more competition), green building can be competitive advantage which is not necessary in times when products are easy to be sold. So, there will be more incentive to go in green building.	23.07%
	Can do more, faster and put more resources in GB/BEE in the economic upturn.	
	More challenges than opportunity in the downturn.	
	During the economic downturn, the developers are more willing to do the energy retrofits, because it's much quicker to get the capital return back.	
	When it is economic downturn, the developers will be more conservative/ reluctant to do any innovative project including green features due to limited budget; while in its upturn, they will be more likely willing to invest in GB/BEE.	15.38%
	Economic downturn is a better chance to further the GB/BEE development, because people expect changes; however, in the economic upturn, everything is prosperous, and why would the developers change their regular earning formula of their investment.	
	The government should take the opportunity in the economic downturn to shout loud to promote BEE/ green, as it is more an opportunity than a challenge to further its development.	
	The previous economic downturns maybe different from the current one, because the green movement is not as heat-up as it is now. Therefore, it more depends on the individual developer and its own capital capacity and business strategy to integrate green features into the practice.	
	Need more government incentives, money wise, to promote GB/BEE to the developer; and education wise, to the whole public.	37.50%
	Both the government and the developers should have long term views and will invest in GB/BEE, even in economic downturn.	47.50%
	At the economic down-turn, the developer would mainly improve the "green" image to add to their brand-name, but the result is not very significant. Because the approach attracts mainly the user-buyers, not the speculators, and developers want to have more speculators for profits than the user-buyers.	2.50%
H2 Changes in economic conditions (upturns and downturns) call for the attention of government to adjust BEE policies as necessary to seize BEE development opportunities.		
Q3 What role should government play and what GB/BEE promotions or incentive could government introduce in times of economic change that would be less upsetting to the market players' normal activities?	During the economic up-turn, government incentives or promotion are less effective than the down-turn, because the property sells well and the buyers are less concern about green features.	20.93%
	During the economic downturn, government incentives are more important, because the developers are more reluctant to invest in green and people who buy also need to be more assured by the cost benefits from green incentives.	41.86%
	During economic upturn, with steady and gentle growth would be the best time to developers to invest in GB/BEE, and the best time for the government to promote GB/BEE, too.	

Table 1: Key Interview Responses on the economic uncertainty of BEE.

Observations on the Findings

Regarding Q1, the respondents had a wide range of viewpoints about new challenges and opportunities at times of changing economic conditions: 25% of the respondents opines that "It all depends on the planning, priorities, and value judgments of the corporation and individual decision-makers", and 23.07% believes that "When products become harder to be sold in the market (less demand, more competition), green building can be competitive advantage. So, there will be more incentive to go in green building". It shows that they collectively agree that the overall decision making depends on the real estate developers business concerns, however, the government incentive has a key role to play. 15.38% of the interviewees thinks that "economic downturn, the developers will be more conservative/

reluctant to do any innovative project including green features due to limited budget; while upturn; they will be more likely willing to invest in GB/BEE", which echoes with only 5.77% agree that "An economic downturn is a better chance to further BEE development", because people expect change; whereas, in economic upturns, everything is prosperous, developers have little reason to change their regular earnings formula to try something new and risky. However, if the government takes the opportunity of the economic downturn to promote BEE vigorously, it is recognized that conditions present more of an opportunity than a challenge (15.38%). A further 9.62% provide a view based on a local example: "In this economic downturn in Malaysia, the developers are more willing to do energy retrofits, because it is much quicker to get the capital return back".

Regarding Q2, 47.5% respond that "Both the government and the developers should have long-term views regarding BEE and will, even in economic downturns". To achieve this, 37.5% think that the BEE market "Need more government incentives, money wise, to promote GB/BEE to the stakeholders; and education wise, to the whole public". In addition, 12.5% express a more conservative view; they think "This economic downturn may be different from earlier ones, because the green movement was not as popular as it is now". However, only 2.5% believe that "In the economic downturn, developers would want improve their reputation for being green to add to their brand name, but the end result would not be very significant, because it attracts mainly the user-buyers, not the speculators. The developers want the speculators for profits more than the user-buyers". Therefore, integrating green features will depend more on the individual developer, its capital capacity, and its business strategy".

Regarding the role that government can play during times of economic change, the majority views (41.86%) agree that basically, "During the economic downturn, government incentives are more important, because the developers are more reluctant to invest in green and end-users who buy also need to be more assured by the cost benefits from green incentives". There is striking alternative views (34.88%), believe that, "Steady and gentle growth, i.e., economic upturn, would be the best time for developers to invest in BEE and the best time for the government to promote BEE, too". 20.93% of the interviewees, however, disagree with the above with the reason that "during the economic up-turn, government incentives or promotion are less effective than the down-turn, because the property sells well and the buyers are less concern about green features".

Discussions and Recommendations

Economic conditions (upturns or downturns) call for attention by the government to adjust BEE policies in order to promote BEE-development. During economic downturn, the developers are less interested in green investment due the limited budget. The government incentives are more important at this situation to boost the BEE market; meanwhile, people who buy also need to be assured of getting the real benefits of green incentives. Although the decision-making on investing in BEE depends on the real estate developers' business concerns, the government incentive has a key role to play. Besides, a long term vision and consistency of both the government and the market regarding the BEE development with a fully deployed package of BEE policies are solid foundation for the healthy BEE market development.

Generally, the study shows that there is a call for turning to stricter guidelines and requirements for BEE, and a variety of policy tools including appliance standards and labeling, building energy code, building energy performance rating and certification, financial incentive, government demonstration, awareness raising etc., are being utilized in Malaysia. Comparing to developed countries, policy for BEE in Malaysia is still at a very early stage of development. A well-established institutional infrastructure to support the implementation of the building energy codes is yet to be established. In the building sector, it needs a more concrete policy and focused strategy to increase the acceptance of energy efficiency measures. The government should have long-term strategies and clear and consistent policy for BEE promotion, to create a positive investment environment and raise the stakeholders' confidence and the market's expectations for business investment in BEE. Market stakeholders want the government to take the lead to try out BEE projects to cut down any uncertainty before they join the investment.

Conclusion

This study examines the stakeholders' concerns on BEE investments due to the economic transition period and has focused on uncertainty in particular. The research employed an interview survey with architects in Malaysia. The data provides a list of findings as they apply to the case of the Malaysia BEE real estate development. The study reveals the market situation and suggests policy solutions to address the concerns of uncertainties to promote BEE.

Acknowledgements

The work described in this paper was supported by research grants provided by The Hong Kong Polytechnic University and University of Technology Malaysia. The authors would like to thank all those who contributed to the interviews and those who contributed in reviewing the manuscript.

References

1. CICA (2002) Confederation of International Contractors Associations: Industry as a partner for sustainable development UK :1-60.

2. Zimmermann M, Althaus HJ, Haas A (2005) Benchmarks for sustainable construction – a contribution to develop a standard. Energy Build 37: 1147–1157.

3. Melcher L (2007) The Dutch sustainable building policy: a model for developing countries? Building and Environment 43: 893-901.

4. Ortiz O, Castells F, Sonnemann G (2009) Review sustainability in the construction industry: a review of recent developments based on LCA. Construction and Building Materials 23: 28-39.

5. Zainordin N, Abdullah SM, Baharum ZA (2012) Users Perception towards Energy Efficient Buildings. Asian Journal of Environment-Behaviour Studies 3: 91-105.

6. ABCSE (2005) Renewable Energy in Asia Malaysia Report. Australia.

7. Atsusaka N (2003) Growing the green building industry in Lane County - program for Watershed and Community Health. Institute for a Sustainable Environment University of Oregon USA.

8. Samari M, Godrati N, Esmaeilifar R, Olfat P, Shafiei MWM (2013) The Investigation of the Barriers in Developing Green Building in Malaysia. Modern Applied Science 7: 1-10.

9. OECD (2003) Environmentally sustainable buildings, challenges and policies. OECD publications Service France.

10. Chan EH, Lau SS (2005) Energy conscious building design for the humid subtropical climate of Southern China, Green Buildings Design: Experiences in Hong Kong and Shanghai. Architecture & Technology Publisher China: 90-113.

11. Zhang QY (2004) Residential Energy Consumption in China and its comparison with Japan Canada and USA. Energy and Buildings 36: 1217-1225.

12. Jaff AB, Stavins RN (1994) The emerge-efficiency gap: what does it mean? Energy Policy 22: 804-810.

13. Bell M, Lowe R, Robert P (1996) Energy efficiency in housing Aldershot Avebury.

14. Finkel G (1997) The economics of the construction industry Sharp London and New York.

15. Qian QK (2012) Barriers to Promote Building Energy Efficiency (BEE)- A Transaction Costs (TCs) Perspective.

16. Staley SR (1998) Ballot-box zoning, transaction costs and land development. Urban Futures Working Paper No: 98-102. Los Angeles CA: Reason Public Policy Institute.

5

Sustainable Solid Waste Collection in Addis Ababa: the Users' Perspective

Mesfin Tilaye[1] and Meine Pieter van Dijk[2*]

[1]*Ethiopian Civil Service University, Addis Ababa, Ethiopia*

[2]*Professor of urban management at ISS of Erasmus University, Netherlands*

*Corresponding author: Meine Pieter van Dijk, Professor of urban management at ISS of Erasmus University, Netherlands
E-mail: mpvandijk@iss.nl

Abstract

Sustainability of solid waste management is high on the agenda of urban managers. Municipalities in developing countries are incapable of meeting the demand for urban services. Some years ago Addis Ababa, the capital of Ethiopia, took the initiative to overcome some of these problems by starting a reform process. It led to a significant shift in the institutional arrangements. Community-based initiatives are becoming increasingly important as a means of addressing the deficiencies of the formal system. This paper analyzes the households' behaviour and their opinions concerning urban solid waste management practices. Sustainability will be considered from the public health, ecological and socio-economic perspective, following the PPP framework: sustainability concerns the people, the planet, and the profit sector.

Primary data consisted of a household survey and interviews of local level officials. Three types of residents were studied: those living in slums, in residential areas, and in a commercial area mixed with houses. 135 households were selected randomly in each condition. The results suggest that from a socio-economic perspective (the profit angle) the service reform sulted the interests of the city community by undertaking the service provision in a more sustainable manner. Regularity, reliability, service coverage and the frequency of service delivery to the households improved. Residents also have a good feeling about cost recovery, though also differing opinions were expressed. With regard to public health (the people's angle), Improvements were observed concerning the cleanliness of the neighborhoods, while the city cleanliness lagged behind. In case of ecological sustainability (the planet perspective) economic incentives played a more important role than ecological concerns in separating and collecting reusable and recyclable items from the waste stream.

Keywords: Environmental; Solid waste management (SWM); People, planet and profit; Household level

Introduction

Sustainability of cities in the developing countries has become a big question mark and has rightly been placed as the focal point of the Sustainable Development Goals, the successor of the millennium development goals. Since the Rio Summit in 1992, the concept of sustainability extends to basic services such as solid waste management (SWM). Many municipalities in developing countries are incapable of meeting the demand for services, resulting in both direct and indirect negative effects on the three dimensions of sustainable development: the people, planet and profit (PPP). This is the framework of the triple bottom line, developed by Sibley et al. [1]. Indicators include among others the area of coverage for a service, cost recovery, regular collection of refuse, dumping and burning of solid waste in open spaces or not.

About 20% to 30% of the waste generated in Addis Ababa remained uncollected and made the city environment aesthetically unpleasant and affected the city's public health. Local initiatives to create sustainable urban solid waste management play a key role. The sustainable urban development programme was initiated by the municipality of Addis Ababa as part of an urban governance strategy in 2004. The strategy was developed to meet the growing need for rendering this service in a sustainable manner. Since then, institutional arrangements have been undergoing significant shifts in Addis Ababa

solid waste collection system. With respect to formal sector service delivery, changes have focused on decentralization of solid waste collection service to local government and the introduction of private sector service delivery. At the neighborhood level, community-based initiatives are also becoming increasingly prominent as a means of addressing the deficiencies of the formal system.

The literature on sustainable development of solid waste management distinguishes a large range of possible relationships between the public and private sector, including public-private partnerships, community-public partnerships, and private-private arrangements [2]. Some studies focused on activities within the relationships in the SWM system notably, separation of waste, and the productive use of waste. There are studies that also deal with small-scale business transactions [3] and the impact of official rules and regulations on private or communal undertakings [4]. Effective provision of services to poor households and the safety and health aspects of activities within the SWM sector were areas of concerns given importance by different authors [5]. Baud and Post tried to connect SWM with sustainable development by operationalizing three broad goals: ecological sustainability, socio-economic equality and improvement of health, which are quite similar to our PPP principles [6]. They argue that there is a gap in the current literature on sustainable SWM in developing countries that the system is rarely investigated in its entirety. Assessments combining ecological, environmental health and socio-economic considerations are still largely absent [6]. Moreover, changing institutional arrangements among the public sector, the private sector and civil society and their implication as a central theme were not sufficiently addressed.

This research focuses on how households (service users) perceive the service provided by the private sector and local government and their readiness to engage in collaborative efforts to make sure that the system is sustainable. The necessity of exploring users' perspectives to consider sustainable solid waste management arises from three factors:

1. The solid waste collection service by the micro-enterprises at the local government level is greatly influenced by households' attitudes and behaviours,

2. Households are recognized as the main solid waste generators.

3. The institutional relationships in the solid waste management system have been heavily influenced by the introduction of a more participatory approach. The objective of this study is to assess the sustainability of solid waste collection system in terms of being beneficial to the society (profit), the city sanitation (the people) and the environment (planet).

Sustainable Solid Waste Management

Tadessa showed that households alone generate about 71% of total waste in Addis Ababa [7]. Solid waste management is an important component of the sustainable development agenda (Agenda 21 in 1992) which is being spearheaded by development partners such as World Bank and the United Nations (UN). During the UN Rio conference a framework on integrated solid waste management was developed, which was eventually transformed into guidelines by Schubeler [8]. Other contributions of solid waste management to socioeconomic, public health and ecologically sustainable development have been made by Sattherthwaite [9].

Sustainability is defined by Van der Klundert and Anschiitz [10] as a system that is: (1) appropriate to the local conditions in which it operates, from a technical, social, economic, financial, institutional, and environmental perspective, and; (2) capable to maintain itself over time without reducing the resources it needs. The problems of solid waste such as inadequate service coverage, irregular waste collection, indiscriminate disposal in unauthorized places, waste spillover from bins and storage containers, and waste littering are common in developing countries [11-13]. These problems eventually lead to public health impact, aesthetic nuisance and environmental pollution.

Van der Klundert and Lardinois [14] highlight the need to review the normal progression of motivations for setting up solid waste management systems. Concerns start with public health and sanitation, develop further to second set of motivations related to quality of life, cleanliness of streets, community appearance, and thirdly again the focus shifts to environmental quality and cost reduction. Finally, the achievement of first-order environmental goals leads to a recognition of the need for sustainable solid waste systems.

The development of 'sustainability indicators', perceived as a first step towards the operationalization of the concept of sustainability, has reflected a pro-active initiative to make a change, itself fired up by a real sense of urgency: "sustainability [becomes] meaningless unless we can do it" [15]. The development of sustainability indicators is now playing an important role in the awakening of new forms of environmental governance [16].

Indicators of sustainability provide a simplified understanding of this concept by providing practical information about the numerous issues encompassed in it. During the past years work was conducted on identifying indicators measuring the level of sustainability reached by solid waste collection systems in selected cities. These normative proxies measures also reflect a trend: they show how far or close we are from being a 'sustainable society' by reflecting the reproducibility of the way a given society utilizes its environment [15].

A major gap in the current literature on SWM in developing countries is that the system is rarely investigated in its entirety, and assessments combining ecological, environmental health and socio-economic considerations are still largely absent. This study addresses the topic of sustainable solid waste collection, following the municipal service reform in Addis Ababa city by considering new governance elements: decentralization and micro-privatization. The study explores the implication of the reform on the sustainability of the city waste collection system. The following hypothesis is tested 'solid waste service reform improved the sustainability of solid waste collection service in Addis Ababa city'. Sustainability is seen from socio-economic, public health and ecological perspectives. The study substantiates the issue of sustainability from users' perspective and takes sustainability as caring for the people, the planet and involving the private sector.

Sustainable Solid Waste Collection from a Socio-economic perspective

Socio-economic sustainability of municipal waste management system depends on many factors: the degree of privatization of waste management services, the extent of public participation, decentralization of responsibilities and tasks related to waste management [16]. Socio-economic sustainability of waste management system indicates its financial viability for households, private enterprises, organizations, and local authorities. The sustainability goal is achieved when the financial costs are balanced with the revenues for all waste managers and consumers paying for the service. The system is not financially viable if one of the partners does not benefit from the existing financial arrangements. From the consumer's point of view "economic affordability" must be taken into account. "Economic affordability" requires that the costs of waste management systems are acceptable to all sectors of the community served (Ibid).

The character of waste management tasks and the technical and organizational nature of appropriate solutions depend a great deal on the economic context of the country and/or city in question and, in fact, on the economic situation in the particular area of a city [17]. Sustainability of a system is achieved when it is able to deliver an appropriate and equitable level of benefits in terms of service quality and affordability over a prolonged period of time without negatively affecting the environment. This implies that the beneficiaries are satisfied with service and the costs are covered through user fees or financial mechanisms [10].

The main effect of micro-privatization has been to improve the existing service. Public service providers are under pressure not so much to reach people whom they are not yet reaching, but to maintain or improve existing services. Micro-privatization achieved both: existing services improved and new clients reached. The services are improved in many different ways; improved the operation, the outreach, the cost effectiveness and the overall quality of the service [18].

Micro-privatization could be a long term and more sustainable solution to the delivery of this public service [2]. Some sort of buyer-seller relationship between clients and micro-enterprises enhances commitment and sustainability. Since micro-enterprises operate at a small scale, and are often based in the neighborhood they serve, they favour community participation and control. All solid waste collection systems require some participation by the residents who receive the service. If a community is aware of the refuse collection service, and knows the workers responsible, they are more likely to be ready to pay for the service, than if the service is impersonal and unseen.

Sustainable Solid Waste Collection from an Environmental health Perspective

The efficient management of urban solid waste in developing countries is vital if environmental cleanliness and public health are not to be compromised. The impact of uncollected waste within cities is enormous. Solid waste accumulation within cities raises public health and environmental concerns because of potential odor from solid waste decomposition and associated insects and rodents. The uncollected solid waste creates conditions favorable for the survival and growth of microbial pathogens. Uncollected solid waste also has aesthetic impact on the environment and also attracts flies, rats, and other creatures that in turn spread disease [19].

Cleanliness should start at household level. The household has the capacity to make its immediate environment healthy and friendly. Household waste is more polluted and miscellaneous, so its collection and sorting costs are quite high. Municipalities face the task of designing schemes for collection of household waste. However, municipal services in developing countries are handicapped by limited finances and an ever-increasing demand on urban services. The failure to provide adequate collection services poses a serious threat to human health and the physical environment [20].

This situation can probably be improved significantly if the inhabitants of low-income communities start assuming the responsibility of handling their own garbage and setting up a system appropriate to their own economic situation. To achieve the objective of sustainability it is necessary to establish systems of solid waste management, which harmonize the technical requirements with the objectives of environmental protection and the needs and interests of different stakeholders especially the urban poor. Many urban poor live in unplanned and unauthorized areas and are, therefore, not eligible for municipal services. Consequently, the solid waste disposal practices of the individual households in un-served high-density areas are mostly detrimental to the living environment of the entire city. To avoid such problems the role played by micro-enterprises is important [18].

Sustainable Solid Waste Collection from Ecological Perspective

Waste management wants to minimize the material resources that leave the production/trade/consumption cycle in the form of refuse instead of as products that could be reintegrated into economic circulation. Although waste generation rates in developing countries are substantially lower than those in industrialized countries, these rates are not proportionally lower relative to income [21].

Recycling only occurs when the commercial value of the recycled materials covers the cost of extracting them from the waste stream. Modernization pushes – even forces – "a marriage" of services and commodities, with different rules and different cultures, to integrate with, or 'marry', each other. From the solid waste system perspective,

the main value of recycling, composting and other commodity-based activities is the reduction in the amount that needs to be moved to a dump, landfill or that is 'lost' and ends up in nature. There is still a big difference between what the principle demands and the social reality in many countries.

Although recycling may be preferred from an environmental perspective, the economic costs involved or the presence of institutional complications may prevent waste recycling from being promoted and implemented in integrated SWM. The actual integration can take place at various levels [22]. These may include waste processors such as formal and informal recyclers, waste generators such as households, industry and agriculture, and government institutions such as waste managers and urban planners.

By working more closely with the communities, micro-enterprises can play a role in public environmental education. Close links with the residents can provide opportunities to introduce separation at source, which would benefit the worker who collects and sells the recyclable materials, and benefit the municipality by reducing the quantity of waste. Community involvement can also improve relationships with service providers, to the benefit of all parties [18].

The pace of development of waste management services heavily depend on the level of public awareness of solid waste-related issues and on participation in making improvements happen at the ground level. The public support for any issue can be greatly increased if the public is fully and well-informed about the reasons behind the actions and the intended benefits [23]. Close links between the micro-enterprises workers and the residents help to develop such cooperation. Recycling is considered to be an indispensable part of solid waste management and many micro-enterprises are involved in this activity. In Bangkok, formally employed refuse collectors are reported as spending 40% of their working time recovering recyclables, thereby doubling their wages (Cointreau 1994).

Solid Waste Management in Addis Ababa City

The city council distinguishes recognizes six major sources of solid waste: households, street, commercial institutes, industries, hotels and hospitals. From total generated solid waste households' account for 71%, street 10%, commercial institutions, 9%, industries 6%, hotels 3% and hospitals 1% [24]. Most of the solid waste materials produced by households are disposed without adequate care. A study made by the Addis Ababa City Administration shows that, the collection coverage has been constantly increasing from 38 per cent in 2000 to 40 in 2001, 53% in 2002, 53.9% in 2004 and 78% in 2005 [25].

According to Dierig inadequate municipal solid waste collection and disposal creates a range of environmental problems in Addis Ababa [26]. Solid waste management is carried out to protect public health; to promote hygiene; recycle materials; avoid waste; reduce waste quantities, and reduce emissions and residuals. KAPB found that there were different types of diseases caused by improper handling of solid waste in the city [27].

The first level for separation at source in the waste recovery system in Addis Ababa is the household. At this level, materials are considered valuable and are therefore usually sorted out for reuse. The relationships between the different actors, which constitute the recovery system, seem to be based on financial benefit from the sale of their materials rather than on environmental awareness. This is mainly attributed to the fact that the majority of people in Addis Ababa are poor. This current state of the recovery system in Addis Ababa is not a perfect example of sustainable recovery of materials [28].

In recent years significant steps have been taken in the appropriate planning of solid waste management to ensure sustainable solid waste management in the city. Partnerships between public, private and community sectors have been established that eventually improved policy statements such as the Waste Strategy developed in 2000 that recognized central principles of sustainable resource management as providing the basis for the development of Addis Ababa Municipal Waste Management. In 2004 the institutional arrangements for the solid waste collection system have been changed. The changes were heavily influenced by decentralization and privatization reforms. Private micro (and small) enterprises are now engaged in collection from households to municipal containers. Simultaneously solid waste collection management has been decentralized to local municipal administration.

Sustainable Solid Waste Collection in Addis Ababa City

This section covers the analysis of sustainable solid waste collection in Addis Ababa using the perception and attitude of citizens. The analysis takes three perspectives:

(a) a socio-economic perspective: reliability of the service, user charge and affordability of service fee, willingness to pay for better service, and customer support service delivery (b) a public health and sanitation perspective: quality of collection service (household solid waste storage and separation at source) and the existence of a clean urban environment (neighborhood cleanliness, cleanliness of the city) and (c) an ecological sustainability perspective: attitude of society in acceptance of separation at source and household participation in recycling and composting and awareness of a change in values in relation to environmental protection.

Socio-economic Sustainability

The analysis in this section covers reliability of the service, user charge and affordability of the service fee, willingness to pay for better service, and customer support service delivery (demand and satisfaction for the service). The reliability of waste collection service was assessed in terms of regularity and frequency of waste collection services. Residents were asked whether the service provision by micro-enterprises to house-to-house collection service was regular.

Is the collection service regular	Residential area			Total
	Urban slum	Residential area	Commercial area with mixed residence	
Yes	92	151	73	316
	81.40%	83.40%	68.90%	79.00%
No	21	30	33	84
	18.60%	16.60%	31.10%	21.00%

Total	113	181	106	400

Table 1: Is the collection service regular in three residential areas?

Table1 presents the residents' response of service reliability of waste collection service. It was evident from the survey that seventy nine (79%) of the respondents said that the service was regular while twenty one (21%) said it was irregular.

Service provision regularity was also looked from the type of settlement perspective (urban slum, residential and residential mixed with commercial areas). Eighty one (81%) of the respondents in slum area said that the service was regular, whilst 19% complained about the regularity of the service. In residential area about eighty three (83%) of respondents said that the service was regular, while 17% denied the regularity of the service. The service regularity in the residential mixed with commercial area was relatively poor. This suggests that the involvement of micro-enterprises improved access and coverage to slum areas for solid waste collection services in the city.

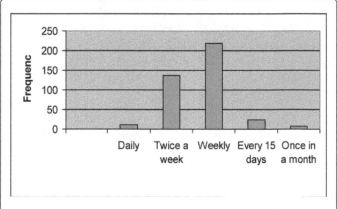

Figure 1: Frequency of waste Collection from Households.

As indicated in Figure 1 the collection frequency of house-to-house collection service rendered by micro-enterprises in Addis Ababa was dominantly once or twice a week. Residents also asked for punctuality of waste collectors in keeping their schedule. The survey results indicate that 67% of the respondents said micro-enterprises do, while 33% said they do not keep their punctuality. Some households complained about the promptness of waste collection and argued that a specific date and time is of paramount importance. Micro-enterprises fail to maintain the schedule all the time, while every household has to make sure that the bins are ready. This is a requirement by the local authorities who are responsible to manage the secondary waste collection system in the city. Hence clients started to put waste on their main gates to be picked by the micro-enterprises in their absence. This created further problem as clients complained they do not find back their waste sacks, especially if the sack is new.

Concerning user charges and affordability, the survey revealed that the mean expenditure of city wide current payment for solid waste management service is 10.42 birr.[1] Solid waste collection service accounts for 30-50% of the expenses of solid waste management in the developing world cities.[2] Households were asked about the affordability. As indicated in Table 2 at the city level 62.9% of the respondents claimed the amount paid was affordable while 23.4% and

---

[1]   It is Ethiopian currency rated 1 birr equivalent to 0.06 USD, September, 2010.

13.7% indicated medium and low level of affordability respectively. Affordability at upper income group is high though there are some variations among middle and low income group.

During the interviews community members mentioned different factors to explain their reluctance of paying the service fee. These include: first the existing procedure of payment for solid waste collection service is based on the amount of water a household consumed. Some households complained about this because they were not clear about the relation between water consumption and waste production; secondly, they use a pit in their compound to prepare compost and sometimes burn it, dispose their waste on municipal skip, and still others argue that they use the waste as fuel to cook food; thirdly they are also worried that water meter readers do not come to read the gauge every month. This could have an effect on the amount of money they paid as water payment was calculated based on a progressive consumption pattern.

Economic status of the household	Household readiness to pay for collection fee			Total
	Low	Medium	High	
Low income	18	24	57	99
	18.20%	24.20%	57.60%	100.00%
Middle income	22	36	99	157
	14.00%	22.90%	63.10%	100.00%
High income	12	29	83	124
	9.70%	23.40%	66.90%	100.00%
Total	52	89	239	380
	13.70%	23.40%	62.90%	100.00%

Table 2: Household affordability to pay for solid waste collection by income.

The affordability can be derived from the payment rate expressing the expenditure for primary solid waste collection as a percentage of the income standard.[3] For the different households (low, middle and high income) it is 0.8%, 0.4% and 0.25% respectively. The result is that low income households are paying more for the solid waste collection service than high income households in the city as a percentage of their income.

Finally concerning the willingness to pay Table 3 shows the responses of the households. City wide about half of the respondents (51.8%) are willing to pay more for the service improvement, while 48.2% were not willing to pay. The amount of money respondents were willing to pay towards improved collection service ranges from 2 to 100 Ethiopian birr with mean of 18.46. Moreover there is clear difference among different economic groups concerning the willingness to pay for improved service; the willingness to pay more is 72%, 51% and 28% for high, middle and low income households respectively.

The underlying factor that affects peoples' willingness to pay for improved service is the low wages of micro-enterprises workers. The payment to micro-enterprises is based on the volume of waste they collected (30 birr/ m3 of waste). Respondents complained that the payments do not serve the mutual interests and benefits of the service provider and service receivers. Moreover households felt that the collected fee was not properly utilized and they consider what they pay is not properly paid to the waste collectors. This notion was substantiated in the community by the idea that workers are working under poor conditions. They are not protected from any exposure of risks and use only elementary equipment.

Economic status of the household	Willingness to pay more for improved service		Total
	Yes	No	
Low income	28	71	99
	28.30%	71.70%	100.00%
Middle income	83	79	162
	51.20%	48.80%	100.00%
High income	88	35	123
	71.50%	28.50%	100.00%
Total	199	185	384
	51.80%	48.20%	100.00%

Table 3: Willingness to pay more for improved service by household income group.

Some households argue that they have paid additional money to part-time workers as the established system is not effective to manage cleanliness of their surroundings. The responsible persons from the local authority do not monitor what is going on at household level and they focus on dealing with matters related to secondary collection only. The poor sanitary conditions of the slum areas also sparked complaints and induced non-payment from households. Some residents argued that they did not sense any significant change in waste collection in terms of service quality that encouraged them to make additional payment.

Effective demand for waste collection services and customers satisfaction were used to assess whether the service provision was oriented to the customer preference variables. Table 4 shows households' demand for solid waste collection service in Addis Ababa. 60% of the households joined micro-enterprises as service provider more than three years ago while 40% less than three years ago. Why they decided to use micro-enterprises as service provider? 42% of the households said that the demand for waste collection was driven by the community. Informal enterprises in collaboration with households initiated a system, which gave rise to formal micro enterprises. 39% of the households said they were advised by the local governments (Kebele) to use micro enterprises for primary solid waste collection service. They claimed that the demand for the service was not governed by the interest of an individual household but the households just do what they are told to do by the local government.

---

[2]   Collection cost from the total solid waste management accounts 30-50%. Collection Cost as percentage of income is 0.9-1.7%, 0.2-0.4% for Low-income, Middle-income and Industrialized Country respectively (Cointreau, 1994).

[3]   In this case the cut-off is less than 600 (low income), 600-2000 (middle income) and greater than 2000 birr(high income).

10% of the respondents said as dumping of waste by households to communal containers was banned they adhered to the new procedure that forced them to use micro-enterprises only.

Some households argue that concerning waste collection service by micro-enterprises there was no discussion and communication with the community. Households were not informed about the system and do not own the process. This made them passive. Households complain that they have lost the power to monitor and control the service providers as the government pays and monitors them.

Period in Service	Frequency	Percent	Valid Percent	Cumulative Percent
Less than a year	61	14.9	15.1	15.1
1-3 years	99	24.1	24.6	39.7
More than three years	243	59.3	60.3	100
Total	403	98.3	100	

Table 4: How long the collection service has been provided to household by the micro-enterprises.

Different views were reflected by the households as to why households preferred the service to be provided by micro-enterprises. Some of the factors raised by the households were: appropriate for the quality of the roads, affordable, involved common people, they were manageable by the community, which is not able to deal with big companies.

Satisfaction Level	Frequency	Valid Percent	Cumulative Percent
Very satisfied	29	7.4	7.4
Reasonably satisfied	291	74	81.4
Not satisfied	64	16.3	97.7
Don't know	9	2.3	100
Total	393	100	

Table 5: Customer Satisfaction Level of primary waste collection Service.

Table 5 shows the rating of the quality of primary solid waste collection service. 74% of the respondents said that they were reasonably satisfied as the service provision is much simpler, manageable and suited the context. 16% are not satisfied with the service while 7% are very satisfied.

Respondents' dissatisfaction was explained by concerns about the interface of primary and secondary collection performed by private micro-enterprises and the public sector, which is not well tuned. The new procedure (contract system) resulted in shifting the problem from the city level to the households. As the respondents explained during the franchise arrangement the waste problem was at city level and this has now become a problem of households since sometimes if the waste is not collected for weeks by micro-enterprises, whenever vehicle and container shortage is encountered, they are instructed to do so by the local government. This aggravated households' dissatisfaction about the collection service. The desired expectation of service quality was not met.

Public Health Sustainability

The analysis in this section covers the attitude and awareness of households of the quality of collection services. It was found during the study that the majority of households were responsible for household waste storage hygiene. That makes the work for collectors easier. In some cases poor storage conditions were found. Various factors such as non-standardization of household storage materials (their receptacles) and poor warranty to get back their original material from micro-enterprises urged them to use less value materials which subsequently affected the quality. Old worn out material is used by many households. Due to poor storage it is not uncommon to see waste littering around the neighborhoods.

The survey revealed that women are mainly responsible for solid waste management at home-level in the city of Addis Ababa. As indicated on Table 6 in most households these tasks are largely allotted to mothers and maid servants accounting for 41% and 39% respectively. They gather the household waste and in some areas push carts are used to deliver it to micro-enterprises.

	Frequency	Valid Percent	Cumulative Percent
Father	5	1.2	4.9
Mother	168	41	45.9
Son	8	2	47.8
Daughter	51	12.4	60.2
Maid	159	38.8	99
Guard	4	1	100
Total	410	100	

Table 6: Who usually manages and throws out the waste at the household level?

The residents were asked whether there is a waste disposal problem in their neighborhood. The survey revealed that 68% of the respondents said there is waste littering in their surrounding while 32% said no. Residents complained about health and safety issues associated with the poor solid waste collection system. Littering was observed during loading of waste on push carts and while transporting. Poor cooperation of residents, poor storage practice of some households resulted in low efficiency of collection and waste remained uncollected on the streets, in open spaces, ditches, roads and rivers and streams. Littering of waste in the neighborhood is common in the city as citizens drop it during the evening illegally. Furthermore it is observed illegal slaughtering and throwing waste in the ditches causes ditches to be clogged and to release an offensive smell. This caused environmental and health hazards, blockage of sewer lines, odour and flies nuisance and aesthetic degradation.

Type of waste related neighborhood environmental problem	Evaluation			
	Yes		No	
	Frequency	%	Frequency	%
Dirty Street	343	86.8	52	13.2
Rubbish heap	75	19	319	81
Rubbish in open drain	339	86	55	14

Rubbish fire	26	6.6	365	93.1
Flies and Vermin nuisance	203	51.5	191	48.5

Table 7: Waste related environmental problems at neighborhoods as rated by the respondents.

Respondents rated waste related environmental problems of the neighborhoods. As shown in Table 7 dirty streets and rubbish in open drains are the major neighborhood waste related problems in the city followed by the nuisance of flies and vermin created by improperly disposed wastes to the surrounding. Households were asked to evaluate their surrounding cleanliness using different items related to the neighborhood waste collection system. Numerical values are assigned to each response in the following manner: A = 4; B = 3; C = 2. Note that the higher the value of the index the cleaner the neighborhood.

Item	Numerical value of rated Mean
The household area looks like pleasant residential area	2.9484
Loose trash are visible in the surrounding area	2.723
Waste is neatly collected and transferred to secondary collection site	3.1642
Neighborhood Cleanliness Index Value	2.95

Table 8: Neighborhood cleanliness Index of Addis Ababa.

The survey indicates that neighborhood cleanliness index of Addis Ababa was 0.74. During field work It was evident that cleanliness variation among neighborhoods exists. There are developed areas whose cleanliness is OK, but also areas whose conditions were very poor.

Solid waste is dumped at different sites and in drains and other unauthorized places. Households were asked to evaluate the city cleanliness. As indicated in Table 9 the survey reveals that the city cleanliness index of Addis Ababa was 0.6. It was much less than the neighborhood cleanliness index and this is due to lack of public awareness of the health implications of unsanitary practices and residents' indifference to the presence of waste. Individuals put waste in plastic bags and throw it in the narrow streets or under the fence of other households. In some crowded and slum neighborhoods there are no toilet facilities nor at individual nor at the communal level. Hence people are compelled to use plastic bags and/or bowls meant for defecation.

Dust bins are availed in most central parts of the city either as standing on the road or mounted on lamp posts or telegraph poles and strategic sites at points where litter was most likely to be thrown such as bus stops, shopping places and walk streets. However, people do not use dust bins properly, instead they throw bus tickets, soft paper, chewing gums, etc. on the street and are not held responsible for their trash. It was not uncommon to see dust bins used for unintended purpose in the city filled with improper materials like bones and leaves.

Item	Numerical value of rated Mean
Loose trash; paper and plastic visible in the city open areas	2.4532
Loose trash; paper and plastic visible in the city street parks	2.6575
Loose trash; paper and plastic visible in the city common areas	2.4701
Solid waste is visible in the city water bodies	2.2125
Solid waste is visible in the city open ditches	2.2857
Solid waste is visible in the city transfer stations	2.3401
Street and sidewalks are dusty and dirty	2.4543
City Cleanliness Index Value	2.41

Table 9: City Cleanliness Index of Solid waste in Addis Ababa.

Some households argue travelers and migrants are negligent about the cleanliness of the city as they lack ownership and awareness. The responsible body in the city did not fulfill the necessary facilities for them that made them not to adhere to the sanitation rules and regulations in place. In some cases it is observed that many of the public facilities are becoming overburdened due to an increasing population or becoming dysfunctional due to a lack of maintenance. The situation is aggravated by the presence of ownerless dogs that disperse the collected waste at the transfer station while searching for food. It was also observed that the wind disperses the dirt especially thin plastic bags from filled bins, which spoils the cleanliness of the city. Municipal skips from every corner of the city travelled long distance to be dumped on the Repi[4]

Some respondents relate the current situation to the structural problems of the city. Although there are modern settlement areas historically it was common to see houses with different qualities stacked together in the same neighborhood of Addis Ababa. This makes it difficult to develop a shared vision among the residents.

Ecological sustainability

Ecological sustainability concerns household participation in recycling and composting and the intention to support recycling by practicing in separation at source. Materials in the household waste stream which are capable of being recycled in Addis Ababa city can be

---

4   The Repi disposal site, nick-named 'Koshé', is the only landfill site to date located 13 km away from the city centre. This site has been in service since 1968. The site has low area capacity (35 ha) with poor road connection. The present method of disposal is crude open dumping; hauling the waste by truck, spreading and levelling by bulldozer and compacting by compactor bulldozer. It has become full to capacity and the identification of another alternative landfill is already too late. The present situation shows that there are settlements clustered around the site and public health is at risk. The newly constructed Ring Road is too close to it, which makes it a nuisance due to bad smell and pollution in general. The gas generated from landfill causes spontaneous combustion and air pollution in the surrounding area. It contributes enormous amounts of methane and carbon dioxide (green house gas) to the atmosphere. There are more than 800 waste scavengers who permanently browse through the heap of waste looking for recyclable and usable items on the site.

categorized as dry recyclables, comprising paper, plastics, metals, glass and textile etc. and organic material, consisting of kitchen (food) and garden waste. Both categories of material are recycled in the city, but the method of collection and treatment are quite different. We observed that households dispose certain items in the waste stream in particular non commodity materials (kitchen waste, garden waste, livestock waste). Items that have a intrinsic commodity value (glasses, cans, rubbers, plastics, low grade paper) are locally traded. These materials are bought by itinerant buyers (korale and liwach). The 'Koryalew'[5] and 'Liwatch'[6] are traders who travel from door-to-door and buy/exchange reusable item (tins, plastics, bottles, nail varnish containers, broken cooking jars, used shoes, old garments, etc.) and supply them to the middle men at Mercato.

These traders have a huge impact on the reduction of solid waste both at household and city level. Although they have a well-established economic system and a market niche in the city they were less recognized by the public sector for their contribution to recycling and reuse. There are industries in Addis Ababa that reuse the recovered materials such as paper, glasses, plastic, iron and steel rod and pieces tin, etc. as raw material. Moreover, a large proportion of reusable materials are transported to small towns and rural areas to be used as household items (plastic bottles for storing water, used cans to store materials, reuse of plastic bags etc.).

Table 10 shows that 80% of the respondents are aware of recycling, while only 13% exercised recycling. Some households' perception of recycling was from the view point of organic wastes only not aware of reuse and recycling activity from the point of view of material recovery. Most households failed to recognize channeling of materials at household level to small scale collectors serves as precursor for supplying of material input to recycling and reuse process.

	HHs have information about recycling		HHs exercising recycling	
	Frequency	Valid Percent	Frequency	Valid Percent
Yes	325	79.5	49	13
No	84	20.5	329	87

Table 10: Households awareness and practice of recycling.

Households reuse/recycle materials for various reasons: economic, environmental concern and both economic and environmental concern. As indicated on Table 11 the survey revealed that 30% of the households recycle waste purely for economic reasons, 28% for environmental concerns and 33% both for economic reason and environmental concern. Interviews with households disclosed that reuse and recycling practice in the city vary with the economic status of the households. There was a growing realization among low income households that household waste should be recycled. The practice of reuse and recycling was attributed to economic and social necessities resulting from poverty and deprivation rather than to environmental considerations.

The positive attitude towards waste re-users and recycling was generally high despite a lack of a formal organized arrangement for source separation and collection of these recyclables. During field

study it was observed that refuse collection workers (micro-enterprise workers) rummaged through refuse for valuables. They collected and sold these recyclable and reusable materials for middle men at the secondary collection site. Buying and selling of recyclable materials also takes place at the transfer stations where micro-enterprises bring all recyclable materials from the households and sort them to sell to traders on the spot.

	For economic interest		From environmental concern		For both economic and environmental concern	
	Frequency	Valid Percent	Frequency	Valid Percent	Frequency	Valid Percent
Yes	121	30	111	27.6	133	33.1
No	283	70	291	72.4	269	66.9

Table 11: Households perspectives of recycling in Addis Ababa.

Organic waste accounts for 70% of the total. The daily generated solid waste is more than 2000 m^3 whereas the annual generated waste amounts to be around 1,000,000 m^3. Despite the huge potential currently only 6% of the solid waste reported used for making compost [27]. Most respondents claimed this to be due to a lack of space to practice gardening. Some said they do not have a market if they produce compost. While others said that they have heard about composting but they did not see anything done so far that would encourage households to participate. Some complained that they neither got support from a concerned body nor have capacity and knowledge to compost.

According to the Refuse Collection and Disposal bye-laws of Addis Ababa city, 2001 Section 4 (1 and 2) and Section 5, all households are required to have two solid waste collection receptacles (one for organic and the other for non-organic waste) of not less than 40 litres fitted with a lid. The two types of waste receptacles are meant to facilitate sorting of waste at the source. It has been noted during the survey that this bye-law has not been adhered to.

90% of the respondents do not practice separation at source. Table 12 shows that 47% of the respondents don't have awareness about waste separation. 23% argue that there is no point of separating waste since as collectors they use only one bag per household. 12% believed that they have no reason to do it as they lack the facility. More education should create the awareness about separation at source.

	Frequency	Valid Percent	Cumulative Percent
It is small	42	10.2	20.7
No facility	47	11.5	32.2
No awareness	183	44.6	76.8
No incentive	95	23.2	100
Total	410	100	

Table 12: Why there is no source separation at Household Level.

---

5   'Koryalew' is an abbreviation for Korkoro-yalew meaning him or her who has tin-made discarded materials for sale.
6   'Liwatch' means old used materials to be swapped with other usable items and includes old clothes, shoes ... etc and in return to take new plates, plastic-made cups, kettles, small containers ... etc.

However there is an effort made by individuals at household level to separate waste for different purposes. There are households that have been trained by government, civil society and community based organization to separate the organic part of the waste and use compost in their compound. Low income households separate waste and use it for different purposes (like energy). Some households separate some types of waste (broken glass, ash) at their own initiative just to ease the task of micro-enterprises as some waste types are difficult to handle with other wastes together.

To understand values and behaviour of residents concerning solid waste management, a factor analysis was done. Table 13 simplified the data by grouping them under two common factors.[7] Factor 1 was designated as 1st and 2nd motivation which include health, sanitation, aesthetic, cleanliness of streets and open areas and city image, since it was generally opted for by certain categories of respondents. Factor 2 was named as 3rd motivation which covered natural resource depletion and trans-boundary effects caused by improper management of solid waste. It became evident that there were certain differences in the perception of these two groups (table 13 and figure 2).

	Component	
	1	**2**
Personal health as most urgent problem related with AA solid waste	0.552	-0.3
Risk of epidemic as most urgent problem related with AA solid waste	0.649	-0.336
Littering of environment and inconvenience due to odour	0.628	-0.067
Reduce aesthetic of the city	0.759	-0.076
Create negative impact to the city image	0.783	-0.093
Affect cleanliness of the city streets and open areas	0.676	-0.07
Causes air pollution	0.641	0.13
Causes surface and underground water pollution	0.643	0.104
Causes natural resource depletion	0.315	0.744
Affect the sustainability of local and regional environment	0.313	0.699

Table 13: Rotated Component Matrix. Extraction Method: Principal Component Analysis. Two components extracted.

The results implied that residents' perception of solid waste problem at the city level was in the transition between first and second order motivation. People are more concerned about second motivation; aesthetic, cleanliness of streets and city image. Indeed the concern given by the residents for first order motivation; health and sanitation was also considerable.

Figure 2 indicates that solid waste minimization as a strategy to reduce natural resource depletion and its impact in lessening local and regional environmental effects are valued differently by the community. Waste minimization/reduction as a principle was

experienced less as the knowledge and support to the system was non-existent at household level in the city.

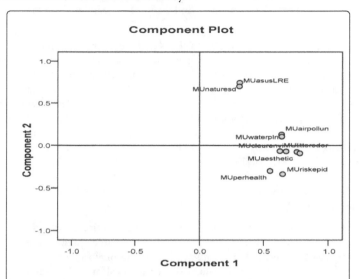

Figure 2: Component plot of solid waste related environmental problems. *MUriskepid: risk of epidemic as most urgent solid waste problem; MUperhealth: personal health as most urgent problem related with solid waste; MUaesthetic: reduction of the city aesthetic due to solid waste; MUlitterodor: littering of the city environment and odour due to solid waste; MUcleurenvi: cleanliness of the city due to solid waste problem; MUwaterpln: water pollution problem due to solid waste; MUairpollun: air pollution problem due to solid waste; MUnaturesd: natural resource depletion due to poor solid waste management; MUasusLRE: effect on sustainability of local and regional environment due to solid waste problem.*

Discussion

The institutional relationships in the solid waste management system have been undergoing significant shifts in Addis Ababa, influenced by decentralization and micro-privatization policies creating an alternative service delivery mechanism to improve municipal solid waste management. Local government, community and private sector are now more involved. This collaboration ensures that each and every move fits into an integrated and sustainable system designed for promoting service satisfaction, public health, environmental protection, and good governance.

Despite private sector involvement in solid waste management in developing countries in the past two decades, there are still problems with solid waste management services [12]. The micro-privatization process in Addis Ababa waste collection strategies using groups of entrepreneurs seems quite effective in maintaining the sustainability of the system from various perspectives: economic status of the residents, the nature and type of waste generated in the city, the capacity of local governments to administer the privatization process, availability of resources and technologies to the public sector should be mentioned. Micro-enterprises operating in a low-income neighborhood have a definite advantage because trucks are unsuited to the hilly, unpaved or

---

[7]   The factor solution was derived from the component analysis with VARIMAX rotation of the 10 environmental problems related with solid waste listed for the purpose of the study. Factor 1 has 8 significant loading while 2 problems listed significantly under factor 2.

narrow streets that are common in the majority of settlement areas. Moreover micro-enterprises are location specific, they are responsible to the area where they function, and keep the cleanliness of the surrounding.

Decentralization of the service to the Kebele level makes it much more manageable and helps to increase efficiency as the Kebele is the smallest unit of the Mass Organizations. It is therefore evident that SWM in Addis Ababa has improved the sustainable provision of solid waste collection service as the reform measures included user financing, proper organization of the responsible bodies, encouraging and promoting of efforts of the other stakeholders that are helpful in alleviating the problems, promoting the SWM Office to become an autonomous player. The Kebeles are acting as bridge connecting the community with the Municipality. Moreover they organize and coordinate the activities of micro-enterprises and community-based organizations (CBOs) pertaining to SWM.

The newly formed institutional arrangements require a different approach to governance to mainstream sustainable solid waste collection. More research is required on institutionalization of the reform (decentralization of service delivery to local authorities and micro-privatization of the service) to ensure the sustainability and appropriateness of the system as an alternative service delivery mechanism. Further research would be necessary to evaluate sustainability from the providers' and regulators' perspectives as this study is limited to users' perspective only.

Conclusions

The study evaluated households' attitude and behaviour concerning sustainable solid waste collection. Sustainability has been considered from the socio-economic, public health and ecological, or the three Ps perspective. Various indicators and the underlying factors were considered under socio-economic, public health and ecological sustainability principles. The research tried to explore the indicators and the underlying factors influencing the sustainability of solid waste collection service provision scheme from the users' perspective.

The study uncovered several important factors that are associated with sustainable solid waste collection. In the case of socio-economic sustainability: reliability, affordability, service coverage, customer support services and willingness to pay for improved service are the variables considered. The results of this study suggest that the service reform suited the interests of the city community in undertaking the service provision function in a sustainable manner. Regularity, reliability, service coverage and frequency of service delivery to the households have improved. Residents have a good feeling about cost recovery, as micro-enterprises require less money and are more cost effective even though there is variation of attitude on cost recovery among different economic groups.

With regard to public health sustainability improvements were observed concerning the cleanliness of the neighborhoods as compared with the city cleanliness. Households are sympathetic to neighborhood level sanitation. The sanitation level of neighborhoods has apparently improved implying that micro-enterprises are taking their responsibility progressively. However, city sanitation suffers largely from the public-private interface in the system. The dreadful attitude and lack of co-operation of inhabitants at the city level account for the irresponsible littering behaviour of transitory people explained by the problem of migration, travelers and traders, the city structure, ownership issues, illegal functions in the city, awareness,

prisoners' dilemma, and local versus city sanitation discourses (procedurally, administratively, and perception wise). Poor monitoring capacity of local administration is also the potential contributor to the problem.

In the case of ecological sustainability even if the awareness level of residents for separation at source is low, economic incentives play a more important role than ecological considerations in separating and channeling reusable and recyclable items from the waste stream at household level. Reuse and recycling of waste in Addis Ababa is mainly undertaken by micro-enterprises. Massive market transactions of recyclable materials take place at the transfer stations, where micro-enterprises bring all recyclable materials from the households and sort them to sell them on the spot. Though there is no legitimate strategy and regulations of methods of collection to be employed, micro-enterprises act informally and offer stable supplies of material for reprocessing. However, the markets for some types of recycled materials are not considered sufficiently, since without this market the recycling loop cannot be closed. The reuse and recycle practice of waste at household level is too low. Only low income households exercise recycling activities because of economic reasons. Waste minimization as a principle is not really adhered to as there is no knowledge of and support for such a system at the household level. These results have implications for further studies. More research is required to consider sustainability not only from the collection perspective but also by looking at solid waste management as a system.

References

1. Sibley J, Hes D, Martin F (2003) A triple helix approach: An inter disciplinary approach to research in to sustainability in outer-suburban housing estates. Methodologies in housing research. Stockholm.

2. Tilaye M, van Dijk MP (2014) Private sector participation in solid waste collection in Addis Ababa (Ethiopia) by involving micro-enterprises. Waste Manag Res 32: 79-87.

3. Van Dijk MP, Tilaye M (2013) Micro-privatization of solid waste collection service, the case of Addis Ababa city. In Water lines 32: 154-162.

4. Oduro-Kwarteng S, van Dijk MP (2013) The effect of increased private sector involvement in solid waste collection in five cities in Ghana. In: Waste management & research 31: 81-93.

5. Huysman M (1994) The Position of Waste Pickers in Solid Waste Management in Bangalore. Rotterdam: IHS of Erasmus university.

6. Baud I, Post J (2002) Between market and partnership: Urban solid waste management and contribution to sustainable development. GBER 3: 46-65.

7. Tadesse K (2004) Dry Waste Management in Addis Ababa City. Accounting for Urban Environment. Ethiopian Development Research Institute, Addis Ababa, Ethiopia.

8. Schubeler (1996) Conceptual Framework for Municipal Solid Waste Management in Low Income Countries.

9. Sattherthwaite D (1997) Sustainable Cities or Cities that Contribute to Sustainable Development? Urban Studies 34: 1667-1691.

10. Van der Klundert A, Anschiitz J (2000) The Sustainability Of Alliances Between Stakeholders In Waste Management.

11. Zurbrugg C (1999) Solid Waste Management in Developing Countries.

12. Onibokun AG, Kumuyi AJ (1999) Governance and Waste Management in Africa.

13. Oduro-Kwarteng S (2011) Private Sector Involvement in Urban Solid Waste Collection: Private Sector Performance, Capacity and Regulation in five Cities in Ghana. Rotterdam: Erasmus University.

14. Van der Klundert A, Lardinois I (1995) Community and Private (formal and informal) Sector involvement in Municipal Solid Waste Management in Developing Countries. Waste.

15. Bell S, Morse S (1999) Sustainability Indicators. Measuring the immeasurable. Earthscan, London.

16. Simon S (2003) Sustainability Indicators. International Society for Ecological Economics.

17. Žickiene S (2005) Municipal Solid Waste Management: Data Analysis and Management Options: Environmental research, engineering and management 3: 47-54.

18. Harper M (2000) Public Services Through Private Enterprise: Micro-privatisation for improved delivery London UK.

19. Sheinberg A (2003) Privatization and the Informal Sector: Thinking locally, acting globally?

20. You N, Allen A (2011) Sustainable Urbanization: Bridging the Green and Brown Agendas, University College London.

21. Tchobanoglous G (1993) Integrated solid waste management: Engineering principles and management issues McGraw-Hill, New York.

22. Lardinois I, Van der Klundert A (1997) Integrated Sustainable Waste Management.

23. Wilson E, McDougall FR, Willmore J (2001) Euro-trash searching Europe for a more sustainable approach to waste management. Resources Conservation and Recycling 31: 327–346.

24. Cointreau-Levine S (1994) Private Sector Participation in Municipal Solid Waste Services in Developing Countries.

25. Addis Ababa Sanitation, Beautification and Parks Development Agency (AASBPDA) (2005) Guideline for Delimiting Service Areas for Micro and Small Enterprises Engaged in Solid Waste Collection and Transportation. Addis Ababa (Amharic Version).

26. Dierig S (1999) Urban Environmental Management in Addis Ababa: Problems Policies Perspectives and the Role of NGOs.

27. Knowledge, Attitude, Perception and Behaviour (KAPB) (2009) Survey of Addis Ababa Sanitation. Addis Ababa.

28. Baudouin A (2009) Waste Management and Governance in Addis Ababa.

6

A Bibliometric Analysis on Acidophilic Microorganism in Recent 30 Years

LI Si yuan[1,2], Hao Chun bo[1,2*], Feng Chuan ping[1,2], Wang Li hua[1,2] and LIU Ying[1,2]

[1]*Key Laboratory of Groundwater Circulation and Evolution of Ministry of Education, China*

[2]*School of Water Resources and Environment, Beijing 100083, China*

*Corresponding author: Hao Chun bo, Key Laboratory of Groundwater Circulation and Evolution of Ministry of Education, China University of Geosciences, Beijing 100083, China
E-mail: mywenyan@gmail.com

Abstract

Objective: To evaluate the global scientific research and the tendencies on acidophilic organism during the past 30 years. Studies in the acidophilic microorganism had significantly increased.

Methods: Articles related with the acidophilic organism were assessed by distribution of countries, institutes, journals using the method of bibliometric analysis.

Results: The results showed seven industrialized countries and four major developing countries were all listed in the top 20 most productive countries, which suggested economic conditions had an important effect on academic development. In addition, researchers in different institutions were more tending to cooperate. However, cooperation always occurred in the interior of the country. Through a synthetic analysis of the paper titles, author keywords and Keywords Plus, it revealed that "resistance to metal" attracted more attentions. Besides, this characteristic was also widely applied in bioleaching. At the level of research environment, "water" was the dominant position, such as acid mine drainage.

Conclusions: Study in the acidophilic microorganism had significantly increased. Cooperation had become the trend. Moreover, economic conditions had an important effect on academic development.

Keywords: Acidophilic microorganism; Bibliometric anaylsis; Resistance to metal; Bioleaching; Diversity

Introduction

Many microorganisms survive in the physically and geo-chemically extreme conditions, which have challenged the limits of life. These conditions include extremes of temperature, pH, pressure, desiccation and others [1-3]. These microorganisms are termed as extremophiles. In recent decades, extremophiles have aroused great interest to researchers. Among them, acidophilic microorganisms are the ones that thrive in acidic environments with pH less than 3.0 [4,5]. They widely exist in acid mine drainage, bioleaching operation, and sulfuric hot spring [6-8]. Acidophilic microorganism not only can adapt to the environment of strong acid, but also can tolerate the high concentration of metal ions. Due to these special characteristics, they have been the hotspot in life science, and lots of microorganisms have been studied deeply, such as the *Acidithiobacillus ferrooxidan* [9]. Besides, acidophilic microorganisms have been widely applied in many respects, such as the bioremediation of heavy metal contaminated soil and water, extraction of enzyme, bio hydrometallurgy and others [10-12].

Scientific articles on acidophilic microorganism have demonstrated a rapid increase over the past several decades. A number of papers presenting the latest research achievements have been published in authoritative scientific journals such as Nature and Science [13,14]. Despite the high growth rate of publications, there have been few attempts to gather systematic data on this special microorganism. A common research tool for this analysis is the bibliometric method [15-17], which has already been widely applied in scientific production and research trends in kinds of topics, for example, global diversity [18], energy efficiency [19], agricultural technology [20], solid waste [21] biotechnology research [22,23]. The Science Citations Index Expanded (SCI-EXPANDED), from the Institute of Scientific Information (ISI) Web of Science databases, is the most important and frequently used source for a broad review of scientific accomplishment in all fields. Traditional bibliometric methods focus on citation and content analysis [24,25]. In recent years, analysis of word distribution of paper titles [26], KeyWords Plus [27], author keywords [28] in different periods has been used widely to get more information related to the research itself.

In this study, a bibliometric analysis of language, source country, institute and research field was performed to describe the importance of research on acidophilic microorganism. Besides, the distributions of keyword were also analyzed to study the research trends during the recent 30 years. Our conclusions not only provided a better understanding of global hotspot for researchers, but also clarified the future research direction on acidophilic microorganism.

Data Sources and Methodology

Data used in this research were based on the online database of the SCI, retrieved from the ISI Web of Science, Philadelphia, USA. According to Journal Citation Reports (JCR), it indexed 7391 major journals with citation references across 173 scientific disciplines in 2011. Besides, the reported impact factor (IF) of each journal was acquired from the 2011 JCR. Here, five search terms, including: "acidophilic organism", "acidophilic microorganism", "acidophilic bacteria" "acidophilic fungi" "acidophilic archaea" were used as keywords to search titles, keywords, document types, addresses and others during 1983-2012. Articles from England, Scotland, Northern Ireland and Wales were reclassified into the UK, and articles originating from Hong Kong and Taiwan were included in China. The collaboration type was determined by the addresses of the authors. The single-country publication was classified if the addresses of authors were in the same country. On the contrary, the internationally collaborative publication was assigned if authors were from different countries. In the same way, the single-institute publication was assigned if the addresses of researchers were from the same institute. The inter-institutionally collaborative publication was assigned if authors were from multiple institutes.

The words in titles were separated, and then conjunctions and prepositions such as "and", "the", "or", "for", "with", "by", and "on" were discarded, as they were meaningless in the further analysis. The ranks and frequencies of keywords between 1983 and 2012 were calculated in order to thoroughly analyze the variations of trends.

Results and Discussion

Characteristics of publication outputs

In this study, 10 document types were contained in the 2199 publications during the 30-year study period, in which article was the most frequent type. Articles contributed a significant portion, 85% or 1873 of the total production. Another two documents were less significant, including proceedings paper (6.5%) and review (4.9%). The devotion of others was rare. As the dominant type of document, articles were used for further analysis. Ninety-eight percent of the journal articles were written in English. Another 12 languages also appeared. However, the proportion was less than 1%. Obviously, English was the dominant language in acidophilic microorganism research.

The number of both of SCI documents and articles was analyzed and performed respectively in Figure 1 to understand the research trend in the 30 years. World academic publications had a notable growth after 1990, and there were two notable increases in the past three decades (1990, 2003). One important reason was that United States proposed the concept of the human genome in 1985, and in 1990 the 3-billion dollars project was formally founded. "Human Genome Project" intended to complete the entire human genome DNA sequence analysis in 15 years [29]. As a result, high-speed DNA sequencing methods came into being, which encouraged the study in microorganism on molecule level. Subsequently, high demand for low-cost sequencing had driven the development of high-throughput sequencing and several new methods for DNA sequencing were developed [30]. These techniques comprised the first of the "next-generation" sequencing methods, which was applied in many studies. Such as massively parallel signature sequencing (MPSS) published and marketed in 2000, 454 Life Sciences marketed in 2004 [31].

Based on the classification of subject categories in JCR 2011, the publication output data of acidophilic microorganism research were distributed into 109 SCI subject categories during 1983-2012. The most significant category was "Microbiology" (20%), followed by "Biotechnology and Applied Microbiology" (13%), "Biochemistry and Molecular Biology" (8.2%) and "Environmental Sciences" (3.9%).

"Microbiology" and "the Biotechnology and Applied Microbiology" were the two fastest growing subject categories, especially after 2001 (Figure 2), suggesting that the value of microorganism had been applied to the practice of production gradually, and molecular technology had been widely used in the study.

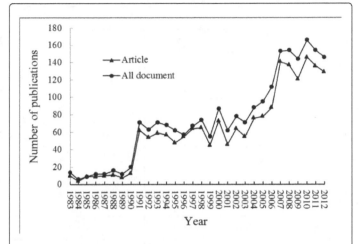

Figure 1: Trends of SCI-EXPANDED publications referring to acidophile during 1983-2012.

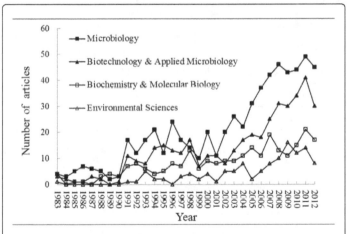

Figure 2: Comparison of the growth trends of the top four productive subject categories.

Articles were published in lots of journals, and the top 20 most productive journals were summarized in Table 1. In this special study field, Applied and Environmental Microbiology published the most articles with (107; 5.7%), followed by International Journal of Systematic and Evolutionary Microbiology with (62; 3.3%), Hydrometallurgy with (48; 2.6%). The average citation rate of journals in acidophilic microorganism was the most direct evidence for indicating the impacts of journal: the higher the citation was, the greater the impact was in this field. Besides, the average cited reference number per article also showed the importance of each article. International Journal of Systematic and Evolutionary Microbiology had the highest citation rate among the 20 journals. And Environmental Microbiology shared the highest average cited reference and the supreme IF. Those manifested the significant position of the two journals in this field.

Journal	TP	TP (%)	IF	TC	TC/TP	NR	NR/TP
Applied and Environmental Microbiology	107	5.7	3.829	3647	34.1	4218	39.4
International Journal of Systematic and Evolutionary Microbiology	62	3.3	2.268	2127	34.3	2158	34.8
Hydrometallurgy	48	2.6	2.027	511	10.6	1427	29.7
Extremophiles	47	2.5	2.941	677	14.4	1647	35
Journal of Bacteriology	35	1.9	3.825	732	20.9	1012	28.9
Microbiology	35	1.9	0.718	99	2.8	657	18.8
FEMS Microbiology Letters	35	1.9	2.044	649	18.5	643	18.4
Bioresource Technology	24	1.3	4.98	189	7.9	777	32.4
Environmental Microbiology	23	1.2	5.843	723	31.4	1210	52.6
Archives of Microbiology	20	1.1	1.431	299	15	623	31.2
Geomicrobiology Journal	20	1.1	2.017	313	15.7	839	42
World Journal of Microbiology & Biotechnology	18	1	1.532	113	6.3	457	25.4
FEMS Microbiology Ecology	18	1	3.408	508	28.2	734	40.8
Systematic and Applied Microbiology	17	0.9	3.366	469	27.6	495	29.1
Canadian Journal of Microbiology	17	0.9	1.363	197	11.6	560	32.9
Minerals Engineering	16	0.9	1.352	227	14.2	449	28.1
Microbiology-Sgm	16	0.9	3.061	496	31	771	48.2
Bioscience Biotechnology and Biochemistry	15	0.8	1.276	312	20.8	454	30.3
Microbial Ecology	14	0.7	2.912	162	11.6	673	48.1
Biotechnology and Bioengineering	14	0.7	3.946	268	19.1	411	29.4

Table 1: The top 20 most productive journals based on the total number of articles. Note: TP: total number of articles, IF: 2011 ISI Impact factor, TC: total citation count, NR: cited reference count, TC/TP: average of citations in a paper, and NR/TP: the average cited reference count per article.

Distribution of country articles

All articles with author addresses could be used to analyze the distribution of country. In 1725 articles with author addresses, 71.6% were single-country publications, and only 28.4% were international. The top 20 most productive countries were summarized in Table 2, ranking with the way of number of journal articles and total citations. Among the 20 countries, the USA was the most productive country, including both single-country articles (214) and internationally collaborative articles (135). Germany ranked second with 201 and Japan ranked third with 190. Economic condition was related to the academic achievement: the seven industrialized countries (G7 group: the USA, Germany, Japan, France, the UK, Canada, and Italy) and four major developing countries ("BRIC": China, India, Brazil, and

Russia) were all included in the top 20 countries [32]. The same phenomenon was revealed in other bibliometric analyses [33,34]. Apart from the economic condition of different countries, one typical acid mine drainage located in USA, Iron Mountain Mine. As one of America's most toxic waste sites, it had been listed as a federal Superfund site since 1983, which made it be one research hot spot [35,36]. Between position of the "G7 group" and "BRIC", Spain was another high-producing country, because another typical acid mine drainage, Tinto River, which was notable for being very acidic (pH 2) and its deep reddish hue, was located in Spain [37,38]. Another important message from these data in Table 2 was that single-country articles were the main trend in the study of acidophilic microorganism, especially in Japan, with a percentage of 90%.

Country	TP	SP	SP (%)	TC	TC/SP	CP	CP (%)	TC	TC/CP
USA	349	214	61.3	7487	35	135	38.7	2649	19.6
Germany	201	104	51.7	2141	20.6	97	48.3	2478	25.5

Japan	190	171	90	2463	14.4	19	10	279	14.7
China	158	118	74.7	676	5.7	40	25.3	459	11.5
UK	152	81	53.3	2175	26.9	71	46.7	1658	23.4
Spain	138	95	68.8	1546	16.3	43	31.2	570	13.3
Russia	97	51	52.6	253	5	46	47.4	1474	32
France	87	48	55.2	1128	23.5	39	44.8	750	19.2
Canada	81	54	66.7	942	17.4	27	33.3	402	14.9
India	81	65	80.2	855	13.2	16	19.8	68	4.3
Italy	62	41	66.1	673	16.4	21	33.9	417	19.9
Chile	54	27	50	391	14.5	27	50	391	14.5
Australia	46	21	45.7	350	16.7	25	54.3	311	12.4
Netherlands	40	20	50	509	25.5	20	50	295	14.8
South Korea	38	18	47.4	158	8.8	20	52.6	154	7.7
Sweden	35	13	37.1	284	21.8	22	62.9	313	14.2
Brazil	33	26	78.8	113	4.3	7	21.2	103	14.7
South Africa	30	19	63.3	305	16.1	11	36.7	205	18.6
Finland	25	10	40	98	9.8	15	60	112	7.5
Belgium	21	11	52.4	107	9.7	10	47.6	130	13

Table 2: Top 20 most productive countries based on the total number of articles. Note: TP: total number of articles, SP: single country articles, CP: internationally collaborative articles, TC: total citation count.

Distribution of institute analysis

The distributions of different institutes were evaluated by the affiliation of at least one author. Of all articles with author addresses, 43.2% were single-institute articles and 56.8% were inter-institutionally collaborative articles, suggesting that study of acidophilic microorganism called for teamwork among institutes. The top 20 most productive institutes were summarized in Table 3, in which the distribution was equal. Three were from China, the USA, the UK, Spain, two were from Russia, and one was from Germany, Japan, France, Sweden, India, Chile. The Russian Academy of Sciences had the most total articles (94), including 12 independent articles and 82 inter-institutionally collaborative articles, followed by Central South University, University of California, Berkeley, University of Chile and two institutes from Spain, Spanish National Research Council and the University of Murcia. It should be noted that

University of California, Berkeley had the highest average citation rate, including independent articles (238) and inter-institutionally collaborative articles (35.6). It suggested that University of California, Berkeley had a fairly high status in this academic field. It should be noted that the university was close to the typical acid mine drainage, Iron Mountain Mine, which was mentioned above. Another observation could be obtained from Table 3 that only four institutes (2 UK, 1 Spain, 1 India) had more independent articles than inter-institutionally collaborative articles. There were also another two institutes, Oak Ridge National Laboratory and Max Planck Institute for Terrestrial Microbiology that only published the inter-institutionally collaborative articles. Moreover, the average citation rate was relatively high. These proved that the academic communities of acidophilic microorganisms were more tending to cooperation.

Institute	TP	SP	SP (%)	TC	TC/SP	CP	CP (%)	TC	TC/CP
Russian Academy of Sciences, Russia	94	12	12.8	118	9.8	82	87.2	1551	18.9
Central South University, China	48	11	22.9	75	6.8	37	77.1	158	4.3
University of California, Berkeley, USA	42	4	9.5	952	238	38	90.5	1351	35.6
University of Chile, Chile	39	7	17.9	107	15.3	32	82.1	505	15.8

Spanish National Research Council, Spain	39	9	23.1	242	26.9	30	76.9	15	0.5
The University of Murcia, Spain	36	20	55.6	578	28.9	16	44.4	164	10.3
Chinese Academy of Sciences, China	36	7	19.4	29	4.1	29	80.6	102	3.5
Okayama University, Japan	27	13	48.1	142	10.9	14	51.9	242	17.3
Autonomous University of Madrid, Spain	26	4	15.4	46	11.5	22	84.6	313	14.2
University of Warwick, UK	25	14	56	491	35.1	11	44	425	38.6
Umea University, Sweden	24	5	20.8	144	28.8	19	79.2	212	11.2
Lomonosov Moscow State University, Russia	24	8	33.3	19	2.4	16	66.7	166	10.4
National Centre for Scientific Research, France	23	8	34.8	199	24.9	15	65.2	323	21.5
The University of Wales, UK	22	16	72.7	440	27.5	6	27.3	228	38
Indian Institute of Chemical Technology, India	22	20	90.9	580	29	2	9.1	16	8
Bangor University, UK	21	7	33.3	71	10.1	14	66.7	134	9.6
The Ohio State University, USA	19	2	10.5	54	27	17	89.5	161	9.5
Chinese Academy of Agricultural Sciences, China	19	1	5.3	12	12	18	94.7	182	10.1
Oak Ridge National Laboratory, USA	18	——	——	0	——	18	100	398	22.1
Max Planck Institute for Terrestrial Microbiology, Germany	16	——	——	0	——	16	100	1181	73.8

Table 3: Top 20 most productive institutes based on the total number of articles. Note: TP: total number of articles, SP: single institute articles, CP: inter-institutionally collaborative articles, TC: total citation count, ——: no articles.

Hot issues

The most important information, which the author expected to express to readers, was presented in the title, keywords and KeyWords Plus. In order to analyze the three separated parts synthetically, the synonymic single words and congeneric phrases were summed and grouped into categories. In this way, the analysis of historical development of science could be more complete and precise, and the new direction about this field could be found. A new method named "word cluster analysis" had been applied to analyze the research trends in risk assessment [39] and nitrate removal. All words included in Figures 3 and 4 contained their plural forms and other transformations, as well as words with the similar meanings.

Research trends in acidophilic microorganism were separated into two categories: research hotspot and research environment. In terms of the research hotspot, the "resistance to metal" was the most predominant for 20 years (Figure 3). Tolerance to various kinds of heavy metal had been analyzed in recent years, such as Fe, Cu, Al, Zn. This characteristic had been widely applied to bioleaching to increase the purity of mixed ore [40-42]. In addition, "diversity" became the second research branch with a fast speed, since the application of cultivation-independent molecular techniques becoming more mature, such as FISH, PCR-DGGE [43,44]. The number of the articles using molecular biology techniques during 2005-2012 was half of the total number in 30 years. On the contrary, because of the limitation of technical means and the difficulty in controlling the micro-world, the development of "enzyme" and "culture" was relatively slow. Therefore, the study in acidophilic microorganism focused on the resistance to

metal and diversity, which had more practical value and substantial benefit.

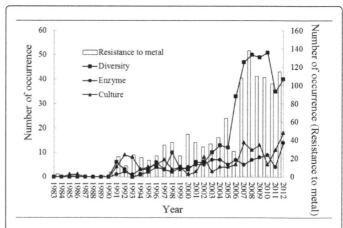

Figure 3: Comparison of the trends of research hotspots, resistance to metal, diversity, enzyme, culture.

Research trend of kinds of environments had been showed in Figure 4, including "water", "soil", "hot spring" and "leaching heap". The difference between four environments became obvious after 2006, because of a rapid increase in the "water" and "bioleaching heap". First of all, "water" was the most wide study environment, such as acid mine drainage. The research in "water" reached a peak point in 2008. Referring to the increasing trend of "bioleaching heap", the main reason was that people realized the value of acidophilic microorganism in industry. France was the earliest one, trying to leaching gold in ores by bacteria in 1964, and achieved encouraging results. Subsequently, bioleaching developed gradually, and became industrialized. The industrialization reached an obvious peak point in 2008. Although "soil" was not main living environment, acidophilic microorganism was abundant in soil which was proved in lots of articles. The field around the acid mine drainage was contaminated by the acid water, and became exclusively suitable for this special microorganism. On the contrary, "hot spring" acted as the original living environment, but articles about "hot spring" were rare.

Another result should be noted that there was a decrease after 2007, including "water", "bioleaching heap", "soil", and even "resistance to metal", "diversity". All of this may be a direct result of economic crisis happened in 2007, which made the fund for scientific research have an obvious shrinkage in the whole world. To a large extent, the economic situation provided a solid foundation to the research.

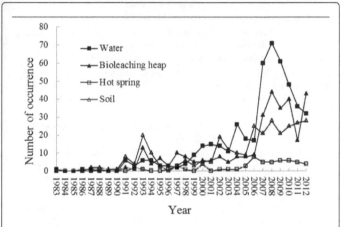

Figure 4: Comparison of the trends of research environment, water, leaching, hot spring, soil.

Conclusions

Based on the databases of acidophilic microorganism listed in SCI-EXPANDED, some significant points on the worldwide research performance were studied using bibliometric analysis during the 30 years. The following conclusions were summarized from this study:

Study in the acidophilic microorganism had significantly increased in the last 30 years, especially in the year 2003-2012.

1873 articles have been distributed into 109 SCI subject categories. "Microbiology" and "the Biotechnology and Applied Microbiology" were the two fastest growing subject categories, suggesting that the value of microorganism had been applied to the practice of production gradually.

At the country level, 71.6% were single-country publications, and 28.4% were international articles. The USA had the most dominant position including the most total articles, single-country articles and internationally collaborative articles. Besides, economic condition was related to the academic achievement. The G7 and the BRIC were listed in the top 20 most productive countries.

In the analysis of distribution of institute, 43.2% of all articles were single-institutional and 56.8% were inter-institutional. In the top 20

most productive institutes, only four institutes had more independent articles than inter-institutionally collaborative articles, and another two institutes only published the inter-institutionally collaborative articles. These proved that the academic communities of acidophilic microorganisms were more tending to cooperation.

Using a new bibliometric method - "word cluster analysis" - found that "resistance to metal" and "diversity" will continue to be the leading research hotspots. Eventually, "water" will be the main research environment.

Acknowledgements

This research was supported by grants from the National Science Foundation of China (40802059), the Fundamental Research Funds for the Central Universities (2010ZD03, 2011YXL035).

References

1. Garcia-Gonzalez L, Rajkovic A, Geeraerd AH, Elst K, Ginneken LV, et al. (2010) The development of Escherichia coli and Listeria monocytogenes variants resistant to high-pressure carbon dioxide inactivation. Letters in Applied Microbiology 50: 653-656.

2. Sharma A, Kawarabayasi Y, Satyanarayana T (2012) Acidophilic bacteria and archaea: acid stable biocatalysts and their potential applications. Extremophiles 16: 1-19.

3. Chen MM, Liu QH, Xin GR, Zhang JG (2013) Characteristics of lactic acid bacteria isolates and their inoculating effects on the silage fermentation at high temperature. Lett Appl Microbiol 56: 71-78.

4. Schleper C, Puehler G, Holz I, Gambacorta A, Janekovic D, et al. (1995) Picrophilus gen nov fam nov: a novel aerobic heterotrophic Thermoacidophilic genus and family comprising archaea capable of growth around pH-0. Journal of Bacteriology 177: 7050-7059.

5. Rothschild LJ, Mancinelli RL (2001) Life in extreme environments. Nature 409: 1092-1101.

6. Hao C, Wang L, Gao Y, Zhang L, Dong H (2010) Microbial diversity in acid mine drainage of Xiang Mountain sulfide mine, Anhui Province, China. Extremophiles 14: 465-474.

7. Ding JN, Zhang R, Yu Y, Jin D, Liang C, et al. (2011) A novel acidophilic, thermophilic iron and sulfur-Oxidizing archaeon Isolated from a hot spring of TengChong YunNan China. Brazilian Journal of Microbiology 42: 514-525.

8. Ruiz LM, Castro M, Barriga A, Jerez CA, Guiliani N (2012) The extremophile Acidithiobacillus ferrooxidans possesses a c-di-GMP signalling pathway that could play a significant role during bioleaching of minerals. Letters in Applied Microbiology 54: 133-139.

9. Ramírez P, Guiliani N, Valenzuela L, Beard S, Jerez CA (2004) Differential protein expression during growth of Acidithiobacillus ferrooxidans on ferrous iron, sulfur compounds, or metal sulfides. Appl Environ Microbiol 70: 4491-4498.

10. Karelová E, Harichová J, Stojnev T, Pangallo D, Ferianc P (2011) The isolation of heavy-metal resistant culturable bacteria and resistance determinants from a heavy-metal-contaminated site. Biologia 66: 18-26.

11. Hallberg KB, Johnson DB (2005) Microbiology of a wetland ecosystem constructed to remediate mine drainage from a heavy metal mine. Sci Total Environ 338: 53-66.

12. Wang H, Luo H, Li J, Bai Y, Huang H, et al. (2010) An alpha-galactosidase from an acidophilic Bispora sp. MEY-1 strain acts synergistically with beta-mannanase. Bioresour Technol 101: 8376-8382.

13. Bräuer SL, Cadillo-Quiroz H, Yashiro E, Yavitt JB, Zinder SH (2006) Isolation of a novel acidiphilic methanogen from an acidic peat bog. Nature 442: 192-194.

14. Baker BJ, Tyson GW, Webb RI, Flanagan J, Hugenholtz P, et al. (2006) Lineages of acidophilic archaea revealed by community genomic analysis. Science 314: 1933-1935.

15. Raan AFJ (1996) Advanced bibliometric methods as quantitative core of peer review based evaluation and foresight exercises. Scientometrics 35: 397-420.

16. Narin F, Hamilton KS (1996) Bibliometric performance measures. Scientometrics 36: 293-310.

17. Wang X, Wang Z, Xu S (2013) Tracing scientist's research trends realtimely. Scientometrics 2013 95: 717-729.

18. Vergidis PI, Karavasiou AI, Paraschakis K, Bliziotis IA, Falagas ME (2005) Bibliometric analysis of global trends for research productivity in microbiology. Eur J Clin Microbiol Infect Dis 24: 342-346.

19. Du HB, Wei LX, Brown MA, Wang YY, Shi Z (2013) A bibliometric analysis of recent energy efficiency literatures: an expanding and shifting focus. Energy Efficiency 6: 177-190.

20. Lee LC, Lee YY, Liaw YC (2012) Bibliometric analysis for development of research strategies in agricultural technology: the case of Taiwan. Scientometrics 93: 813-830.

21. Fu HZ, Ho YS, Sui YM, Li ZS (2010) A bibliometric analysis of solid waste research during the period 1993-2008. Waste Manag 30: 2410-2417.

22. Bajwa RS, Yaldram K (2013) Bibliometric analysis of biotechnology research in Pakistan. Scientometrics 95: 529-540.

23. Costa BMG, Pedro EDS, Macedo GRD (2013) Scientific collaboration in biotechnology: the case of the northeast region of Brazil. Scientometrics 95: 571-592.

24. Khan MA, Ho YS (2012) Top-cited articles in environmental sciences: merits and demerits of citation analysis. Sci Total Environ 431: 122-127.

25. Ahmed T, Johnson B, Oppenheim C, Peck C (2004) Highly cited old papers and the reasons why they continue to be cited. Part II. The 1953 Watson and Crick article on the structure of DNA. Scientometrics 61: 147-156.

26. Li J, Wang MH, Ho YS (2011) Trends in research on global climate change: a Science Citation Index Expanded-based analysis. Global and Planetary Change 77: 13-20.

27. Qin J (2000) Semantic similarities between a keyword database and a controlled vocabulary database: An investigation in the antibiotic resistance literature. Journal of the American Society for Information Science 51: 166-180.

28. Ugolini D, Cimmino MA, Casilli C, Mela GS (2001) How the European Union writes about ophthalmology. Scientometrics 52: 45-58.

29. DeLisi C (2008) Meetings that changed the world: Santa Fe 1986: Human genome baby-steps. Nature 455: 876-877.

30. Hall N (2007) Advanced sequencing technologies and their wider impact in microbiology. J Exp Biol 210: 1518-1525.

31. Mardis ER (2008) Next-generation DNA sequencing methods. Annu Rev Genomics Hum Genet 9: 387-402.

32. Yang LY, Yue T, Ding JL, Han T (2012) A comparison of disciplinary structure in science between the G7 and the BRIC countries by bibliometric methods. Scientometrics 93: 497-516.

33. Xie SD, Zhang J, Ho YS (2008) Assessment of world aerosol research trends by bibliometric analysis. Scientometrics 77: 113-130.

34. Huang WL, Zhang BG, Feng CP, Li M, Zhang J (2012) Research trends on nitrate removal: a bibliometric analysis. Desalination and Water Treatment 50: 67-77.

35. Bond PL, Smriga SP, Banfield JF (2000) Phylogeny of microorganisms populating a thick subaerial predominantly lithotrophic biofilm at an extreme acid mine drainage Site. Applied and Environmental Microbiology 66: 3842–3849.

36. Edwards KJ, Bond PL, Gihring TM, Banfield JF (2000) An archaeal iron-oxidizing extreme acidophile important in acid mine drainage. Science 287: 1796-1799.

37. Sanchez-Andrea I, Knittel K, Amann R, Amils R,Sanz JL (2012) Quantification of Tinto River sediment microbial communities: importance of sulfate-reducing bacteria and their role in attenuating acid mine drainage. Applied Environmental Microbiology 78: 4638-4645.

38. Cánovas CR, Olias M, Vazquez-Suñé E, Ayora C, Miguel Nieto J (2012) Influence of releases from a fresh water reservoir on the hydrochemistry of the Tinto River (SW Spain). Sci Total Environ 416: 418-428.

39. Mao N, Wang MH, Ho YS (2010) A bibliometric study of the trend in articles related to risk assessment published in Science Citation Index. Human and Ecological Risk Assessment: An International Journal 16: 801-824.

40. Küsel K, Dorsch T (2000) Effect of Supplemental Electron Donors on the Microbial Reduction of Fe(III), Sulfate, and CO(2) in Coal Mining-Impacted Freshwater Lake Sediments. Microb Ecol 40: 238-249.

41. Fischer J, Quentmeier A, Gansel S, Sabados V, Friedrich CG (2002) Inducible aluminum resistance of Acidiphilium cryptum and aluminum tolerance of other acidophilic bacteria. Arch Microbiol 178: 554-558.

42. Mangold S, Potrykus J, Björn E, Lövgren L, Dopson M (2013) Extreme zinc tolerance in acidophilic microorganisms from the bacterial and archaeal domains. Extremophiles 17: 75-85.

43. Sun FL, Wang YS, Wu ML, Wang YT, Li QP (2011) Spatial heterogeneity of bacterial community structure in the sediments of the Pearl River estuary. Biologia 66: 574-584.

44. Lakatošová M, Holecková B (2007) Fluorescence in situ hybridisation. Biologia 62: 243-250.

7

Recycling of Local Qatar's Steel Slag and Gravel Deposits in Road Construction

Ramzi Taha*, Okan Sirin and Husam Sadek

Department of Civil and Architectural Engineering College of Engineering Qatar University, Qatar

*Corresponding author: Ramzi Taha, Department of Civil and Architectural Engineering College of Engineering, Qatar University, Qatar
E-mail: ramzitaha@qu.edu.qa

Abstract

Every year, the State of Qatar generates about 400,000 tons of steel slag and another 500,000 tons of gravel as a result of steel manufacturing and washing sand, respectively. The two materials (by-products) are not fully utilized to their best market values. At the same time, infrastructural renewal will take place in Qatar over the next ten years, and there will be a greater demand for aggregates and other construction materials as the country suffers from the availability of good aggregates. This paper presents results obtained on the use of steel slag, gravel and gabbro (control) in hot mix asphalt concrete (HMAC) paving mixtures and road bases and sub-bases. Tests were conducted in accordance with Qatar Construction Specifications (QCS-2010) and results were compared with QCS requirements for aggregates used in these applications. Based on the data obtained in this work, steel slag and gravel aggregates have a promising potential to be used in hot mix asphalt concrete paving mixtures on Qatar's roads, whether in asphalt base and asphalt wearing courses or as unbound aggregates in the base and sub-base pavement structure.

Keywords: Steel slag; Gravel; By-products; Asphalt concrete; Sub-base; Qatar

Introduction

In the State of Qatar, steel slag, a by-product of steel manufacturing, is generated in large quantities. In fact, it is estimated that more than 400,000 tons (1 ton = 1016.05kg) of steel slag are generated annually and they are not efficiently utilized in construction [1]. The disposal of such quantities poses a great burden on Qatar's Steel. In addition, gravel, resulting from washing sand, is also produced at more than 500,000 tons per year in Qatar [1]. Simultaneously, infrastructural renewal (roads, bridges, metro, railways, new airport, deep-water seaport, hotels, stadiums, etc.) will take place in the State of Qatar over the next ten years and there will be a greater demand for aggregates and other construction materials. Qatar suffers from the availability of good aggregates that could be utilized in roads, parking, buildings and other construction. In fact, Qatar imports most of its aggregates needs from neighboring countries. Thus, our environmental responsibilities and potential economic benefits that might be realized dictate that steel slag and other discarded materials should be utilized in the construction sector.

Research is thus needed to promote and investigate, where possible, the recycling of steel slag and gravel deposits in Qatar's construction industry. This paper presents results obtained on the use of steel slag, gravel and gabbro in hot mix asphalt concrete (HMAC) paving mixtures in addition to road bases and sub-bases. All tests were conducted in accordance with Qatar Construction Specifications (QCS-2010).

Literature Review

Waste is an unavoidable by-product of most human activities. Economic growth and rising living standards in many parts of the world have led to an increase in the quantities of generated wastes.

Steel slag, a by-product of steel manufacturing, is no exception. It is produced either from the conversion of iron to steel in a Basic Oxygen Furnace (BOF) or by the melting of scarp to make steel in the Electric Arc Furnace (EAF). The slag is produced as a molten liquid melt and it is a complex solution of silicates and oxides that solidifies upon cooling. The American Society for Testing and Materials (ASTM) defines steel slag as "a non-metallic product, consisting essentially of calcium silicates and ferrites combined with fused oxides of iron, aluminium, manganese, calcium and magnesium that are developed simultaneously with steel in basic oxygen, electric furnace, or open hearth furnaces [2]."

Chemical composition, mechanical and environmental properties in addition to some undesirable characteristics of steel slag were presented in a previous work [1].

Hot mix asphalt concrete

Steel slag has been successfully used in asphalt paving mixtures in the United States, Canada, Europe and Japan [3,4,6-9]. It is used as an aggregate in hot mix asphalt wearing courses and surface treatments, including chip seals. Positive properties of steel slag aggregates, when used in asphalt paving, include high stability, excellent stripping and skid resistance, and resistance to rutting. Proper processing of steel slag and special quality control procedures should be in-place when selecting steel slag for use in asphalt paving. ASTM D5106 [2] is the Standard Specification for Steel Slag Aggregates for Bituminous Paving Mixtures. BS 4987: Part I [10] specifies the use of steel slag as one of the aggregate types in coated macadam for roads and other paved areas.

However, the literature [5,6] cautions against using 100% steel slag aggregate in asphalt paving as such hot mix asphalt concrete mixes might be susceptible to high void space and bulking problems due to the angularity of the steel slag. Hot mix asphalt with 100% steel aggregates might be prone to over asphalting during production, which could lead to subsequent flushing during in-service traffic compaction. Thus, steel slag aggregates' use in asphalt paving should be restricted to either the fine or coarse aggregate fraction, but not both, for better aggregate interlocking, lower void space and higher frictional resistance. This can be achieved by blending the coarse or fine steel slag aggregates with conventional natural materials such as gravels (more rounded) for better compatibility of the final asphalt mix.

Granular bases and sub-bases

Steel slag aggregates can be used in the construction of unpaved parking lots, as railroad ballast, as a shoulder material, and also in the construction of berms and embankment. Experience in many countries including the United States, Belgium, Japan, The Netherlands, Germany, and Saudi Arabia [4,5,11-13] has referred that steel slag aggregates, when properly selected, processed, aged, and tested, can be used as granular base for roads in above-grade applications. Positive properties of steel slag include very high stability and good soundness.

S. Aiban [12] examined the use of steel slag aggregates generated in Saudi Arabia in road bases and he asserted that "laboratory and field data have shown the superior performance of steel slag aggregates over the locally available calcareous sediments. The resulting California Bearing Ratio values are doubled and the water sensitivity is much less when using steel slag aggregates instead of the local calcareous material."

Though, the literature [5,14] cautions against using steel slag aggregates in confined applications, such as backfill behind structures, granular bases and sub-bases confined by curb and gutter, and trenches. This has primarily to do with the potential for volumetric expansion of steel slag due to free lime hydration. Also, concerns have risen over the formation of tufa-like precipitates that might clog sub-drains and drain outlets [11,15].

Objective of the Study

The main objective of this paper is to present the results obtained on the feasibility of using steel slag, gravel and gabbro in hot mix asphalt concrete mixtures and in road base and sub-bases according to Qatar Construction Specifications (QCS-2010). The conventional aggregate "gabbro" was used as the control mixture.

Experimental Program

Steel slag, gabbro and gravel aggregates used in the study were supplied by Slag Aggregates Producer (SAP) from their plant in Mesaieed City in Qatar. A total of 15 bags were delivered to the asphalt laboratory at Qatar University (QU), including 6 tons of slag, 2 tons of gabbro and 2 tons of gravel of different sizes. Gabbro is an igneous rock that has been used in road construction for a long time in this region. The steel slag aggregate was aged at the SAP Plant between 1 and 8 years. SAP delivered all three aggregates in compliance with QCS gradations' requirements.

Hot mix asphalt concrete

All hot mix asphalt concrete paving mixtures were designed using the Marshall Mix design method (ASTM D6926 and ASTM D6927), which is the standard method specified for use in Qatar. The design of an asphalt concrete mixture includes:

1. Selection of best aggregate blend.
2. Determination of the optimum asphalt content.

Finally, the mix should meet specifications' requirements and be economical at the same time.

A total of three aggregate types (steel slag, gravel and gabbro), supplied in different sizes, and were used in this study. Aggregate blend gradation was prepared to satisfy the gradations' requirements for an Asphaltic Concrete Wearing Course (SC-TYPE 1) in QCS-2010. Maximum aggregate size was 25mm. The blend gradation and specification limits are plotted in Figure 1.

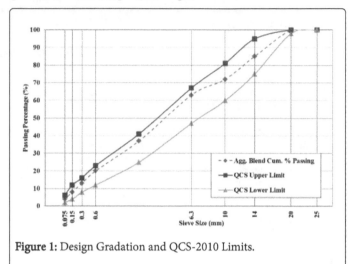

Figure 1: Design Gradation and QCS-2010 Limits.

A total of six different aggregate combinations were used to prepare asphalt mixtures in the laboratory. For the 50% steel slag with 50% gravel, 50% steel slag with 50% gabbro, and 50% gravel with 50% gabbro mixtures, the quantity of every aggregate size of the blend was divided by two. The conventional unmodified Pen 60/70 bitumen, obtained from the Woqod Company, was used to prepare all Marshall Mixes.

A 102mm diameter by 64mm high cylindrical bituminous mixture samples were prepared and compacted in the laboratory according to ASTM D6926. In order to determine the optimum asphalt content for each mix, a series of test samples was prepared for a range of asphalt contents (from 3 to 7%) in 0.5% increments. Three replicate samples were prepared at each asphalt content. All samples were compacted using 75 blows on each side and designed for heavy traffic conditions.

Road base and sub-base

On the other hand, different sizes of aggregates were blended to meet the gradation of a sub-base course (Class C) as given in QCS-2010. The gradation of the sub-base material used in this project along with the upper and lower % passing requirements given in QCS-2010 can be seen in Table 1.

For this purpose, different amounts of water were added to sub-base mixes composed of 100% steel slag, 100% gabbro and 100% gravel to determine the optimum moisture content (OMC) using the modified Proctor compaction test (ASTM D1557-12). Samples were compacted using 25 blows in 5 layers using a 44.48N rammer dropped from a distance of 457.2mm. Corrected moisture contents were determined after drying samples at an oven temperature of $110 \pm 5°C$ for 24 hours.

B.S. Sieve Size	Cumulative Passing (%)	QCS 2010 Limits	
25mm	100	100	100
20.0mm	95	90	100
10.0mm	67	50	85
5.0mm	50	35	65
2.36mm	37	25	50
425μm	22	15	30
75μm	10	5	15

Table 1: Gradation of sub-base aggregate blend.

Then, samples were compacted manually by 62 blows in 5 layers using a 4.5kg rammer at the OMC in accordance with BS 1377-4: 1990. Compacted samples were soaked in water for 96 hours before the California Bearing Ratio (CBR) test is conducted. Two replicate samples were used for each aggregate type to determine the CBR values.

Discussion of Results

Physical properties

First, physical properties of the steel slag, gabbro and gravel aggregates were investigated. All tests were conducted in accordance with ASTM standards. The steel slag aggregate shown in Figure 2 has the following general properties:

- Physical state: solid.
- Color: dark gray.
- Shape and characteristics: very rough and porous surface with high angularity.
- Odor: odorless.
- Solubility: insoluble in water, oil and solvents.

Figure 2: Steel slag aggregates.

The steel slag, gravel and gabbro aggregates failed the liquid limit and plastic limit tests (i.e. the aggregates are non-plastic). In addition, the bulk, saturated surface-dry (SSD) and apparent (APP) specific gravity tests were performed on the steel slag, gravel and gabbro aggregates. Table 2 indicates that all specific gravity values for steel slag (fine and coarse) are greater than those of gabbro and gravel.

Specific gravity	Steel Slag	Gabbro	Gravel
Bulk SG for coarse aggregates	3.39	2.95	2.62
Bulk SG for fine aggregates	3.54	2.86	2.59
SSD SG1 for coarse aggregates	3.43	2.96	2.64
SSD SG for fine aggregates	3.58	2.9	2.64
APP SG2 for coarse aggregates	3.52	2.98	2.69
APP SG for fine aggregates	3.69	2.98	2.72

Table 2: Specific gravity results of steel slag, gabbro and gravel. ([1]SSD: Saturated Surface Dry; [2]APP: Apparent Specific Gravity).

Then, the unit weight values for all aggregate types used in this study were determined as shown in Table 3. As a result of its high specific gravity, the steel slag had the highest unit weight compared to other aggregate types. Also, the average absorption values for the steel slag aggregates were 1.06% for coarse and 1.13% for fine aggregates (Table 3). These results are acceptable according to the QCS-2010specifications. However, the coarse aggregates of gabbro had the lowest water absorption percentage and this is due to the strong inert structure of gabbro aggregates that have the least voids.

On the other hand, the toughness results of steel slag, gabbro and gravel were obtained from the Los Angeles abrasion Test. The average L.A. abrasion for steel slag was 14.9% as presented in Table 3. This was less than the 25-30% limit established for coarse aggregates and it was less than that of gravel deposits (22.7%), but more than that of gabbro (8.1%). Table 3 presents also the sand equivalent results for the steel slag, gravel and gabbro aggregates. All samples had high percentage values, which were above the minimum threshold specified by the QCS-2010 specifications' requirements. Gabbro had the highest sand equivalent value.

To determine the flakiness and elongation indices, 2 kg of each aggregate type was tested according to BS 812: Sections 105.1 and 105.2, respectively. The results presented in Table 3 alluded that all aggregate types are below the acceptable limit. However, gravel had the highest flakiness and elongation indices. In the soundness test, two sizes from coarse and fine aggregates were tested and the results are presented in Table 3. The data indicate that steel slag had the lowest soundness value, while gravel had the highest one. The soundness value for fine gravel aggregates (30.3%) failed the QCS-2010 requirement of ≤ 18% for asphalt works.

In general, the steel slag aggregates and gravel test results presented in Tables 2 and3 were comparable to typical values reported in the literature, and they met the QCS-2010 requirements.

Property	ASTM Standard	QCS-2010 Specifications		Steel Slag	Gabbro	Gravel
		Unbound Materials	Asphalt Works			
Unit weight, (kg/m³)	C29	-	-	2595	2169	2338
Water absorption for coarse aggregates, (%)	C128	-	≤ 1.5	1.06	0.34	1.12
Water absorption for fine aggregates, (%)	C128	-	-	1.13	1.38	1.93
L.A. abrasion, (%)	C131/C535	≤ 40	≤ 25-30	14.9	8.1	22.7
Sand equivalent for fine aggregates, (%)	D2419	≥ 25	> 30	41	47	33
Flakiness index, (%)	BS 812	≤ 35	≤ 25-30	1	8	17
Elongation index, (%)	BS 812	≤ 40	≤ 25	13	24	26
Soundness for coarse aggregates, (%)	C88	≤ 20	≤ 10-15	1	2	7
Soundness for fine aggregates, (%)	C88	-	≤ 18	4.2	16	30.3

Table 3: Physical properties of steel slag, gabbro and gravel.

Radiological properties

A total of 11 samples of asphalt concrete cylinders made of 100% steel slag, were received by the Nuclear Laboratory at QU. Samples were measured directly in plastic bags on a HPGe detector in order to determine which radionuclides had activity concentrations significantly higher than the background level. Each sample was assessed for the following naturally occurring radionuclides from the Uranium and Thorium decay series, as well as ^{40}K.

Each sample was prepared directly into a 200 ml PET container, filling the container as much as possible. No measures were taken to further homogenize the samples. The samples were measured directly on a 50% p-type HPGe detector for a time period of between two hours and one day, depending on the activity level in the sample. This detector has been efficiency calibrated using a radioactivity standard from NPL, UK. Spectrum acquisition and analysis was carried out using Ortec GammaVision software. All activities were decay corrected to the actual measurement date. In order to correct for disparities between the sample matrix and standard matrix, a post-adjustment geometry correction of the measurement result was performed using efficiency transfer methods. For this geometry correction, slag is assumed to mainly be composed of $CaSiO_3$, while limestone and river stone is assumed to be composed mainly of $CaMg(CO_3)_2$. Gravel is assumed to be composed of 50% $CaSiO_3$ and 50% $CaMg(CO_3)_2$, while Gabbro is simulated as basalt. In practice, this chemical composition will have only a slight impact on the final measurement results compared to the effect of the sample density and volume.

For the natural decay series, secular radioactive equilibrium is assumed between ^{226}Ra and daughters as well as ^{228}Ra and daughters. Often, ^{228}Ra can further be assumed to be in equilibrium with ^{232}Th. Activity concentrations have been calculated as follows:

- ^{40}K – directly from 1460keV peak
- ^{226}Ra – from daughter nuclides ^{214}Pb and ^{214}Bi
- ^{228}Ra – from daughter nuclide ^{228}Ac
- ^{228}Th – from daughter nuclides ^{208}Tl and ^{212}Pb
- ^{238}U – from daughter nuclide ^{234m}Pa

Table 4 below presents the nuclides of interest and their calculated activity concentrations for raw materials and asphalt concrete made of 100% steel slag aggregates. In the cases where the activity concentration of the sample falls below the minimum detectable activity (MDA) of the measurement, the result is reported as less than (<) a value. MDA's and uncertainties are reported at a 95% confidence level.

Sample Type	Activity Concentration (Bq/kg)				
	^{40}K	^{226}Ra	^{228}Ra	^{228}Th	^{238}U
Crushed slag 3/4"	23 ± 14	252 ± 25	144 ± 15	151 ± 15	290 ± 140
Slag raw	< 21	212 ± 22	167 ± 18	166 ± 17	270 ± 150
River stone raw	98 ± 10	8.0 ± 0.8	4.8 ± 0.8	5.1 ± 0.5	< 36
Gravel 5mm	244 ± 27	15.9 ± 1.6	8.3 ± 2.0	9.2 ± 1.1	< 95
Limestone 5mm	12.5 ± 2.6	4.7 ± 0.5	2.2 ± 0.8	2.3 ± 0.4	< 28
Gabbro 7mm	< 7.7	< 1.5	< 2.6	< 1.6	< 56
Slag 5mm	22 ± 11	237 ± 18	152 ± 16	158 ± 16	300 ± 120
Slag powder 0-1mm	62 ± 15	167 ± 17	128 ± 15	136 ± 14	350 ± 150
Slag powder 0-5mm	36 ± 12	218 ± 16	156 ± 16	172 ± 17	210 ± 110
Slag raw (new)	< 17	184 ± 19	182 ± 19	195 ± 20	270 ± 150
Slag raw (old)	25 ± 12	213 ± 22	156 ± 16	171 ± 17	280 ± 100

Asphalt concrete	< 8.0	182 ± 18	175 ± 23	195 ± 20	200 ± 90

Table 4: Nuclides' activity concentration results.

Table 5 presents the activity concentration indices for different combinations of gabbro/gravel/slag compared against standard values (H1and H2) recommended by the Qatari Ministry of Environment (MOE). These standard values were specified by MOE-Laboratories and Standardization Affairs as part of a memorandum issued to a local materials company operating in Qatar that allowed the use of a maximum 20% steel slag in certain asphalt concrete and Portland cement concrete road applications. The memorandum specifies other requirements as part of the whole package. H1 is the activity concentration index standard for the use of steel slag in certain asphalt concrete and Portland cement concrete road applications close to populated and residential areas, while H2 is the activity concentration index standard for the use of steel slag in certain asphalt concrete road applications for freeways and roads outside Doha, Capital of Qatar. It should be noted that all activity concentration indices were calculated based on the values measured in the Nuclear Laboratory at Qatar University and presented earlier in Table 4.

Mix Type	H1	H2	MOE Standard
G20-100 (100% Gabbro)	0.0205	0.0083	≤1.0
S20-100 (100% Slag)	1.7706	0.7317	≤1.0
GL20-100 (100% Gravel)	0.1758	0.0698	≤1.0
S20-75 (75% Slag+ 25% Gabbro)	1.3333	0.5508	≤1.0
GL20-75 (75% Gravel+ 25% Gabbro)	0.1369	0.0544	≤1.0
S20-50 (50% Slag+50% Gabbro)	0.8956	0.37	≤1.0
GL20-50 (50% Gravel+50% Gabbro)	0.0982	0.039	≤1.0
S20-25 (25% Slag+75% Gabbro)	0.458	0.1891	≤1.0
GL20-25 (25% Gravel+ 75% Gabbro)	0.0594	0.0236	≤1.0
Asphalt Concrete (100% Slag)	1.4843	0.611	≤1.0
Portland Cement Concrete (100% Slag)	1.2163	0.4992	≤1.0

Table 5: Activity concentration indices for different gabbro/gravel/slag mixtures.

Note:

$$H_1 = \frac{CTH}{200} + \frac{CRa}{300} + \frac{CK}{3000} + \ldots$$

$$H_2 = \frac{CTH}{500} + \frac{CRa}{700} + \frac{CK}{8000} + \frac{CCs}{2000} + \ldots$$

C_{TH} = Activity concentration value of ^{232}Th (assumed in equilibrium with ^{228}Ra) in Bq/kg

C_{Ra} = Activity concentration value of ^{226}Ra in Bq/kg

C_k = Activity concentration value of ^{40}K in Bq/kg

C_{Cs} = Activity concentration value of ^{137}Cs in Bq/kg

Table 5 indicates that the use of higher percentages of steel slag, beyond the 20% specified by MOE, will easily meet the H2

requirement of ≤ 1.0 for asphalt concrete road applications such as freeways and roads outside Doha. Even for steel slag usage in certain asphalt concrete road applications close to populated and residential areas, it will be possible to meet the H1 requirement of ≤ 1.0 if other mix design requirements and MOE recommendations are met.

Hot mix asphalt concrete results

After the samples were compacted using the Marshall compactor, the bulk specific gravity and density of specimens were determined in accordance with ASTM D2726. Prior to the stability and flow tests, compacted asphalt samples were immersed in a water bath at 60°C for 30-40 minutes. All samples were tested in the Marshall Test Apparatus.

Using the graphs of air voids (%AV) vs. asphalt content (%AC), voids in mineral aggregate (%VMA) vs. %AC and stability vs. %AC, the optimum asphalt content (OAC) can be determined by averaging the % AC determined for each of three parameters. Examples of these graphs are shown in Figures 3 and 4.

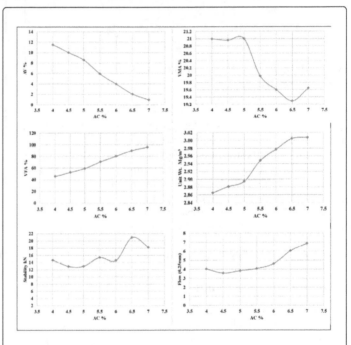

Figure 3: Marshall Test graphs for 100% steel slag asphalt concrete mixture.

After determining the OAC for each mixture, three samples were prepared at this optimum and mixture properties were compared to the mix design criteria given in Table 6.

As presented in Table 6, the 100% steel slag aggregate mixture had the highest optimum asphalt content of 6.39%, which is attributed to the high void space and bulking of the material. However, mixtures prepared using 50% steel slag and 50% gabbro or gravel aggregates produced better stability, lower flow, and lower optimum asphalt content.

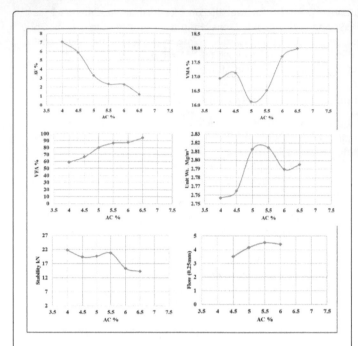

Figure 4: Marshall test graphs for 50% steel slag + 50% gabbro asphalt concrete mixture.

In general, Table 6 indicates that all six mixtures have met stability, flow and Marshall Quotient/Stiffness criteria as specified in QCS-2010. VMA values for all mixtures, except for the 100% gravel aggregates mixture, were also satisfied. However, all mixtures resulted in lower air voids contents than that of a minimum value requirement of 5%. All mixtures had higher VFA values than that of a maximum value requirement of 75%. The most probable reason for this is the lack of sand in the mixtures. In this study, it was only attempted to maximize the use of steel slag and gravel aggregates in the mixtures. Aggregate gradations used were within the QCS-2010 lower and upper limits.

Road base and sub-base results

Optimum moisture content: The compaction curves for each aggregate type were established after a sufficient number of water contents were used. The relationship between the dry unit weight and water content for the three aggregates are shown in Figure 5.

Mix Type	Optimum Asphalt Content (%)	Stability (N)	Flow (mm)	Marshall Quotient/ Stiffness, (kN/mm)	AV (%)	VMA (%)	VFA (%)	Filler/ Asphalt Ratio
100% Slag	6.39	22,080	3.53	6.25	2.67	19.5	86	0.63
100% Gravel	5.4	15,550	3.1	5.02	2.13	13	83	0.74
100% Gabbro	5.35	16,640	3.82	4.36	3.26	15.3	79	0.75
50% Slag + 50% Gabbro	5.19	23,880	3.02	7.91	2.68	16.1	83	0.77
50% Slag + 50% Gravel	5.22	22,200	2.75	8.07	2.96	16.5	82	0.77
50% Gravel + 50% Gabbro	5.08	22,900	3.07	7.46	2.1	13	84	0.79
QCS-2010 (75 blows compaction) Min	Min	10,000	2	4	5	15	50	0.75
Max	Max	-	4	-	8	-	75	1.35

Table 6: Marshall Mix design results for all mixes.

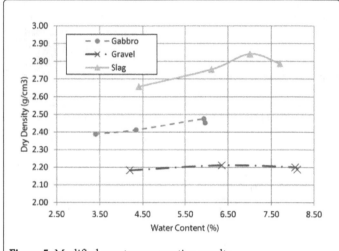

Figure 5: Modified proctor compaction results.

Figure 5 indicates that the optimum moisture contents for steel slag, gravel and gabbro aggregates are 7.0%, 6.4% and 5.9%, respectively. The largest dry density was achieved in the steel slag aggregates.

CBR results: After determining the optimum moisture content for each aggregate type, the California Bearing Ratio (CBR) test was conducted. Two replicate samples were used for each aggregate type to determine the CBR values. No swelling was observed in the soaked samples. Table 7 presents the CBR values for different aggregate types used in this study. All aggregates satisfied the minimum CBR requirement of 80% specified in QCS-2010 for base and sub-base materials.

	Steel Slag	Gravel	Gabbro	QCS-2010
CBR (%)	239	143	129	≥ 80

Table 7: CBR results for soaked samples.

Conclusions and Recommendations

The main objective of this research work was to produce the fundamental data needed to establish the suitability of using local discarded materials, such as steel slag and gravel in roads construction (bases/sub-bases and hot mix asphalt concrete mixtures).

Physical properties, such as Los Angeles abrasion, flakiness and elongation indices, soundness and sand equivalency values for steel slag, gravel and gabbro aggregates satisfied all criteria set forth in QCS-2010. Nevertheless, fine aggregates of gravel did not satisfy the QCS soundness requirement. Based on the mixtures prepared and the data obtained in this work, steel slag and gravel aggregates have a promising potential to be used in hot mix asphalt concrete paving mixtures on Qatar highways, whether as an asphalt base course or as an asphalt wearing course.

In addition radiological properties were investigated and the results indicated that the use of high percentages of steel slag (>20%) will easily meet the H2 requirement of ≤1.0 for asphalt concrete road applications. Even for steel slag usage in certain asphalt concrete road applications close to populated and residential areas, it will be possible to meet the H1 requirement of ≤1.0 if other mix design requirements and MOE recommendations are met.

It is also worth mentioning here that hot mix asphalt concrete (HMAC) mixtures can be designed using a variety of aggregate structures and blends that may result in different optimum asphalt contents and volumetric properties. Therefore, the results that were obtained here are meant to be an evidence of the possibility of using steel slag and gravel in HMAC mixtures and should not be treated as standard recipes for routine applications.

Steel slag and gravel physical properties, determined in the research project, met the QCS-2010 specifications' requirements for unbound materials. Also, based on compaction and California Bearing Ratio (CBR) test, steel slag and gravel aggregates have high CBR values that qualify their use in the base and sub-base layers.

Recommendations for further work include the construction of pilot field studies to establish the final validity for the construction use of steel slag and gravel. Such studies might encompass the construction of short road test sections, where steel slag and/or gravel could be used in the asphalt concrete base course, in the wearing (surface) course layer or as unbound aggregates in the base and sub-base layers. Short- and long-term monitoring of such sections will be critical to establish construction practices as well as field performance under actual traffic and environmental (temperature, rain, humidity, etc.) conditions.

Acknowledgments

The research team gratefully acknowledges the financial support provided by Slag Aggregate Producer, under Qatar University (QU) Research Grant No. QUEX-CENG-SAP-12/13-1. In particular, special thanks are due to Mr. Devassy Baby of SAP, who was very supportive of the proposed research.

References

1. Taha RA, Sirin O, Al-Nuaimi N, Benyahia F (2013) Beneficial Utilization of Steel Slag and Gravel in Construction. Final Report, Slag Aggregates Producer.

2. ASTM D5106 (1991) Standard specification for steel aggregates for bituminous paving mixtures. ASTM International, USA.

3. Emery J (1984) Steel Slag Utilization in Asphalt Mixes. Proceedings. Canadian Technical Asphalt Association.

4. Collins RJ, Ciesielski SK (1994) Recycling and use of waste materials and by-products in highway construction. National Cooperative Highway Research Program, Synthesis of Highway Practice 199, Transportation Research Board, Washington DC, USA.

5. Recycled materials resource center (2013) User guidelines for by-products and secondary use materials in pavement construction. Federal Highway Administration, USA.

6. Kandahl PS, Hoffman GL (1982) The Use of steel slag as bituminous concrete fine aggregate. Final report, Research project No. 79-26, Pennsylvania Department of Transportation, USA.

7. Hunt L, Boyle G, (2000) Steel slag in hot mix asphalt concrete. Final report, Report No. OR-RD-00-09, Oregon Department of Transportation, USA.

8. Xue Y, Wu S, Hou H, Zha J (2006) Experimental investigation of basic oxygen furnace slag used as aggregate in asphalt mixture. J Hazard Mater 138: 261-268.

9. Ahmedzade P, Sengoz B (2009) Evaluation of Steel Slag Coarse Aggregate in Hot Mix Asphalt Concrete. J Hazard Mater 165: 300-305.

10. BS 4987: Part 1 (1993) Coated macadam for roads and other paved areas. Part 1: Specification for constituent materials and for mixtures, British Standard, UK.

11. Gupta JD, Kneller WA, Tamirisa R, Skrzypezak-Jankun E (1994) Characterization of base and subbase iron and steel aggregates causing deposition of calcareous tufa in drains. Transportation research record, Washington DC, USA 1434: 8-16.

12. Aiba SA (2006) Utilization of steel slag aggregate for road bases. Journal of Testing and Evaluation 34.

13. Shen W, Zhou M, Ma W, Hu J, Cai Z (2009) Investigation on the application of steel slag-fly ash-phosphogypsum solidified material as road base material. J Hazard Mater 164: 99-104.

14. Carl BC, Burn KN (1969) Building damage from expansive steel slag backfill. Journal of the Soil Mechanics and Foundations Division 95: 1325-1334.

15. Gupta JD, Kneller WA (1993) Precipitate potential of highway sub base aggregates. Report No. FHWA/OH-94/004, Ohio Department of Transportation, USA.

8

Where Did That Pollution Come From? A Review of Chemical and Microbial Markers of Organic Pollution

John J Harwood*

¹Department of Chemistry, Tennessee Technological University, Cookeville, TN, USA

*Corresponding author: John J Harwood, Department of Chemistry, Tennessee Technological University, Cookeville, TN, USA
E-mail: jharwood@tntech.edu

Abstract

This review presents the current range of analytical markers which can be used to identify the sources of organic matter in environmental waters. Both chemical and microbial markers are presented. Applications which have been developed include identification of pollution input from human domestic sources, agriculture, landfills, and urban runoff. DNA and RNA characteristic of individual animal species, fecal sterols and bile acids, artificial sweeteners, commercial chemicals, and isotope ratios of carbon and hydrogen have been found useful as pollution markers. The review is intended to both introduce the reader to this vital area of research, and to present an overview which can be useful to practitioners in the field.

Keywords: Organic pollution; Molecular markers; Microbial markers; Microbial source tracking; MST; fecal sterols; Sucralose; Acesulfame-K; Bacteriodes

Introduction

While much research in recent years has focused on determining the extent of pollution by pharmaceuticals and personal care products in environmental waters, the traditional challenges of gross organic pollution (dissolved organic carbon, DOC) and contamination by pathogens in sewage remain. Reduction of these pollutants requires identifying the sources. One means of this is through the use of "markers". Both chemical and microbial markers are used.

Impacts of DOC

The impacts of DOC in waters are several. In consuming organic matter, bacteria deplete the dissolved oxygen content in water on which dersirable aquatic organisms, fish and macroinvertebrates, depend. The importance of this impact is shown by the ubiquitous use of the measure biochemical oxygen demand (BOD), the oldest and the most common method for assessing biodegradable organic contaminants in water and wastewater [1]. In unpolluted waters, DOC can range from 1 mg/L to greater than 20 mg/L [2-4]. Without reaeration, just 3.1 mg/L DOC will deplete oxygen in water [5]. Depleted oxygen can result in a buildup of toxic ammonia and hydrogen sulphide [1]. Degradation of DOC may release nutrient phosphorus to water, and lowers pH through a build-up of CO_2. Lower pH and oxygen lowers redox potential, facilitating desorption of toxic metal ions from sediments. These physico-chemical impacts reduce biodiversity in aquatic systems.

DOC is also the most important factor influencing drinking water treatment costs [4]. Degradation of DOC in source water through chlorination produces hazardous disinfection byproducts, trihalomethanes and haloacetic acids [6,7].

Depending on the source, DOC in water may include priority pollutants [8]. The Nationwide Urban Runoff Program conducted by the US Environmental Protection Agency in the early 1980's found 14 toxic organic compounds, predominantly polycyclic aromatic hydrocarbons (PAH), in more than 10% of samples [9]. These results continue to be observed in United States stormwater runoff [10].

Pollution from pulp and paper mills has a large variety of toxic effects on aquatic communities [11]. A significant number of these pollutants are classified as carcinogenic, mutagenic or endocrine disrupting. This pollution can both cause fish kills and affect the reproductive physiology of fish. Coking processes, oil production, pulp mills, textile production, and tanneries all significantly elevate BOD in receiving waters [3,11-14]. These sources also release many toxic compounds including PAH, organic solvents, phenolic compounds, and dioxin and other chlorination products. Landfills contribute BOD and toxic organic compounds as well, including benzene and other volatile organic compounds [8,15].

Even large rivers can be heavily degraded by uncontrolled organic pollution, as is documented by Kumar and Wata in the case of the Ganga River [3]. Before efforts to reduce organic pollution from industry and domestic sources in Kanpur, India, in 1987 this river had nearly been reduced to "an ecological desert". A "much reduced" fish production in Mumbai Harbor and the Uhas River estuary has been attributed to organic pollution [14]. In rural areas, DOC from agricultural fields strongly impacts receiving water quality [16].

The US EPA National Water Quality Inventory in 2004 found that of 16% of total river and stream miles reported on in the United States, 44% were impaired or not clean enough to support their designated uses [17]. Organic enrichment/oxygen depletion was a source of impairment in 17% of these impaired stream miles. The top sources of impairment included agricultural activities (38%) and municipal discharges/sewage (15%). This report also found 30% of the 29% of total area of U.S. bays and estuaries assessed were reported to be impaired. Organic enrichment/oxygen depletion was again a major cause of pollution, impairing 29% of impaired stream miles.

Agricultural activities (11%) and municipal discharges/sewage (33%) again accounted for most of this source of impairment.

Applications of Source Markers

Zgheib et al. present a very innovative example of using chemical source markers to study organic matter in combined wet weather flows containing domestic wastewater and runoff storm water in Paris [18]. This water contains organic particles from three main sources: runoff water, domestic wastewater and erosion of an organic in-sewer deposit. Vegetable and animal sterols (stigmastanol, β-sitosterol and cholestanol) can be found in runoff water, whereas different fecal sterols are expected to be found in in-sewer stocks and wastewater. Eight sterols were selected to distinguish between the sources of organic matter: cholesterol, cholestanol, fecal sterols (coprostanol, epicoprostanol, coprostanone), and phytosterols (β-sitosterol, stigmasterol, stigmastanol).

The sterol profile of wastewater was found to be slightly different from that of in-sewer deposits [18]. It is characterized by similar abundances of cholesterol (30-49%) and coprostanol (34-52%), and a low percentage of other sterols. Biofilm is distinguished from other sewer deposits by the highest abundance of β-sitosterol (13-20%), coprostanol between 30 and 35%, and cholesterol 27-38%. Roof and street runoff were found to contain the same individual sterols, especially cholesterol, β-sitosterol and stigmasterol, with a total absence of fecal sterols.

As in other studies utilizing these markers, principal component analysis (PCA) was applied to help interpret the rather complex results of the sterol profiles [18]. This analysis found both sterol profiles to be similar between the combined wet weather flows, wastewater and the organic layer deposits, while gross bed sediment and runoff clusters are far apart.

As discussed below, researchers have explored using molecular markers to source pollution from treated and untreated domestic wastewater, agricultural runoff, and urban runoff and landfill leachate. Microbial markers are used to determine sewage impacts on recreational waters and shellfish production.

Detecting contamination by sewage and treated wastewater

Organic contamination from human and animal sources is often accompanied by microbial and viral contamination, constituting a public health concern. Both biological and chemical markers have been employed to detect this contamination with a series of analytical methods referred to as "Microbial Source Tracking" (MST) [19]. Traditionally, the presence of indicator bacteria in water, *Escherichia coli (E. coli)* or fecal coliform, and enterococci, has been tested to determine possible contamination by human sewage. However, these indicators require time-consuming bacterial culturing and are found to not be specific to human sewage [20]. Indeed, one river basin-wide study which compared different fecal indicators found a "tighter linkage between E. coli counts and ruminant source markers than between *E. coli* counts and human markers" [21]. More advanced techniques have been developed for water analysis, most commonly based on DNA or RNA analysis utilizing polymerase chain reaction (PCR) technology, and also on antibiotic resistance profiling.

Antibiotic resistance analysis (ARA) consists of culturing the bacteria in a water sample on a series of media containing a variety of antibiotics. A phenotype library must be assembled prior to application of the test. This library documents the resistance of bacteria which may present in waste to the different antibiotitics. These bacteria are characteristic of different animal species which may release waste to the water. The antiobiotic resistance is encoded in plasmids, which may be lost or uptaken by bacteria, which puts in question the stability of the libraries [19]. ARA has been observed to produce a high level of false-positive results [22].

The most common PCR target for identification of bacterial sources in water has been 16S ribosomal DNA and RNA [23]. The bacteria probed are the genus Bacteroides, which are obligate anaerobes and are among the most numerous bacteria in human and animal intestines [19]. The PCR technique is relatively rapid as it does not require cell culturing. This technique can be used to detect human fecal sources in fresh and marine waters, and it can identify other fecal sources such as ruminants, dogs, pigs, and horses [23]. PCR is labor intensive, however, and a finger-print library is again required [19]. Liu et al. have suggested that the observed lack of specificity, leading to frequent false positives, is due to the inability of PCR to discriminate between sequences having "single internal mismatches" [24]. This group designed primers with more pronounced differences in sequence between bacteria species. A study of potential regional variability in applying PCR assays in MST tested 280 samples from 16 countries across six continents [25]. Ruminant associated marker concentrations correlated strongly with total intestinal Bacteroidetes populations and with each other, indicating that the detected ruminant-associated populations seem to be part of the intestinal core microbiome of ruminants worldwide. However, human-targeted assays were of relatively low specificity, indicating strong need for improved human-targeted assays.

Chemical markers of human sewage have also been studied as indicators of the potential presence of human pathogens. Murtaugh and Bunch first suggested coprostanol (5β(H)-cholestan-3β-ol), a product of the bacterial degradation of cholesterol in the human gut, be used as an indicator of human pollution [26]. Many studies have since been conducted associating coprostanol in water with sewage contamination. Other fecal sterols as well as bile acids have been enlisted in attempting to source human sewage input into waters [27].

Isobe et al. used sterol markers to perform regional monitoring of sewage impact in Malaysia and Vietnam [28]. This study demonstrated two great advantages of sterols as molecular markers, stability and transportability. The compounds are more stable in storage than are bacteria. Also, sterols concentrate in particulate matter, and so are collected from water samples on filters which can be easily transported. In the study gas chromatography-mass spectrometry was used to analyze for ten sterols at 59 sampling stations. The results showed C_{27} sterols to be abundant in urban areas, with coprostanol and cholesterol predominating, and significantly C_{29} sterols, indicative of ruminants, were depleted. These sterol profiles were typical of those previously observed in areas with heavy sewage impact. In rural areas, C_{29} sterols, including β-sitosterol and stigmastanol, were dominant while only trace amounts of coprostanol were found; this indicates contamination by non-human sources. A strong linear relationship was observed between concentrations of coprostanol and *E. coli* in both Malaysia and Vietnam.

Laundry detergent components find their way into environmental waters and serve as good markers of wastewater. Linear alkylbenzenes (LABs) were identified in the 1990's to be useful in this application. LABs are manufactured for the production of the linear alkylbenzenesulfonate surfactants used in commercial detergent formulations, and are present as minor components in detergents. LABs concentrate in suspended particulate matter and sediment, where they are found generally in sub-µg/L or in µg/g concentrations (dry weight), respectively [29]. In secondary water treatment, microbial alteration results in depletion of the "external isomers," where the benzene ring is attached near the end of the alkyl chain, with enrichment of "internal isomers." The ratio of these isomers can be used to indicate whether contamination is from secondary treatment (high "I/E") or untreated or primary treatment discharge (low "I/E") [30,31].

Takada et al. studied LABs and other markers in the Tokyo Bay and in a deep ocean dumpsite off the eastern seaboard of the United States [32]. The markers showed where sewage residue concentrates in the bay. Also, sedimentation rates can be estimated using the correlation between alkylbenzene production and sediment alkylbenzene concentration with depth [33-35]. Isobe et al. performed a large-scale monitoring project using LABs to show the wide-spread contamination by untreated sewage of canals and rivers in Southeast Asia [31].

Laundry fluorescent whitening agents (FWA) are also molecular markers for municipal and domestic wastewaters. The two most used detergent fluorescent whiteners are DSBP, a distyrylbiphenyl FWA, and DAS 1, a diaminostilbene [36]. Whiteners are subject to "photofading," isomerization to non-fluorescent geometric isomers, and photolysis. Hyashi et al. used these makers to track wastewater effluents from sewage treatment through rivers and estuaries and into Tokyo Bay [37]. Apparently due to a difference in photodegradation rate, the ratio DSBP/DAS1 was found to decrease as the water travelled. This ratio can be used to indicate fresh verses aged sediments [38], and to make inferences concerning sedimentation processes [39].

Artificial weeteners used in beverages and other products are now found widespread in environmental waters and can serve as markers of wastewater. Sucralose and acesulfame-K are both very stable and water soluble, resisting metabolism, decomposition and loss through adsorption in wastewater treatment and in the environment. In a large sampling of Swiss wastewater treatment plants, rivers, lakes and groundwater, acesulfame-K was consistently detected in untreated and treated wastewater (12-46 µg/L), in most surface waters, in 65% of investigated groundwater samples, and even in several tap water samples (up to 2.6 µg/L) [40]. Sucralose has been measured in surface waters of 27 European countries [41] and in fresh and marine waters of the United States [42,43].

Distinguishing agricultural pollution

Analysis of viruses has been explored in sourcing animal waste. Adenoviruses have been suggested as indicating contamination of human origin [44]. Formiga-Cruz et al. used PCR to analyze for adenoviruses, enteroviruses and hepatitis A virus in shellfish and sewage samples. The method was found to be as or more sensitive than testing for fecal coliforms or bacteriophages. As in previous studies, all samples testing positive for enteroviruses or hepatitis A virus were also positive for human adenoviruses, hence the latter viruses may simplify future pollution assays.

Gourmelon et al. have combined biological and chemical markers to produce a "MST Toolbox" to trace the origin of the fecal pollution [45,46]. The group recommends for distinguishing human pollution the markers caffeine, TCEP and benzophenone and the steroid ratios sitostanol/coprostanol and coprostanol/(coprostanol + 24-

ethylcoprostanol). PCR analysis of bacterial markers HF183 and *Bifidobacterium adolescentis* and genotype II of FRNAPH were determined to be indicators of human pollution. For porcine and ruminant pollution, the use of the same steroid ratios and PCR bacterial markers Pig-2-Bac and *Lactobacillus amylovorus* (porcine) and Rum-2-Bac (ruminants) was found adequate. The markers function best when the level of coliform in water is above 500 cfu.

Significant progress has been achieved in distinguishing pollution from agricultural production of different animals through analysis of profiles of fecal sterols, which vary between animal species. Recently, bile acids have been combined with fecal sterols to more definitively identify waste from individual animal species.

Leeming et al. published a groundbreaking analysis of feces from humans and 14 species of animals in which 17 sterols were identified [47]. Principal component analysis (PCA) was applied to analyze sterols contents. Cows and sheep, humans, and hens, which group with dogs and cats, are clearly distinguished from each other by this analysis. Distinguishing sterol markers include coprostanol (~60% of the total sterols in human feces), C29 sterols and 5β-stanols, which dominate the sterol profiles of herbivores, and 5β- and 5α- stanols, which were found in very low occurrence in birds and dogs feces. The sterol content of bird feces was found to be extremely variable and largely dependent on the diet of the animals. Only cats and pigs were found to have fecal sterol profiles similar to that of humans. While analysis of feces from individual animals produces clear distinctions between sources, mixed source samples can produce "confounding results" in source identification [48].

Jardé et al. used the sterol ratios $(C_{29}+C_{28})/C_{27}$ and $5β/C_{27}$ to distinguish animal sources in five rivers in Brittany, the principal animal production and dairy region of France [49]. Cross-plots of the sterol ratios contained in river particulate against type of manure spread in the watersheds clearly indicate a systematic relationship between sterol ratios and the type of animal breeding or animal manure spreading on their catchments. This work was continued to find that PCA analysis of six stanols can clearly distinguish bovine, porcine, and human feces [50]. The analysis successfully identified these individual sources of contamination in three river basins in France, with the results being verified through microbial analysis.

Building on previous work with molecular markers in archeology studies, Bull et al. present a multiple biomarker analysis flowchart for identifying individual animal sources using 5β-stanols and bile acids [27]. Like the sterols, bile acids concentrate in particulates and sediment. Clearly, adding a second class of compounds can increase the power of discrimination in delineating sources of fecal pollution. For instance, the presences of hyocholic acids in porcine fecal material enable it to be distinguished from human and canine contamination.

To assess the contribution of domesticated animal sources of fecal pollution, Tyagi et al. derived a multiple regression model with selected fecal sterols and bile acids [51]. Five compounds were found to determine to identify pollution sources efficiently: epicoprostanol, cholesterol, cholestanol, chenodeoxycholic acid, and hyodeoxycholic acid. Almost 100% accuracy was obtained in identifying sources when measuring compounds from runoff from test plots.

Organic Pollution from Non-Animal Sources

Landfill leachate

Municipal landfill leachate can add signicant organic matter to receiving waters [15]. Trace but potentially ecologically harmful components of landfill leachate include phthalates, alkylphenols and polynuclear aromatic hydrocarbons, persistent organic pollutants, including polychlorinated biphenyls and chlorinated pesticides, and volatile organic compounds, including chlorinated solvents [52-55]. Even as some hazardous materials used in commerce are phased out, new ones, such as brominated and fluorinated compounds are introduced [54]. These compounds may be so stable as to outlast integrity of engineered landfills, so may leach and impact environmental waters well into the future. While many of the components have low water solubility, the compounds are solubilized through association with soluble organic colloids in the leachates [52].

Fourie has noted that a unique feature of landfill leachate is that both inorganic and organic contaminants have high concentrations [56]. He suggests two ratios may serve as a "first pass testing protocol": A - COD x sulphate, and B - chloride x iron x ammonia; COD is chemical oxygen demand. In landfill leachate, both these ratios will be typically greater than 105 and 104, respectively.

In a relatively early detailed study of organic constituents leaching from a Barcelona, Spain sanitary landfill, seventeen carboxylic acids and a host of phenolics, alcohols and phthalates were identified [57]. These researchers suggested the gas chromatographic profile of short-chain organic acids, derivatized with pentafluorobenzyl bromide to allow selective detection with the electron capture detector, could be used to identify leachate pollution in groundwater. The series of C4 - C7 carboxylic acids was clearly recognizable in sampled groundwater and exhibited patterns similar to those of leachate. The predominance of the even carbon numbered acids, with a greater abundance of hexanoic acid, "may reflect" β-oxidation sequence characteristic of long-chain fatty acid oxidation.

Schwarzbauer et al. used gas chromatography-mass spectrometry to analyze seepage from a municipal landfill in Germany [58]. Given the high degree of stability and similarly high concentrations of some compounds measured in the landfill seepage and leakage waters, these authors suggested compounds which might serve as markers of water pollution from municipal landfills: the pharmaceutical propyphenazone, the plasticizer N-butyl benzene sulfonamide, and the insecticide N,N-diethyl toluamide (DEET). Clofibric acid, a plasticizer, and the herbicide mecoprop were also commonly detected. A continuation of this work used these five compounds to study lateral and vertical distribution of contamination as well as the long-term emission from the landfill [59].

Alkyl organotin compounds might serve as unique makers of landfill leachate. These compounds are used as PVC stabilizers, wood preservatives, pesticides, fungicides, and polyurethane and silicone catalysts [60,61]. While not found at levels considered hazardous to the environment, the compounds are found in landfill leachates at levels as high as 229 ng/L. Analysis can be achieved by gas chromatography-inductively coupled plasma mass spectrometry.

Bergström et al. have developed a comprehensive scheme for evaluation of landfill leachate treatment processes which may suggest useful markers [62]. The "LAQUA protocol" includes analysis of organic compounds, metals, inorganic ions, water-quality parameters, and toxicity. Polar organic compounds are represented by phenol and several phenolic compounds including 3-methyl-4-chlorophenol. A polychlorinated biphenyl reference standard was chosen to monitor removal of these compounds during treatment.

To characterize the biogeochemical evolution of DOC in landfill leachates, and to interpret the origin of DOC in groundwater, Mohammadzadeh et al. developed a technique to distinguish carbon isotope ratios in small organic acids [63]. The analytical technique, which interfaces a high performance liquid chromatograph to an isotope ratio mass spectrometer, was used to analyze leachate from a municipal landfill near Ottawa, Canada. The group found a difference in $\delta^{13}C$ values for leachate acetate (-10.7‰ to -16.9‰ VPDB) and the "precursor" DOC within the landfill (-24.7‰). The enrichment of $\delta^{13}C$ in the acetate suggests secondary biogeochemical reaction, likely methanogenesis, removes the lighter ^{12}C from the leachate.

This approach was expanded to include potential changes in nitrogen isotope composition in leachates [64]. Leachate from a municipal solid waste landfill in New Zealand were analyzed for changes in ^{13}C composition in dissolved inorganic carbon (DIC), ^{15}N in nitrate and ammonia nitrogen, and isotopic changes in particulate organic matter. Landfill leachate was found to have a distinct isotopic signature characterized by highly enriched 13C-DIC and highly enriched ^{15}N-NH_4^+. Downstream sites had ^{15}N values approaching those of the leachate, elevated above both upstream and reference stream site waters. The contrasting decrease in ^{15}N-NO_3^- values may be explained by isotope selection during nitrification. ^{13}C- DIC was enriched in landfill leachate and downstream sites; this enrichment is explained by isotope selection during bacterial methanogenises.

Tritium levels may also be indicative of landfill leachate [56]. Luminescent paint used in watches and other devices is apparently the source. In leachates from landfills throughout the world tritium has been measured at levels from 10 to 20,000 times background levels. A significant advantage of this tracer is that it is conservative, unaffected by processes such as ion exchange and adsorption that can affect concentrations of other contaminants moving through subsurface geology.

Road runoff

Organic pollution from roads is a major source of PAH among other toxic pollutants [65]. However, using PAH as road runoff markers is complicated by the presence of the compounds in waters from other sources including power, residential heating, industrial activities, by the changing composition of fuels, and by evolving emission controls used with these sources [66]. Building on previous work of Spies et al. [67], researchers have analyzed tire rubber components 2-(4-morpholinyl)benzothiazole (24MoBT) and N-cyclohexyl-2-benzothiazolamine (NCBA), benzothiazolamines present in different vulcanization accelerators, as markers of road runoff in urban sediments in Japan [68]. Sediment cores were dated using Cs-137 and tetrapropylene-based alkylbenzenes. "Changeovers" in the concentrations of the markers in sediments coincide well with changes in the production history of vulcanization accelerators. The dated downcore profile of 24MoBT and NCBA show positive correlation with the traffic data in the Tokyo Metropolitan Area.

Nitro-PAHs and triphenylene have been compared with sulfur-PAHs as markers of urban stormwater road runoff [69]. Based on abundance, source specificity, and persistence, dibenzothiophene and triphenylene were judged the most promising among the candidate markers. This study found 24MoBT to photodegrade rapidly in the aqueous phase and to not concentrate in sediment, limiting the potential usefulness of this marker. The ratio of 1-nitropyrene to total PAH has been found to be a useful indicator of road runoff from diesel fueled as compared with gasoline fueled vehicles [70]. Cluster analysis of tri-terpenes can distinguish atmospheric dust from road dust [71].

Conclusion

This review has focused on chemical and microbial markers used in determining sources of pollution. The underlying story is how the development of increasingly powerful analytical tools has allowed progress in this area. Originally pollution was monitored visually, then through laborious bacterial culturing techniques. In the 1990's, high resolution gas chromatography-mass spectrometry could be applied to analyze for steroid markers and xenobiotic organic compounds. A decade later, high performance liquid chromatography-mass spectrometry has added the capability to readily analyze for polar compounds, including artificial sweeteners. PCR analysis of nucleic acids is being applied to identify bacteria in water corresponding to pollution from specific animal hosts. This technical innovation has greatly expanded the potential applications of pollution source identification through analysis of chemical and microbial markers. From the large array of potential markers, the work of many researchers has refined the science to the point that a relatively short list of the most useful markers may be selected for common use. As more studies are performed, quantitative relationships between marker concentrations and associated pollution may be developed. Rapid, definitive identification and delineation of pollution sources through analysis of source markers seems very possible in coming years.

References:

1. PK Goel (2009) Water Pollution: Causes, Effects and Control. (Revised 2ndedn), New Age International Pvt Ltd New Delhi.

2. Carvalho P, Thomaz S M, Bini LM (2003) Effects of water level, abiotic and biotic factors on bacterioplankton abundance in lagoons of a tropical floodplain (Paraná River, Brazil). Hydrobiologia 510: 67–74.

3. Kumar, Amit, Wata (2003) Impact of Manmade Environmental Degradations on the Chemical Composition and Organic Production in River Ganga at Kanpur. Research Journal of Chemistry and Environment 7: 63-66.

4. Cox, JW, Ashley R (2000) Water quality of gully drainage from texture-contrast soils in the Adelaide Hills in low rainfall years. Australian Journal of Soil Research 38: 959-972.

5. Langmuir, Donald (1997) Aqueous Environmental Geochemistry. Prentice Hall, Upper Saddle River, New Jersey.

6. Wassink JK, Andrews RC, Peiris RH, Legge RL (2011) Evaluation of fluorescence excitation–emission and LC-OCD as methods of detecting removal of NOM and DBP precursors by enhanced coagulation. Water Science & Technology: Water Supply 115: 621–630.

7. Kraus, Tamara EC, Anderson, Chauncey A Morgenstern, Karl, Downing, Bryan D, Pellerin, Brian A, Bergamaschi, Brian A (2010) Determining Sources of Dissolved Organic Carbon and Disinfection Byproduct Precursors to the McKenzie River, Oregon. Journal of Environmental Quality 39: 2100–2112.

8. Hallbourg, Robin R, Delfino, Joseph J, Miller WL (1992) Organic priority pollutants in groundwater and surface water at three landfills in north central Florida. Water, Air, and Soil Pollution 65: 307-22.

9. Boving, Thomas B, Stolt, Mark H, Augenstern, Janelle, Brosnan, Brian (2008) Potential for localized groundwater contamination in a porous

10. Bathi JR, Pitt R, Clark SE (2009) Associations of PAHs with size fractionated sediment particles. 2009 World Environmental and Water Resources Congress Proceedings, Kansas City, MO. 18-22.

11. Karrasch B, Parra O, Cid H, Mehrens M, Pacheco P (2006) Effects of pulp and paper mill effluents on the microplankton and microbial self-purification capabilities of the Biobío River, Chile. Science of the Total Environment 359: 194–208.

12. Mrkva, Miroslav (1975) Automatic UV-control system for relative evaluation of organic water pollution. Water Research 9: 587-589.

13. Lundegard, Paul D, Jeffrey RK (2001) Polar Organics in Crude Oil and Their Potential Impacts on Water Quality. Petroleum Hydrocarbons and Organic Chemicals in Ground Water: Prevention, Detection, and Remediation, Conference and Exposition, Houston, TX.

14. Chavan RP, Lokhande RS, Rajput SI (2005) Monitoring of organic pollutants in Thane Creek water. Nature Environment and Pollution Technology 4: 633-636.

15. Castrillón L1, Fernández-Nava Y, Ulmanu M, Anger I, Marañón E (2010) Physico-chemical and biological treatment of MSW landfill leachate. Waste Manag 30: 228-235.

16. Udeigwe, Theophilus K, Wang, Jim J (2010) Biochemical Oxygen Demand Relationships in Typical Agricultural Effluents. Water, Air, and Soil Pollution 213: 237–249.

17. USEPA (2009) National Water Quality Inventory: Report to Congress 2004 Reporting Cycle EPA 841-R-08-001. United States Environmental Protection Agency Office of Water Washington, DC.

18. Zgheib S1, Gromaire MC, Lorgeoux C, Saad M, Chebbo G (2008) Sterols: a tracer of organic matter in combined sewers. Water Sci Technol 57: 1705-1712.

19. Seurinck, Sylvie, Verstraete, Willy, Siciliano, Steven D (2005) Microbial source tracking for identification of fecal pollution. Reviews in Environmental Science and Bio/Technology. 4: 19–37.

20. Shanks, Orin C, Nietch, Christopher, Simonich, Michael, Younger, Melissa, Reynolds, Don Field, Katharine G (2006) Basin-Wide Analysis of the Dynamics of Fecal Contamination and Fecal Source Identification in Tillamook Bay, Oregon. Applied and Environmental Microbiology 72: 5537–5546.

21. Glassmeyer, Susan T, Urlong, Edward TF, Kolpin, Dana W, Cahill, Jeffery D, Zaugg, Steven D et al. (2005) Transport of Chemical and Microbial Compounds from Known Wastewater Discharges: Potential for Use as Indicators of Human Fecal Contamination. Environmental Science & Technolology. 39: 5157-5169.

22. Harwood VJ, Wiggins BA, Hagedorn C, Ellender RD, Gooch J (2003) Phenotypic library-based microbial source tracking methods: Efficacy in the California collaborative study. Journal of Water and Health 1: 153-166.

23. Bernhard AE1, Goyard T, Simonich MT, Field KG (2003) Application of a rapid method for identifying fecal pollution sources in a multi-use estuary. Water Res 37: 909-913.

24. Liu R1, Chan CF, Lun CH, Lau SC (2012) Improving the performance of an end-point PCR assay commonly used for the detection of Bacteroidales pertaining to cow feces. Appl Microbiol Biotechnol 93: 1703-1713.

25. Reischer GH1, Ebdon JE, Bauer JM, Schuster N, Ahmed W, et al. (2013) Performance characteristics of qPCR assays targeting human- and ruminant-associated bacteroidetes for microbial source tracking across sixteen countries on six continents. Environ Sci Technol 47: 8548-8556.

26. Georg HR, James EE, Johanna MB, Nathalie S, Warish A, Johan Å, Anicet RB, Günter B, Denis B, Tricia C, Christobel F, Goraw G, GwangPyo K, Ana Maria de RH, Douglas M, Ramiro P, Bandana P, Veronica R, Margit AS, Regina S, Huw T, Erika MT , Virgil V, Stefan W, Robert LM, Andreas HF (2013) Performance Characteristics of qPCR Assays Targeting Human- and Ruminant-Associated Bacteroidetes for Microbial Source Tracking across Sixteen Countries on Six Continents. Environmental Science & Technology. 47: 8548-8556.

27. Murtaugh JJ, Bunch RL (1967) Sterols as a measure of fecal pollution. J Water Pollut Control Fed 39: 404-409.

28. Bull ID1, Lockheart MJ, Elhmmali MM, Roberts DJ, Evershed RP (2002) The origin of faeces by means of biomarker detection. Environ Int 27: 647-654.

29. Isobe, Kei O, Tarao, Mitsunori, Zakaria, Mohamad P, Chiem, Nguyen H, Minh, Le Y (2002) Quantitative Application of Fecal Sterols Using Gas Chromatography-Mass Spectrometry to Investigate Fecal Pollution in Tropical Waters: Western Malaysia and Mekong Delta, Vietnam Environmental Science & Technolology 36: 4497-4507.

30. Takada, Hideshige, Eganhouse, Robert P (1998) Molecular Markers of Anthropogenic Waste, in: Robert A Meyers, The Encyclopedia of Environmental Analysis and Remediation, Vol 5 John Wiley & Sons, Inc, Hoboken, New Jersey.

31. Tsutsumi S1, Yamaguchi Y, Nishida I, Akiyama K, Zakaria MP, et al. (2002) Alkylbenzenes in mussels from South and South East Asian coasts as a molecular tool to assess sewage impact. Mar Pollut Bull 45: 325-331.

32. Isobe KO1, Zakaria MP, Chiem NH, Minh le Y, Prudente M, et al. (2004) Distribution of linear alkylbenzenes (LABs) in riverine and coastal environments in South and Southeast Asia. Water Res 38: 2448-2458.

33. Takada, Hideshige, Satoh, Futoshi, Bothner, Michael H, Tripp, Bruce W, Johnson, Carl G (1997) Anthropogenic Molecular Markers: Tools to Identify the Sources and Transport Pathways of Pollutants. In: Molecular Markers in Environmental Geochemistry ACS Symposium Series Vol 671, American Chemical Society, Washington, DC.

34. Eganhouse RP, Pontolillo J, Leiker TJ (2000) Diagenetic fate of organic contaminants on the Palos Verdes Shelf, California. Marine Chemistry 70: 289-315.

35. Eganhouse RP, Pontolillo J (2000) Depositional history of organic contaminants on the Palos Verdes Shelf, California. Marine Chemistry 70: 317-338.

36. Boonyatumanond R, Wattayakorn G, Amano A, Inouchi Y, Takada H (2007) Reconstruction of pollution history of organic contaminants in the upper Gulf of Thailand by using sediment cores: First report from Tropical Asia Core (TACO) project. Marine Pollution Bullutin 54: 554-565.

37. Poiger, Thomas, Karim Franz Günter, and Giger, Walter (1999) Fate of Fluorescent Whitening Agents in the River Glatt. Environmental Science & Technology 33: 533-539.

38. Hayashi Y, Managaki S, Takada H (2002) Fluorescent Whitening Agents (FWAs) in Tokyo Bay and adjacent rivers: their application as anthropogenic molecular markers in coastal environments. Environmental Science & Technology. 36: 3556-3563.

39. Managaki S, Takada H (2005) Fluorescent Whitening Agents in Tokyo Bay sediments: Molecular evidence of lateral transport of land-derived particulate matter. Marine Chemistry 95: 113-127.

40. Buerge Ignaz J, Buser, Hans-Rudolf, Kahle, Maren, Muller, Markus D, Poiger, Thomas (2009) Ubiquitous Occurrence of the Artificial Sweetener Acesulfame in the Aquatic Environment: An Ideal Chemical Marker of Domestic Wastewater in Groundwater. Environmental Science & Technology. 43: 4381-4385.

41. Loos R, Gawlik B, Boettcher K, Locoro G, Contini S, Bidoglio G (2009) Sucralose Screening in European Surface Waters Using a Solid-phase Extraction-liquid Chromatography–triple Quadrupole Mass Spectrometry Method. Journal of Chromatography. 1216: 1126-131.

42. Mead, Ralph N, Morgan, Jeremy B, Avery Jr, Brooks, G, Kieber, Robert J, Kirk, Aleksandra M (2009) Occurrence of the artificial sweetener sucralose in coastal and marine waters of the United States. Marine Chemistry. 116: 13-17.

43. Mawhinney, Douglas B, Young, Robert B, Vanderford, Brett J, Borch, Thomas, Snyder, Shane A (2011) Artificial Sweetener Sucralose in U.S. Drinking Water Systems. Environmental Science & Technology 45: 8716-8722.

44. Formiga-Cruz, Meritxell, Hundesa, Ayalkibet, Clemente-Casares, Pilar, Albinana-Gimenez, Nestor, Allard, Annika (2005) Nested multiplex PCR assay for detection of human enteric viruses in shellfish and sewage. Journal of Virological Methods 125: 111-118.

45. Mieszkin S1, Furet JP, Corthier G, Gourmelon M (2009) Estimation of pig fecal contamination in a river catchment by real-time PCR using two pig-specific Bacteroidales 16S rRNA genetic markers. Appl Environ Microbiol 75: 3045-3054.

46. Gourmelon M1, Caprais MP, Mieszkin S, Marti R, Wéry N, et al. (2010) Development of microbial and chemical MST tools to identify the origin of the faecal pollution in bathing and shellfish harvesting waters in France. Water Res 44: 4812-4824.

47. Leeming R, Ball A, Ashbolt N, Nichols P (1996) Using Faecal Sterols from Humans And Animals to Distinguish Faecal Pollution in Receiving Waters. Water Research. 30: 2893-2900.

48. Shah, Vikaskumar G, Dunstan, R Hugh, Geary, Phillip M, Coombes, Peter, Roberts, Timothy K (2007) Evaluating potential applications of faecal sterols in distinguishing sources of faecal contamination from mixed faecal samples. Water Research 41: 3691-3700.

49. Jardé, Emilie, Gruau, Gérard, Mansuy-Huault, Laurence (2007) Detection of manure-derived organic compounds in rivers draining agricultural areas of intensive manure spreading. Applied Geochemistry 22: 1814-1824.

50. Derrien M1, Jardé E, Gruau G, Pourcher AM, Gourmelon M, et al. (2012) Origin of fecal contamination in waters from contrasted areas: stanols as Microbial Source Tracking markers. Water Res 46: 4009-4016.

51. Tyagi, Punam, Edwards, Dwayne R, Coyne, Mark S (2007) Use of selected chemical markers in combination with a multiple regression model to assess the contribution of domesticated animal sources of fecal pollution in the environment. Chemosphere. 69: 1617-1624.

52. Kalmykova, Yuliya, Björklund, Karin, Strömvall, Ann-Margret, Blom, Lena (2013) Partitioning of polycyclic aromatic hydrocarbons, alkylphenols, bisphenol A and phthalates in landfill leachates and storm water. Water Research 47: 1317-1328.

53. Castillo M, Barceló D (2001) Characterization of organic pollutants in textile wastewaters and landfill leachate by using toxicity-based fractionation methods followed by liquid and gas chromatography coupled to mass spectrometric detection. Analytica Chimica Acta 426: 253-264.

54. Weber, Roland, Watson, Alan, Forter, Martin, Oliaei, Fardin (2011) Persistent organic pollutants and landfills – A review of past experiences and future challenges. Waste Management & Research. 29: 107-121.

55. Sabel, Gretchen V, Clark, Thomas P (1984) Volatile organic compounds as indicators of municipal solid waste leachate contamination. Waste Management & Research. 2: 119-130.

56. Fourie ABA, De Mello, Luiz Guilherme, Almeida, Marcio (2002) Strategy to determine if a landfill is the source of detected water pollution. In: Environmental Geotechnics, Proceedings of the International Congress on Environmental Geotechnics, 4th, Rio de Janeiro, Brazil.

57. Albaiges J, Casado F, Ventura F (1986) Organic indicators of groundwater pollution by a sanitary landfill. Water Research 20: 1153-1159.

58. Schwarzbauer J1, Heim S, Brinker S, Littke R (2002) Occurrence and alteration of organic contaminants in seepage and leakage water from a waste deposit landfill. Water Res 36: 2275-2287.

59. Heim, Sabine, Schwarzbauer, Jan, Littke, Ralf (2004) Monitoring of waste deposit derived groundwater contamination with organic tracers. Environ Chem Lett 2: 21-25.

60. Vahcic, Mitja, Milacic, Radmila, Scancar, Janez (2011) Development of analytical procedure for the determination of methyltin, butyltin, phenyltin and octyltin compounds in landfill leachates by gas chromatography-inductively coupled plasma mass spectrometry. Analytica Chimica Acta. 694: 21-30.

61. Bjoern, Annika (2010) Leaching of organotin stabilizers from PVC under prevailing landfill conditions. Advances in Chemistry Research 3: 669-673.

62. Bergstroem, Staffan, Svensson, Britt-Marie, Maartensson, Lennart, Mathiasson, Lennart (2007) Development and application of an

analytical protocol for evaluation of treatment processes for landfill leachates I Development of an analytical protocol for handling organic compounds in complex leachate samples. International Journal of Environmental Analytical Chemistry 87: 1-15.

63.   Mohammadzadeh, Hossein, Clarka, Ian, Marschnera, Mark, St-Jean, Gilles (2005) Compound Specific Isotopic Analysis (CSIA) of landfill leachate DOC components. Chemical Geology 218: 3–13.

64.   North, Jessica C, Frew, Russell D, Peake, Barrie M (2004) The use of carbon and nitrogen isotope ratios to identify landfill leachate contamination: Green Island Landfill, Dunedin, New Zealand/ Environment International 30: 631-637.

65.   Mangani, Giovanna, Berloni, Arnaldo, Bellucci, Francesca, Tatano, Fabio, Maione, Michela, (2005) Evaluation of the pollutant content in road runoff first flush water. Water, Air, Soil Pollutution 160: 213-228.

66.   Kumata, Hidetoshi, Sanada, Yukihisa, Takada, Hideshige, Ueno, Takashi (2000) Historical Trends of N-Cyclohexyl-2-benzothiazolamine, 2-(4-Morpholinyl) benzothiazol and Other Anthropogenic Contaminants in the Urban Reservoir Sediment Core. Environmental Science & Technology 34: 246-253.

67.   Kumata H1, Takada H, Ogura N (1996) Determination of 2-(4-Morpholinyl)benzothiazole in Environmental Samples by a Gas Chromatograph Equipped with a Flame Photometric Detector. Anal Chem 68: 1976-1981.

68.   Kumata, Hidetoshi, Yamada, Junya, Masuda, Kouji, Takada, Hideshige, Sato, Yukio (2002) Benzothiazolamines as Tire-Derived Molecular Markers: Sorptive Behavior in Street Runoff and Application to Source Apportioning. Environmental Science & Technology 36: 702-708.

69.   Zeng EY1, Tran K, Young D (2004) Evaluation of potential molecular markers for urban stormwater runoff. Environ Monit Assess 90: 23-43.

70.   Murakami, Michio, Yamada, Junya, Kumata, Hidetoshi, Takada, Hideshige (2008) Sorptive Behavior of Nitro-PAHs in Street Runoff and Their Potential as Indicators of Diesel Vehicle Exhaust Particles. Environmental Science & Technology 42: 1144–1150.

71.   Kose, T, Yamamoto, T, Anegawa, A, Mohri, S, Ono, Y (2008) Source analysis for polycyclic aromatic hydrocarbon in road dust and urban runoff using marker compounds. Desalination 226: 151-159.

9

Extended Producer Responsibility and Product Stewardship for Tobacco Product Waste

Clifton Curtis[1], Susan Collins[2], Shea Cunningham[3], Paula Stigler[4] and Thomas E Novotny[5*]

[1]*Director, The Varda Group; and Policy Director, Cigarette Butt Pollution Project, USA*

[2]*President, Container Recycling Institute, USA*

[3]*Sustainability Policy, Research & Planning Consultant, Container Recycling Institute, USA*

[4]*Assistant Professor, University of Texas Health Sciences, San Antonio Regional Campus, USA*

[5]*Chief Executive Officer, Cigarette Butt Pollution Project and Professor of Epidemiology, Graduate School of Public Health, San Diego State University, USA*

*Corresponding author: Thomas E Novotny, Chief Executive Officer, Cigarette Butt Pollution Project and Professor of Epidemiology, Graduate School of Public Health, San Diego State University, USA
E-mail: tnovotny@cigwaste.org

Abstract

This paper reviews several environmental principles, including Extended Producer Responsibility (EPR), Product Stewardship (PS), the Polluter Pays Principle (PPP), and the Precautionary Principle, as they may apply to tobacco product waste (TPW). The review addresses specific criteria that apply in deciding whether a particular toxic product should adhere to these principles; presents three case studies of similar approaches to other toxic and/or environmentally harmful products; and describes 10 possible interventions or policy actions that may help prevent, reduce, and mitigate the effects of TPW. EPR promotes total lifecycle environmental improvements, placing economic, physical, and informational responsibilities onto the tobacco industry, while PS complements EPR, but with responsibility shared by all parties involved in the tobacco product lifecycle. Both principles focus on toxic source reduction, post-consumer take-back, and final disposal of consumer products. These principles when applied to TPW have the potential to substantially decrease the environmental and public health harms of cigarette butts and other TPW throughout the world. TPW is the most commonly littered item picked up during environmental, urban, and coastal cleanups globally.

Keywords: Tobacco control; Tobacco product waste; Cigarette butts; Producer responsibility; Product stewardship

Introduction

The human health effects of smoking are well known, but far less is known about the environmental impacts of tobacco product waste (TPW), especially cigarette butts. This paper addresses the environmental concerns regarding TPW throughout its lifecycle, with special emphasis on cigarette butt waste. The lifecycle environmental issues for tobacco include the growing process (with concerns for heavy pesticide and petroleum-based fertilizer use, land degradation, and deforestation) [1,2], as well as production (manufacturing, packaging, and distribution wastes)[3]; and consumer use (including CO_2 production, methane release, second hand smoke exposure, and third-hand smoke effects[4]), and finally, disposal of cigarette butts and packaging as TPW[5,6].

There were an estimated 5.5 trillion cigarettes sold globally in 2011, with approximately 293 billion sold in the United States [7,8]. By some estimates, at least one-third of all cigarettes smoked are tossed into the environment, comprising by far the largest single type of litter by count, about 30-40% of all items picked up, in coastal and urban cleanups dating back to the 1980s[9].

Beginning in the 1950s, the tobacco industry shifted production of manufactured cigarettes from unfiltered to filtered, using a variety of different components. The filtered cigarettes were marketed as being "healthier" in response to the new concerns for the health risks of smoking [10]. Since at least the 1990s, over 98% of all cigarettes sold in the United States are filtered, and nearly all of the filters sold are made of cellulose acetate, a separately manufactured plastic element that is attached to the tobacco product [11,12]. The increase in production and the fraudulent marketing of filtered cigarettes as a healthier option for smokers over the last 60 years presents us with not only a public health problem due to the filter fraud, but also an environmental concern with the non-biodegradable filters that are the primary component of discarded cigarette butts.

The US National Cancer Institute reviewed the changing cigarette product, in particular 'light' and 'low-tar' designations, and concluded that "Epidemiological and other scientific evidence, including patterns of mortality from smoking-caused diseases, does not indicate a benefit to public health from changes in cigarette design and manufacturing over the last fifty years"[13]. This design specifically refers to the filtered cigarette, and thus discarded cigarette butts, especially the plastic filters, may be considered a dispersed source of non-biodegradable, toxic environmental waste that could be subject to elimination without concern for the health effects of this product change [14]. Filters are still believed by many smokers and non-smokers to be health-protective devices, but there have been no benefits to public health from filters, and in fact the risks for lung cancer and chronic pulmonary disease due to smoking have actually increased since becoming widely used by uninformed smokers.

The cost to municipalities to clean up TPW is substantial. The City and County of San Francisco studied the costs of litter cleanup and disposal in 2007-2009 and estimated the costs attributed to TPW to be

$22 million annually [14]. A separate study, funded by the US Environmental Protection Agency (EPA), estimated total cleanup, prevention and disposal costs of all sources of litter (including TPW) at over $500 million for West Coast communities [15]. From an environmental perspective, aquatic ecosystems, such as shorelines and waterways, may be very vulnerable to the environmental impact of TPW, as so much of this waste is deposited on land and ultimately flows downstream via storm drains, rivers, creeks and other pathways to those environments [16].

Under specific circumstances of sunlight and moisture, the filter component of cigarette butts may be broken into smaller plastic pieces that also contain and leach out some of the seven thousand chemicals contained in a cigarette [17]. Many of these chemicals, such as ethyl phenol, heavy metals, and nicotine, are in themselves environmentally toxic, and at least 50 are known human carcinogens [18]. TPW leachates have in fact been shown to be of environmental concern, with measureable amounts of heavy metals such as cadmium, arsenic, and lead in laboratory analyses [19]. They have been found to be acutely toxic to freshwater micro-organisms, with the main lethal chemicals being nicotine and ethyl phenol [20]. Recent studies using standardized EPA toxicity assessment protocols have shown that cigarette butts soaked in either fresh or salt water for 96 hours have a Lethal Concentration 50 (killing half the exposed test fish) of about one cigarette butt per liter [21].

In a May 2011 editorial in the international public health journal, Tobacco Control, tobacco control advocates and scientists urged key stakeholders "to join forces and find solutions for eliminating this especially toxic form of [cigarette butt-related] trash"[22]." The Washington, DC-based Legacy Foundation, which helped fund that special journal supplement, then convened a national webcast focusing on how public health experts, policy leaders, environmental, and community leaders can eliminate toxic TPW [23].

This paper reviews Extended Producer Responsibility (EPR), Product Stewardship (PS), and two additional related environmental principles as possible approaches to TPW prevention, reduction, and mitigation. We will also review criteria that may apply in deciding whether TPW may adhere to EPR/PS. We then present three case study summaries of EPR/PS approaches that have been used with other environmentally harmful products. Finally, we propose ten policy actions that can help prevent, reduce, and mitigate the potential environmental impacts of TPW.

Review of Extended Producer Responsibility, Product Stewardship, and Other Related Environmental Principles

Extended Producer Responsibility

The EPR concept dates to the early 1990's when Thomas Lindhqvist, a Swedish graduate student, prepared a report for Sweden's Ministry of the Environment that called for making manufacturers of products responsible for the entire lifecycle of the products they produce [24]. Lindhqvist defined EPR as "an environmental policy protection strategy to reach an environmental objective of a decreased total environmental impact from a product, by making the manufacturer of the product responsible for the entire life-cycle of the product and especially for take-back, recycling and final disposal of the product." Three central tenets embedded in this concept were:

To internalize the environmental cost of products into their retail price.

To shift the economic burden of managing toxicity and other environmental harm associated with post-consumer waste away from local governments and taxpayers and on to producers.

To provide incentives to producers to incorporate environmental considerations in the design of their products.

Lindhqvist's focus included those three tenets, as well as four specific categories of responsibility (Figure 1).

EPR-based laws have been enacted in more than 20 US states, with legally binding features requiring manufacturers of products containing toxic or environmentally unsustainable materials to take responsibility for management throughout key parts of their lifecycle, especially for management of post-consumer waste [25]. The products addressed are diverse, including: paint, batteries, beverage containers, pesticide containers, electronics, packaging, cell phones, sharps, carpets, fluorescent lighting, mercury thermostats, radioactive devices, motor oils, mattresses, plastic bags, photographic film, smoke detectors, and auto switches.

Internationally, EPR laws and regulatory systems have been implemented in several countries, including Canada, the European Union member states, Australia, New Zealand, Japan, Sweden and Norway [26]. As with U.S. states, international approaches vary widely with respect to specific producer, consumer, retailer, and government responsibilities for end-of-life product management.

Liability refers to responsibility for proved environmental damages caused by the product in question. The extent of the liability is determined by legislation and may embrace different parts of the lifecycle of the product, including usage and final disposal.

1. *Economic responsibility* means that the producer will cover all or part of the costs for the collection, recycling, or final disposal of the products manufactured; these expenses could be paid directly by the producer or by a special fee.
2. *Physical responsibility* is used to characterize systems in which the manufacturer is involved in the physical management of the products or in managing the adverse effects of the products throughout their life cycle; here, the manufacturer may be linked to environmental impacts of the product.
3. *Informative responsibility* requires the manufacturer to supply information on the environmental risks of the products manufactured.

Figure 1: Categories of Extended Producer Responsibilities [24].

There are three reasons why producers should assume EPR for TPW management at the end of tobacco product life [27]:

Shifts waste management responsibilities and costs from local governments and taxpayers back to the polluter/producer, which in most instances is in the private sector;

Economic costs of TPW management may encourage manufacturers to design non-toxic, non-hazardous products; and

Internalizing the costs of waste product management to the producer will be fairer overall when net external costs to the public and communities are taken into account.

While all of these rationales are understandable with respect to shifting economic responsibility to producers of most consumer products, pursuing tobacco product design changes to reduce TPW toxicity are unlikely to be effective. The tobacco product is inherently hazardous to human health and contains many chemicals that are on the Agency for Toxic Substances and Disease registry priority list of hazardous substances as well as the State of California Proposition 65 list of chemicals known to cause cancer [24,28]. Chemicals covered by these lists are those that cause one or more of the following: cancer or other chronic human health effects, significant adverse acute human health effects, or significant adverse environmental effects. TPW will remain filled with these chemicals, no matter how the tobacco product or filter is altered.

Product Stewardship

PS contrasts with EPR in that PS may involve other actors along the supply and retail chain, whereas EPR focuses all the responsibility for waste management onto manufacturers [29,30]. During the early-to-mid 1990s, the idea of shared responsibility, also referred to as "product responsibility," began generating attention. PS was possibly introduced by industry as a way to dilute the EPR concept and share responsibility rather than have all responsibility fall to the producers [31]. PS is usually designed as a voluntary system that shares responsibility for the adverse environmental effects of products by all parties involved in the lifecycle [32]. PS principles therefore require much wider and more diverse involvement of parties than does an EPR-only-based approach (Figure 2) [33].

PS is the act of minimizing health, safety, environmental and social impacts, and maximizing economic benefits of a product and its packaging throughout all lifecycle stages. The producer of the product has the greatest ability to minimize adverse impacts, but other stake holders, such as suppliers, retailer, and consumers, also play a role. Stewardship can be either voluntary or required by law.

Figure 2: Joint Statement on Product Stewardship Principals by the Product Policy Institute, PS Institute, and California PS Council, 2012.

A key variable determining whether a product management system may be eligible for EPR and/or PS involves the funding system that is adopted for these approaches. With EPR, costs are to be paid by the producer; when the program is a cost-sharing arrangement between producers and other stakeholders, it would be thought of as a PS approach. To date, the tobacco industry has denied any form of producer responsibility for TPW, shifting almost the entire focus onto the consumer. For example, in formerly secret tobacco industry documents, Philip Morris companies, Inc. (1998) described their position on EPR as follows:

"The Company opposes the concept of manufacturer/producer responsibility when defined to mean that the manufacturer/producer must accept sole and complete responsibility for a product/package throughout its life cycle. Specifically, it's the Company's position that these waste management practices….create a highly inefficient system due to the fact that responsibility continues even after the manufacturer has relinquished control of the product and/or package. Consistent with our Environmental Principles, we commit to provide consumers with appropriate and useful information on the environment and their role, as well as that of communities and business, in becoming part of environmental solutions, including those which affect solid waste management" [34]."

Other Environmental Principles that May be Applicable to TPW:

The Polluter Pays Principle (PPP) was introduced in the 1970s by the Organization for Economic Cooperation and Development (OECD) in consideration of the economic costs associated with protection of the environment. As framed, the PPP "[meant] that the polluter should bear the expenses of carrying out [pollution prevention and control] measures decided by public authorities to ensure that the environment is in an acceptable state [35]." A 1989 OECD initiative, dealing with accidental pollution, made specific reference to "hazardous" components when invoking the use of PPP [36]. The PPP integrates environmental protection, social development, and economic activities by using market and/or regulatory instruments to ensure that persons or organizations responsible for pollution bear the full environmental and social costs of their activities, and that those costs are reflected in the market price for goods and services. Over time, the PPP has become a generally accepted principle of international environmental law and policy, perhaps most advanced in its application within the EU, in focusing attention and responsibility on polluting sources. In line with resistance to EPR, there is no evidence that PPP has applied to tobacco industry responsibility for TPW.

The Precautionary Principle is based on the caution that governs many aspects of daily life, and responds to the complexity of environmental risks to health and the often indeterminate nature of cause-and-effect relationships between potentially hazardous waste products and health effects. This principle first appeared in the 1970s as a basis for water protection policies in Germany. Over the years it has provided an overarching framework for addressing threats from toxic chemicals involving a wide range of exposures [37]. At its core, this principle calls for preventive, anticipatory measures to be taken when an activity raises threats of harm to the environment, wildlife, or human health, even if cause-and-effect relationships are not fully established. The principle is instructive with regard to TPW, given the evidence that this waste stream has toxic, carcinogenic, and otherwise harmful chemicals derived from tobacco products and the attached cellulose acetate filters. As such, prevention, reduction and mitigation efforts involving TPW could be undertaken to help prevent potential TPW-related harm to humans, animals, and ecosystems before it is evident. Moreover, application of this principle would include not only current TPW prevention and reduction, but would apply to past polluting practices that have produced environmentally persistent TPW such as cigarette filters and plastic packaging.

Review of Criteria for Applying EPR/PS

A variety of criteria may be applicable in determining whether TPW should adhere to EPR and complementary PS principles and standards. The six criteria mentioned below are framed as questions for which the answers provide a sense of whether any consumer product waste should qualify for an EPR/PS based policy, legal, regulatory, or voluntary regime (Figure 3) [38,39].

Criteria similar to these have been used in several states, including California, Oregon, and Washington, to determine whether consumer product wastes should be managed through EPR and/or PS approaches [40,41,42]. With regard to TPW, the end-of-life tobacco product phase is a strong candidate for use of EPR/PS. As noted earlier in relation to the core tenets of EPR framed by Lindhvqist, TPW will not be amenable to resource recovery/conservation or environmental design as described in the final criterion in Figure 3. Although detoxification, biodegradation, and disposal strategies may be among the best options for EPR/PS approaches to many other consumer products, TPW may need more novel approaches, given its ubiquity and the specific toxic, harmful environmental contaminants it produces.

Relevant Case Studies

EPR/PS have been applied to a variety of products, including beverage containers, paint, and batteries. Many other products are also candidates for application of those environmental principles, especially for products with similar toxic characteristics. We present three case study summaries that may inform EPR/PS-related strategies for TPW (full case studies are available from the authors).

The Oregon PaintCare Stewardship Program

Leftover paint is the largest component of household hazardous waste (HHW) in the United States. The EPA estimates that about 10% of all paint purchased each year (approximately 64 million gallons) goes unused [43] According to the EPA, municipal governments, which bear the managerial burden of leftover paint collection, could avoid more than a half billion dollars annually in mitigation costs with a paint stewardship program managed by the paint industry and funded by consumers.

Oregon enacted the first paint stewardship law in the United States in 2009 and followed up with strengthening amendments in 2013 [44]. Its PaintCare program requires paint manufacturers to implement a cost-effective and environmentally sound program for managing left-over paint. The program mandates a recycling fee at the point of sale for paint in five-gallon or less containers; these fees must be sufficient to cover all program administrative and recycling costs. At the end of the first two years of the program, collection sites numbered more than 100, most of which are located in retail paint stores. During that same time, the quantity of paint collected and reprocessed increased by 34%. Oregon's PaintCare program uses a solid-waste management hierarchy similar to that followed by the EPA, which focuses sequentially on source reduction, reuse, recycling, recovery, and biodegradation [45].

While the Paintcare program overall has received high approval ratings, several challenges required amendments to the program; the program in Oregon was made permanent in June 2013. PaintCare programs have now been established in California and Connecticut, and four other states have passed similar legislation (Rhode Island, Vermont, Minnesota and Maine).

Stewardship-related elements of PaintCare in four areas are very relevant and worthy of replication in addressing TPW issues:

Creation of a stewardship organization: Similar to features in Oregon's state law addressing leftover paint, a TPW stewardship entity could be established at the state level, as a corporation or nonprofit, created by the tobacco industry. This would be accorded the legal mandate and responsibility to implement an industry-sponsored program, with oversight provided by the state's environmental quality/ protection department or agency. Given the health-related issues involved with TPW, consultation with a state's health department or agency would be recommended.

Access to convenient collection sites: Establishment of Hazardous Household Waste and retail collection sites for TPW. Given the toxic,

poisonous nature of the substantial quantities of filters and remnant tobacco and paper discarded randomly, though, TPW protective recovery paraphernalia would likely be needed in gathering and returning TPW to collection sites.

Educational and outreach activities: A TPW stewardship organization could promote activities, including but not limited to signage, written materials and templates for reproduction, shared with retailers for distribution to the consumer at time of sale. Also, they could identify collection opportunities, and promote waste management hierarchy, especially technologies and management of biodegradation and safe, secure disposal, consistent with other applicable laws.

Plans, annual reports and program/budgeting: Documentation tasks involving required review and approval by an oversight department should be part and parcel of the responsibilities associated with any TPW program. Having in place supervisory review, guidance and sign-off, independent yet deeply familiar with and interested in the operations of the program, would help significantly to facilitate and ensure its accountability and success.

- Does the product create/cause adverse risks to the environment or to public health and safety?
  - Are toxic and hazardous constituents present?
  - Are there adequate, mandatory contaminate controls in place to address those risks?
- Does the product's post-consumer waste significantly burden government solid waste or other cleanup and disposal programs?
  - Do local governments and taxpayers bear most or all of the management costs?
  - Does the waste have the potential to act as a contaminant in those programs, in relation to storm water contamination or other diffuse, uncontrolled disposal?
- Are existing, voluntary cleanup/disposal programs effective?
  - Is the cleanup/disposal of the waste dealt with in a safe, responsible manner?
  - Is most or all of the waste that is created by the product part of those programs?
- Are there examples of success in collecting and processing other toxic or non-toxic products in other states or countries?
  - Do those other examples involve products that result in environmental or public health or safety risks similar to those associated with the product at issue?
  - Do other states or countries have problems applying similar rules to the product at issue, relative to programs for other products addressing comparable risks?
- Are existing EPR or PS programs for the product at issue viewed as not being effective?
  - Do existing programs lack mandatory measures necessary to achieving a sufficient degree of compliance and success?
  - Has the producer been uncooperative in trying to find effective solutions in addressing the necessary safe cleanup and disposal of the product?
- Does the product have potential for enhanced resource conservation?
  - Potential resource recovery and material conservation?
  - Are there opportunities for environmental design/increased reuse or recycling?

Figure 3: Criteria for EPR/PS Approaches to Consumer Product Waste

British Columbia Beverage Container Recycling Program

The Litter Act of 1970 in British Columbia was the first beverage container deposit law, and the first EPR law in North America; the law is now called the Recycling Regulation, and litter concerns were a primary reason for passage of this law [46]. This container recycling program now requires a mandatory deposit on every beverage container offered for sale (with minor exceptions). Consumers pay the deposit at the time of purchase, and the deposit amount appears on their receipt. Consumers return empty beverage containers to retail stores or special take-back locations ("depots"), and they receive the full amount of the deposit in return. This program achieves a recycling rate of 80% or more.

Such a container deposit law places a monetary value on beverage containers, and this value then reduces litter in two ways: (1) people are less likely to litter, because the container can be returned for a refund, and (2) if they do litter, another person is likely to pick up the container and return it to receive the refund. Data from the Great Canadian Shoreline Cleanup indicate that beverage container litter is about 30% lower in British Columbia than in the provinces of Manitoba and Ontario, where there are no mandatory container deposit-refund laws [47].

Recycling centers, where consumers return beverage containers to the depots for recycling and refilling, are licensed for a specific geographic area. Consumers may also return limited numbers of containers to grocery stores and liquor stores. The beverage producers operate the deposit-refund system in British Columbia, and there are no statutory fees or charges remitted to government under the system. To carry out deposit-refund obligations within a province-wide system, beverage producers formed two stewardship organizations to manage program operations. These stewardship organizations are answerable to the Provincial Ministry of Environment, the agency authorized to carry out the Recycling Regulation. This Ministry approves the stewardship organization program plans as well as annual reports and five-year updates.

Among the 20 states with EPR-based laws, the beverage container deposit laws adopted first in Oregon and Vermont in the 1970s to reduce container litter are noteworthy. Prior to this, the vast majority of packaged beverages were sold in refillable bottles, and consumers returned those containers to retrieve deposits. In the 1960s, the ownership and distribution streams for beverage companies were consolidated, and these companies almost completely embraced single-use containers. There are now beverage container deposit laws in 10 U.S. states and the Territory of Guam, in 10 Canadian Provinces, and in more than 20 other countries worldwide.

A deposit-return scheme for TPW may only be feasible if the toxicity of the returned TPW can be managed. TPW differs significantly from beverage containers because of esthetics (odor), toxicity of the chemicals exuded from TPW, and the special care that may be necessary with regard to disposal and transport of this toxic waste material. Thus, other models of product stewardship for toxic and/or hazardous waste products may be more applicable.

Recycling Household Batteries in Canada

Many household batteries are classified as hazardous waste and contain a number of heavy metals and toxic chemicals such as mercury, lead, and cadmium. When these batteries are incinerated or deposited in landfills, they can contaminate soil and waterways, and they may present a risk to human health [48]. Battery recycling aims to reduce the number of batteries disposed of as municipal solid waste.

Canada's legislation and management of used batteries is conducted on a province-by-province basis. However, the responsible parties for collecting and recycling used batteries are the manufacturers, brand owners, or first importers. These businesses joined collectives and established two non-profit stewardship organizations: US-based Call2Recycle (operating in all provinces) and Stewardship Ontario (Ontario only) [49]. These organizations are financed by manufacturers on a market share-based-reimbursement arrangement. Neither organization receives funding from the government. There are now approximately 7,000 used battery drop-off sites at retail centers,

public agencies, community centers, and businesses in British Columbia, Manitoba, Ontario, and Quebec.

Ontario, in fact, has an extensive list of HHW's that must be recycled or taken back. These include: antifreeze, lubricating oil and filters, fertilizers, paints, solvents, and single-use dry batteries. The recycling drop-off service (known as Orange Drop) is free to consumers (www.makethedrop.ca). In addition to the drop off service, Stewardship Ontario funds that province's Battery Incentive Program (BIP), which pays transporters for returning recycled batteries. Ontario's battery take-back system has resulted in battery recycling rates of 12%, and these are the highest in North America [50].

Call2Recycle also recycles rechargeable batteries in the province [51]. Contractors receive the used batteries at their warehouses, record details about the weight and battery types of the shipment, and then separate the batteries by chemical content. The batteries are then shipped to the appropriate specialty processors by chemical type. The processors extract usable chemicals and metals to be used in the manufacture of new products. Waste products are disposed of according to Responsible Recycling and Basel Action Network standards [52].

Like spent batteries, TPW contains toxic components (ethyl phenol, nicotine, heavy metals, and many carcinogens) that can have detrimental effects to human health and the environment. Thus, a take-back program, as demonstrated in Canada for spent batteries, may be applied to TPW.

EPR/PS-Related Policy Actions to Prevent, Reduce, and Mitigate TPW

To expand on these findings, we next present ten policy approaches based on the weight of evidence regarding the environmental impact of TPW thus far.

Extended producer responsibility and product stewardship laws and programs

TPW prevention, reduction, and mitigation could be made the responsibility of the tobacco industry as well as other parties in the lifecycle of tobacco product sales and usage through EPR/PS. These could be legally binding and/or voluntary programs for cleanup, take-back, and final disposal. In addition, public agencies tasked to regulate the stewardship agencies, similar to those involved in the battery recycling case study, must ensure follow-through on obligations by benchmarking, setting financial and operational reporting standards, requiring transparency, requiring annual reporting and third party audits, and establishing mechanisms for public input and continual improvement.

Bans of single use, disposable filters

Sales of some products known to be hazardous or prone to improper disposal have simply been banned by state-level authorities (e.g., pop-tops on aluminum cans, plastic tampon applicators, etc.). Given that cellulose acetate cigarette filters are not biodegradable and may cause significant environmental degradation, and given their toxicity, persistence, and ubiquity, a sales ban on single-use filters on cigarettes would reduce a significant portion of TPW. Unfortunately, tobacco remnants from unfiltered cigarette butt waste will still leach out some toxicants, but the removal of the plastic filter will reduce a significant volume of TPW while reducing the time needed for any biodegradability of the tobacco remnant.

Bans on outdoor public smoking

Laws that ban smoking vary widely across the United States, with some states banning it in certain areas and others banning it nearly everywhere. According to a 2013 report of the American Nonsmokers' Rights Foundation [53], more than 80% of the U.S. population now lives under a ban on smoking in "workplaces, restaurants, and/or bars, by a state, commonwealth, or local law," though only 48.7% live under a ban covering all workplaces, restaurants, and bars. In addition, as of April 5, 2013, at least 1,159 U.S. colleges or universities have adopted 100% smoke free campus policies [54]. Overall, these restrictions on smoking serve to change the social norm against public smoking and may also reduce the burden of TPW in outdoor environments if properly enforced. On the other hand, as smokers must move outdoors to smoke, more TPW is deposited onto streets, parks, and other public outdoor spaces, and this is then more likely to wash into storm drains and aquatic environments.

Product Labeling

Some products carry warnings not to litter the product or packages, but this intervention has never been used to inform smokers about the non-biodegradability of filters or tobacco packaging waste. Under the 2009 Family Smoking Prevention and Control Act [55]the Food and Drug Administration could require a label of sufficient size that simply states: "Cigarette filters are non-biodegradable toxic waste. Safe disposal should be required in accordance with state law." Additional information could also describe potential toxicity of TPW, methods for safe handling, and applicable fines for littering.

Litigation against the Tobacco Industry

To date, most litigation against the tobacco industry has focused on health care costs [56]. Similarly, the industry could be held responsible for the environmental costs associated TPW cleanup. Litigation has been pursued against manufacturers of products that damage the environment, with those lawsuits typically based on negligence and nuisance-related legal theories involving proof of the defendant's wrongful conduct, for failure to take reasonable steps to prevent harm, or for protecting someone's right to use and enjoy real property. Given the accumulating evidence for the toxicity of TPW, the tobacco industry may be considered a toxic waste generator, and thus they may be liable for the costs of safe clean-up, take-back, or disposal of their products.

Litter fees

TPW cleanup and disposal costs are substantial at local municipal levels. As noted earlier, local authorities may apply litter fees as part of a program framework to recover cleanup and abatement costs, to conduct public education, and to administer the program [13].

Deposit/Return

Similar to beverage container deposit laws, cigarettes could be sold with a "butt deposit" to be refunded when the cigarette butts are returned to the vender. The challenge in such a program would be to develop safe transport and destruction mechanisms for TPW as part of a take back and disposal regime.

Waste fees

Concern about toxic waste resulting from contaminated products has given rise to consumer-funded Advanced Recycling Fees (ARF) [57]. Assessed at the point of purchase, such fees can help cover the costs of recycling the item and properly disposing of non-recyclable material. ARFs differ importantly from EPR approaches in that ARFs are set by the government as a fixed fee paid when products are purchased and are used to manage a governmental program. EPR would involve a variable fee set by producers based on the true cost of recovery, with their programs financed and managed by producers. Fees typically fund recycling collection systems and provide no economic incentives for the consumer or for system efficiency. EPR, on the other hand, shifts the focus upstream, providing an incentive to manufacturers to reduce recycling costs and to improve product/packaging design for source reduction and increased recyclability. If applied to TPW, the fee could potentially contribute to butt collection and transfer centers, as well as to the establishment of monitored, hazardous waste storage sites for TPW.

Fines for Littering

Fines are levied by state and local communities for littering on roadways, beaches, parks, and other public spaces [58]. Fines could also be levied against cigarette manufacturers based on the quantity of brand-specific cigarette waste found on cleanups or as improperly disposed waste from ashtrays, cigarette butt receptacles, or other sources. The fines would at least partially compensate taxpayers for clean-up, collecting, and disposing of cigarette waste. At a national level, the Comprehensive Environmental Resource Compensation and Liability Act of 1980 (Superfund Program), provides a broad framework for requiring companies or other parties to clean-up pollution activities for which they are responsible and/or to pay fines and damages associated with the pollution being cleaned up by others [59].

Changing social norms

Changing the perceptions about TPW as harmless litter will involve extraordinary social normative changes in the smoking ritual itself [60].Smokers and non-smokers alike must recognize the externalities of discarding TPW. Cigarette butts are not simply a minor littering problem but rather an externality burdening non-smokers and communities. TPW is the most common waste product (by count) globally. The policy interventions listed above will all contribute to a changing social norm about smoking. Smoking itself has become less and less socially acceptable; TPW disposal into the environment should also become less and less socially acceptable, and its differential impact on poor and minority communities may also classify it as a social justice issue.

Conclusion

This review suggests that there is precedent for enacting local, state, and national laws, regulations and other mandatory or voluntary interventions to protect the environment from toxic and non-biodegradable solid TPW through EPR/PS. TPW has not as yet been subject to any systematic take back or safe disposal regulations that create EPR for the tobacco industry; nor has it been subject to PS, whereby others along the supply and retail chain, including distributors, retailers, employers, governments, or other parties may share responsibility to prevent TPW contamination of the environment. Despite EPR and PS taking different approaches for responsibility, the two principles are best viewed as complementary in that they can work in tandem to prevent, reduce, and mitigate TPW's environmental effects.

The first tenet of EPR calls for internalizing the environmental cost of products into their retail price, and the second tenet calls for shifting the economic burden of managing toxicity and other environmental harm associated with post-consumer waste away from local governments and taxpayers and on to producers. For TPW, they are both very applicable, and very appropriate. Regrettably, the third tenet, providing incentives to producers to incorporate environmental considerations in the design of the product, is unachievable, given the toxic, hazardous chemicals permanently embedded in the tobacco product. Nonetheless, a specific sales ban on single use filters, which are not a health-protective device, may reduce the non-biodegradable portion of TPW. In 2014, a bill was introduced in the California Assembly to ban the sale of single-use cigarette filters for environmental reasons; states have the authority to restrict the sales but not the manufacturing of tobacco products. While the bill did not emerge from committee deliberations, this novel approach is very likely to be considered again in California and other jurisdictions [61].

Based on application of EPR and PS principles to other products, PS is more likely to be the operative system for TPW prevention, reduction, and mitigation, though steps may be taken to place financial responsibilities on to the tobacco industry. While safe cleanup and disposal approaches to TPW would benefit greatly from an EPR and/or PS regime, the tobacco industry is likely to fiercely resist any measures that would shift responsibility directly back to the industry, or to other parties involved in the lifecycle, for the environmental costs or impacts of TPW [62].

The Polluter Pays and Precautionary Principles may apply to TPW. The PPP supports the view that the commercial polluter should bear the environmental and social costs of its activities, with those costs reflected in the market price for goods and services. The PPP, as well as the Precautionary Principle, support EPR/PS application, given the evidence that the TPW waste stream has toxic, carcinogenic, and otherwise harmful chemicals derived from tobacco products and the attached cellulose acetate filters. As such, prevention, reduction and mitigation efforts involving TPW should be undertaken to help prevent potential harm to humans, animals, and ecosystems before it is evident.

For some of ten policy actions we have reviewed for EPR/PW application to TPW, the connection is direct while for others it is indirect. The 1st policy action, calling for laws and programs that mandate EPR and/or PS as well as voluntary actions by the tobacco industry can be compared to policies applied to toxic as well as non-toxic products as suggested by our three case studies. The 5th policy action, litigation, places responsibility via negligence, nuisance, product liability, and other legal theories on to the tobacco industry for failure to take reasonable steps to prevent harm or protect rights to use and enjoy real property (e.g., the beach environment). Evidence of those types of harms may be sufficient to make the industry liable for safe clean-up, take-back and/or disposal of their products under an EPR/PS system. Four of the other ten policy actions focus on litter fees, deposit-return refunds, waste fees and fines, which involve the exchange of money among parties who are stakeholders in the life cycle of tobacco products. In that context, PS applies, as these parties share responsibility for managing the cleanup and disposal of TPW. The remaining four policy actions focus on bans of disposable filters,

bans on public smoking, mandates for labeling, and efforts to change social norms. All ten policy options contribute to changing of social norms about TPW and may help frame new channels through which society may achieve the end of the tobacco use epidemic.

This review is limited by the existing disconnect between the perception of TPW as harmless waste and the growing recognition that it is toxic and hazardous. Consumers, environmental policymakers, and even smokers do not fully recognize the environmental issues around TPW, and hence, EPR/PS strategies have not been considered for TPW. The focus of the review is further limited to the post-consumer, downstream, end-of-life management of TPW. As referenced in the Introduction, however, there are also environmental impacts at or near the upstream, front-end of the product life cycle, involving the growing process, manufacturing processes, and product design. Ideally, an integrated, comprehensive system of EPR/PS-related management is needed to prevent, reduce and mitigate TPW throughout its life cycle.

Nevertheless, while strategies to reduce smoking and mitigate TPW may vary significantly in their methods and aspirations, they share two core goals: 1) the status quo is unacceptable, and 2) reducing TPW, its environmental impacts, and smoking overall, will require bold, new, and fundamentally different strategies to assure success. These will require a diverse mix of ideas for achieving the goal of a TPW-free environment and better understanding of the life-cycle environmental hazards of tobacco productions, marketing, and consumption. We have asserted that EPR/PS may provide important pathways to achieve these goals.

Acknowledgement

Research reported in this publication was supported by the US National Cancer Institute of the US National Institutes of Health under award number 2R01CA091021-10A1 and by University of California Office of the President's Tobacco-related Disease Research Program, under award number 21XT-0030.

References

1. Lecours N, Almeida GE, Abdallah JM, Novotny TE (2012) Environmental health impacts of tobacco farming: a review of the literature. Tob Control 21: 191-196.

2. Geist HJ (1999) Global assessment of deforestation related to tobacco farming. Tob Control 8: 18-28.

3. Novotny TE, Zhao F (1999) Consumption and production waste: another externality of tobacco use. Tob Control 8: 75-80.

4. Matt GE, Quintana PJE, Destaillats H, Gundel LA, Sleiman M, et al. (2011) Thirdhand tobacco smoke: Emerging evidence and arguments for a multidisciplinary research agenda. Environ Health Perspec 119: 1218-1226.

5. Novotny TE, Lum K, Smith E, Wang V, Barnes R (2009) Cigarettes butts and the case for an environmental policy on hazardous cigarette waste. Int J Environ Res Public Health 6: 1691-1705.

6. Action on Smoking and Health (2009) Tobacco and the Environment.

7. U.S. Department of Agriculture (2007) Tobacco Outlook. Electronic Outlook Report from the Electronic Research Service TBS -263 Washington DC: US Department of Agriculture.

8. Tynan MA, McAfee T, Promoff G (2012) Consumption of Cigarettes and Combustible Tobacco-United States, 2000–2011. Morbidity and Mortality Weekly Report (MMWR) 61:565-569.

9. Ocean Conservancy (2012) International Coastal Cleanup 2012 Data Release.

10. Pollay RW, Dewhirst T (2002) The dark side of marketing seemingly "Light" cigarettes: successful images and failed fact. Tob Control 11 Suppl 1: 18-31.

11. Zipser A (1954) Cigarette industry convalescing; filter prescription seems to help: cigarette output survives a crisis. New York Times.

12. Harris B (2011) The intractable cigarette 'filter problem'. Tob Control 20 Suppl 1: i10-16.

13. US National Cancer Institute (2001) Risks Associated with Smoking Cigarettes with Low Machine-Measured Yields of Tar and Nicotine. Tobacco Control Monograph.

14. Schneider JE, Peterson NA, Kiss N, Ebeid O, Doyle AS (2011) Tobacco litter costs and public policy: a framework and methodology for considering the use of fees to offset abatement costs. Tob Control 20 Suppl 1: i36-41.

15. Stickel BH, Jahn A, Kier W (2012) The Cost to West Coast Communities of Dealing with Trash, Reducing Marine Debris.

16. KAB-MSW Consultants (2009) National Visible Litter Survey and Litter Cost Study. New Market.

17. Moerman JW, Potts GE (2011) Analysis of metals leached from smoked cigarette litter. Tob Control 20 Suppl 1: i30-35.

18. US (2014) Department of Health and Human Services. The Health Consequences of Smoking—50 Years of Progress: A Report of the Surgeon General.

19. Moerman JW, Potts GE (2011) Analysis of metals leached from smoked cigarette litter. Tob Control 20 Suppl 1: i30-35.

20. Warne MStJ, Patra RW, Cole B (2002) Toxicity and a hazard assessment of cigarette butts to aquatic organisms.

21. Slaughter E, Gersberg RM, Watanabe K, Rudolph J, Stransky C, et al. (2011) Toxicity of cigarette butts, and their chemical components, to marine and freshwater fish. Tob Control 20 Suppl 1: i25-29.

22. Healton CG, Cummings KM, O'Connor RJ, Novotny TE (2011) Butt really? The environmental impact of cigarettes. Tob Control 20 Suppl 1: i1.

23. Legacy (2014) Environmental Impact of Cigarettes.

24. Lindhqvist T (2000) Extended Producer Responsibility in Cleaner Production Policy Principle to Promote Environmental Improvements of Product Systems.

25. Product Stewardship Institute (2014) Extended Producer Responsibility State Laws as of January 2014.

26. Cal Recycle (2014) Product Stewardship and Extended Producer Responsibility (EPR).

27. New South Wales Environmental Protection Agency (2013) Extended Producer Responsibility Clean Production.

28. Ostrowski SR, Wilbur S, Chou CHSJ, Pohl HR, Stevens YW, et al. (1999) Agency for Toxic Substances and Disease Registry's 1997 priority list of hazardous substances. Latent effects—carcinogenesis, neurotoxicology, and developmental deficits in humans and animals. Toxicology and Industrial Health 15: 602-644.

29. Snir EM (2001) Liability as a catalyst for product stewardship. Production and Operations Management 10: 190-206.

30. Nicol S, Thompson S (2007) Policy options to reduce consumer waste to zero: comparing product stewardship and extended producer responsibility for refrigerator waste. Waste management & research 25: 227-233.

31. Sheehan, Bill, and Helen Spiegelman. (2006) Extended producer responsibility policies in the United States and Canada: history and status.

32. President's Council on Sustainable Development and US Environmental Protection Agency. (1997) Proceedings of the Workshop on Extended Producer Responsibility.

33. Product Policy Institute (2014) Product Stewardship Institute California Product Stewardship Council. Product Stewardship and Extended Producer Responsibility: Definitions and Principles.

34. Morris P (1998) Cigarettes as litter.

35. Organisation for Economic Co-operation and Development Council (1972) Recommendations of the Council on Guiding Principles Concerning International Economic Aspects of Environmental Policies.

36. Organisation for Economic Co-operation and Development Council (1989) Recommendation Concerning the Application of the Polluter-Pays Principle to Accidental Pollution.

37. Boehmer-Christiansen S (1994) The precautionary principle in Germany-enabling government. In: Interpreting the Precautionary Principle.

38. Frevert K (2014) ASTSWMO Product Stewardship Framework Policy Document. Association of State and Territorial Solid Waste Management Officials.

39. State of Maine (2010) An Act to Provide Leadership Regarding the Responsible Recycling of Consumer Products. Maine Public Law.

40. CalRecycle Extended Producer Responsibility (EPR) Legislation Checklist, Discussion Draft.

41. Doppelt B, Nelson H (2001) Extended Producer Responsibility and product take-back: Applications for the Pacific Northwest.

42. Toffel MW, Stein A, Lee KL (2008) Extending producer responsibility: An evaluation framework for product take-back policies. Harvard Business School Working Paper.

43. Oregon Paint Stewardship Program Evaluation-PaintCare.

44. Paintcare (2014) About PaintCare. Drop-Off Sites.

45. US Environmental Protection Agency (2014) Non-Hazardous Waste Management Hierarchy.

46. British Columbia Ministry of the Environment (2004) Bottle Bill Resource Guide.

47. CM Consulting (2012) Who Pays What: An Analysis of Beverage Container Collection and Costs in Canada.

48. Shapek RA (1995) Local government household battery collection programs: costs and benefits. Resources Conserv Recycl 15: 1-19.

49. Orange Drop (2013) Program Guide. Curbside Battery Collection Program.

50. CM Consulting (2014) Battery Recycling Boost – CM Consulting releases study reviewing battery recycling programs across Canada.

51. Ontario Industry Stewardship Plan

52. Basel Action Network.

53. Americans for Nonsmokers' Rights Foundation (2012) Overview List – How many Smokefree Laws.

54. Americans for Nonsmokers' Rights Foundation (2013) US Colleges and Universities with Smokefree and Tobacco-Free Policies.

55. US Food and Drug Administration (2009) Family Smoking Prevention and Tobacco Control Act, Public Law.

56. Nolo Law for All (2014) Tobacco Litigation: History & Recent Developments.

57. Cal Recycle (2008) Overall Framework for an EPR System in CA.

58. Litterbutt (2003) Litter Laws by State.

59. Environmental Protection Agency (2014) CERCLA Overview.

60. Warner KE (2013) An endgame for tobacco? Tob Control 22 Suppl 1: 3-5.

61. Stone M (2014) Preventing Toxic Cigarette Waste.

62. Smith EA, Novotny TE (2011) Whose butt is it? tobacco industry research about smokers and cigarette butt waste. Tob Control 20 Suppl 1: i2-9.

10

Environmental Study of the Release of BTEX from Asphalt Modified with Used Motor Oil and Crumb Rubber Modifier

Daniel Bergerson*, Magdy Abdelrahman and Mohyeldin Ragab

Department of Civil and Environmental Engineering, North Dakota State University, North Dakota State University, CIE 201, Fargo, USA

*Corresponding author: Daniel Bergerson, Graduate Student, Department of Civil and Environmental Engineering, North Dakota State University, North Dakota State University, CIE 201, Fargo, USA
E-mail: daniel.bergerson@ndsu.edu

Abstract

The need to be more environmentally conscious has recently shifted toward the forefront of society. With this new focus on environmentally responsible behavior comes the practice of using recycled materials in construction when possible. Therefore it is necessary to carefully evaluate the safety of any recycled materials being used in pavement materials. Under this study, a reference of testing is presented and implemented to test leachate of samples of two different asphalt binder types containing used motor oil (UMO) and/or crumb rubber modifier (CRM) for the presence of benzene, toluene, ethylbenzene, and xylenes (BTEX). Alongside batch leaching tests, air testing was also conducted. Air testing was carried out for select samples to evaluate BTEX content of air above the interactions between the asphalt binder and UMO. Air testing was also completed for asphalt binders interacted with both UMO and CRM. It was found that asphalt binders containing UMO have the potential to leach benzene at concentrations above the national drinking water limits. Results also show that interaction temperature, interaction time, and binder grade affect the amount of BTEX leached from modified asphalt binder. It was also found that binders modified with both CRM and UMO released less BTEX to both leachate and the air at the end of interaction time, meaning it is likely that CRM retains BTEX and prevents it from being released into the environment when used in conjunction with UMO.

Keywords: Used motor oil; Crumb rubber modifier; Benzene; Toluene; Ethylbenzene; Xylene; Batch Leaching; Air testing

Abbreviations:

UMO: Used Motor Oil; CRM: Crumb Rubber Modifier; BTEX: Benzene, Toluene, Ethylbenzene and Xylene; PAH: Polynuclear Aromatic Hydrocarbons; VOC: Volatile Organic Compounds; RAP: Reclaimed Asphalt Pavement

Introduction

In recent years, attention has been brought to the need to preserve the environment and its resources for future generations. This can be achieved through the utilization of waste materials, in addition to limiting the use of virgin products. However, one of the drawbacks that arise from the utilization of waste products is the fact that there is no control over the constituents of such products. This leads to the inability to forecast the environmental and safety aspects that result from utilizing such products as raw materials for other industries. Used motor oil (UMO) and crumb rubber modifier (CRM) are both waste materials that can be implemented in the paving industry. Although both materials have been investigated separately, up to this point no research has been dedicated to investigating the combined effect of such binder modifiers on the environment.

An asphalt pavement can have recycled materials within it that have to be taken into consideration when trying to account for all environmental outputs. CRM is an example of a recycled material that is incorporated into asphalt pavement. CRM is made from recycled tires, which in some cases contain carbon black and some oils.

Research by Thayumanavan et al. states that the NCHRP evaluation methodology developed in project 25-9 is to help transportation agencies make prudent decisions about the reuse of waste materials and by-products in highway construction. Both scrap tires and reclaimed asphalt pavement (RAP) were tested in this evaluation, and the primary objective was to assess the potential impact on surface water and groundwater of constituents released from these materials. Other materials that were tested include coal fly ash, bottom ash, blast furnace slags, and foundry sand [1].

Results of screening tests showed that crumb rubber does have toxicity impact. Crumb rubber was also shown to contain potential organic toxicants. Most concern for these toxic organics can be eased by the volatilization that is likely to occur in most scenarios. Testing needs to be done on rejuvenated mixes that contain crumb rubber despite the likely occurrence of volatilization. These screening tests have shown the potential danger of crumb rubber, and therefore it needs to be evaluated in all potential uses for it [1].

Dedene et al. explored the usability of UMO in the rejuvenation of RAP and found that it might serve as a rejuvenator; however the environmental aspect of utilizing UMO pavements was not investigated [2]. In work by Villanueva et al., the authors confirmed that the addition of UMO to asphalt enhances the low temperature behavior of asphalt [3].

In the work by Duhalt, the author indicates that UMO contains high concentrations of polynuclear aromatic hydrocarbons (PAH), such as benzo[a]pyrene, which are carcinogenic compounds. These PAH accumulate in motor oil during engine operation. The PAH content of used motor oil can be 670 times greater than that of new motor oil. In addition, the lubrication additives present in the oils are, in some cases, pollutants [4].

Jia et al.[5] investigated the utilization of UMO in asphalt and concluded that the inclusion of UMO in asphalt binder should generally be avoided due to some critical negative effects on the performance of asphalt binder. Jia et al. also stated that if the inclusion of UMO becomes necessary for any reason, the amount should be strictly limited. The authors found that the inclusion of UMO in asphalt binder generally reduced the high temperature grade of the binder by $6°C$ on average at 5% oil concentration and $12°C$ at 10% oil content, depending on binder type.

Hidayah et al. [6] investigated the addition of UMO to RAP. They stated that the appropriate amount of UMO depends on the constituent of aged mixture material. The addition of UMO to RAP offered stiffness reduction and therefore improved resistance to cracking.

In the work of Hesp et al., the authors indicated that physical hardening and losses in strain tolerance due to the presence of UMO in asphalt are largely to blame for observed premature and excessive failures in pavement [7].

In another work by Hesp et al. [8] the authors indicated that Paraffins in UMO sludge precipitate asphaltenes from the base asphalt cement. This premature precipitation leads to increased pavement hardening. In addition, Paraffins in UMO sludge promote additional physical hardening to asphalt during cold storage.

Dedene [9] stated that UMO has the ability to counteract the stiffening from incorporating RAP into pavement and restore the performance grade (PG) to that of virgin binder. UMO is also capable of restoring maltenes to rubber asphalt binder (RAB), which will improve the structure of the asphalt binder. Mixture testing with RAP and UMO was shown to soften the asphalt specimens, both by increasing the amount of rutting and lowering the samples indirect tensile strength.

Benzene, toluene, ethylbenzene, and the ortho, para, and meta xylenes, are the four Alkyl Benzene volatile organic compounds (VOC) that represent common constituents of petroleum fuels. In gasoline, for example, these compounds may contain approximately 2% ethylbenzene, 5% Benzene, and 11-12% Toluene and Xylenes depending on the formulation. The term BTEX is commonly used by the petroleum industry in measuring the quality of fuels as it represents the sum of the concentrations of these four compounds. EPA and state agencies have adapted this parameter for use by to serve as a measure of effluent quality of these contaminants in water and to serve as an "indicator" parameter representing the wide variety of compounds found in petroleum products [10].

Several factors determine whether harmful health effects will occur or not, as well as the type and severity of such health effects after exposure to BTEX. The amount of BTEX to which a person is exposed and the length of time of the exposure, as well as which BTEX compound a person is exposed to, are all factors included. Of the four BTEX compounds, benzene is the most toxic [11].

As an example, the result of exposure to very high concentrations of benzene in air (10,000,000 ppb and above) can cause death [12]. Lower levels (700,000 – 3,000,000 ppb) can cause drowsiness, dizziness, rapid heart rate, headaches, tremors, confusion, and unconsciousness [11].

The public water systems have supplied U.S. EPA established permissible levels for chemical contaminants in drinking water. These levels are called maximum contaminant levels (MCLs). The US EPA uses a number of conservative assumptions to derive these MCLs, thereby ensuring adequate protection of the public. In the case of known or suspected carcinogens, such as benzene, the MCL is calculated based on assumption that the average adult weighs 154 lbs and drinks approximately 2 quarts of water per day over a lifetime (70 years). The MCL is set so that a lifetime exposure to the contaminant at the MCL concentration would result in no more than 1 to 100 (depending on the chemical) excess cases of cancer per million people exposed [13].

Materials and Methods

Used motor oil

Ten samples of UMO were collected from auto shops, labeled with numbers for reference, and analyzed for BTEX content according to the procedure utilized by Wang et al. in EPA report 600/R-03/072 [14]. This procedure involves diluting the UMO in pentane and analyzing with gas chromatography-mass spectrometry (GC/MS). All samples were kept in sealed containers from the time of sampling to the time of analysis. All samples were initially analyzed within seven days of being obtained. Sample 6 was chosen to be used in all interactions as a binder modifier. Repeated analysis of the BTEX content of sample 6 was continued over an extended period of time in order to determine the effect of time on BTEX content of UMO. In particular, this was investigated to determine how much, if any loss of BTEX compounds occurs in short term sealed storage of UMO. Sample 6 contained the most benzene, while also having relatively high concentrations of toluene, ethylbenzene, and xylenes. It was therefore chosen for use in asphalt applications as a worst-case scenario.

Asphalt interactions

Asphalt interactions were conducted using either PG 64-22 or PG 52-34 asphalt binder in combination with UMO ranging from 0% to 9% by final weight of modified binder as well as CRM ranging from 0% to 20% of initial binder weight. The CRM was prepared from a mixed source of scrap tires using the cryogenic method with a particle size controlled to be smaller than mesh #30 and larger than mesh #40, according to US standard system. The interaction was conducted in a 1 gallon can and a heating mantle (Glos-col 100B – 618) connected to a bench type controller was utilized. A long temperature probe (TJC 36 – 12") was used to control the interaction temperature. A high shear mixer (HSM-100LM-2) was used to mix the binder, UMO, and CRM. All interactions were conducted under a nitrogen blanket to prevent aging of the material. Interactions for which air samples were taken from the enclosed area above the modified binder were enclosed in an airtight fashion using a thermal plastic bag, thermal rubber, and thermal caulking. Table 1 illustrates the list of interaction conditions utilized in this research. In Table 1, only one source of UMO (sample 6) was used for consistency, as will be explained in the BTEX content of UMO section. A specific coding for the samples was adopted in the current work, starting with the asphalt type, HU-52 or HU-64, then followed by the interaction temperature, interaction speed, with additional information about additives and time following.

Binder	Interaction Temperature (°C)	Interaction Speed (Hz)	% UMO	UMO Source	% CRM	CRM Source
PG 64-22	160	30	0	-	0	-
PG 52-34	160	30	0	-	0	-
PG 64-22	190	30	0	-	0	-
PG 52-34	190	30	0	-	0	-
PG 64-22	160	30	3		0	-
PG 52-34	160	30	3		0	-
PG 64-22	160	30	3		10	CRM 30-40
PG 64-22	190	30	3		0	-
PG 52-34	190	30	3		0	-
PG 64-22	190	30	3	Sample 6	10	CRM 30-40
PG 64-22	160	30	9		0	-
PG 52-34	160	30	9		0	-
PG 64-22	160	30	9		20	CRM 30-40
PG 64-22	190	30	9		0	-
PG 52-34	190	30	9		0	-
PG 64-22	220	30	9		0	-

Table 1: List of Interaction Conditions.

Dynamic mechanical analysis

A dynamic shear rheometer (DSR) was used for dynamic mechanical analysis of modified asphalt. All tests were conducted at 64°C and 10 rad/sec. Samples were tested on 25 mm parallel plates with 2 mm gap for samples containing CRM, or a 1 mm gap for binder containing UMO and unmodified binder. The utilization of a 2 mm gap to test crumb rubber modified asphalt (CRMA) by a DSR was carried out by various researchers [15].

Flash Point of asphalt binder containing UMO

Flash point tests were conducted on binders modified with high percentages of UMO to ensure safety. Both PG 52-34 and PG 64-22 binders containing 15% UMO by final weight were evaluated, along with a control PG 52-34 binder sample. Flash point tests were conducted courtesy of the North Dakota Department of Transportation (NDDOT).

Leaching procedure

Leaching tests were carried out according to the NCHRP 24-hour batch leaching test as described in NCHRP report 25-9 [16-20]. A sample size of 100 grams of solid material (95% glass beads, 5% binder consisting of various UMO percentages by weight) and 400 mL of deionized (DI) water was employed. Glass beads were used to simulate aggregate without affecting sample composition. They effectively increase surface area of the asphalt binder in samples while not interacting with the binder. Samples were prepared at times of 2, 30,

and 120 minutes into the interaction process. After the 24-hour leaching process, pentane extraction was used to extract any BTEX present in a sample of the leachate. The process of mixing a leachate sample with saturated salt solution as well as n-pentane showed to have very good extraction efficiency. The pentane was then carefully removed and analyzed for BTEX content using GC/MS. Five BTEX standards were analyzed in conjunction with each set of samples to ensure accuracy and reduce variability of results and interpretation.

Air testing

Air testing was performed utilizing the National Institute for Occupational Safety and Health (NIOSH) 1501 method [21]. In this method, a solid sorbent tube (coconut shell charcoal, 100mg/50mg) is attached to an air pump with a 0.2 L/min. flow rate. The asphalt interaction was completely enclosed and airtight when air sampling was conducted. Samples were taken from 2 minutes to 30 minutes and from 30 minutes to 120 minutes into the interaction time. This configuration results in sampling times of 28 minutes and 90 minutes respectively. Samples were not taken from 0 to 2 minutes because the resulting sample volume would not be sufficient. The carrier gas for the sampling was nitrogen. As soon as the sorbent tubes were collected, they were covered tightly at both ends and stored at -10°C until they were analyzed.

Results

Dynamic mechanical analysis

Figure 1 illustrates the rheological properties of the binders modified with only UMO, only CRM, and UMO together with CRM. As can be seen in Figure 1, the addition of 9% UMO resulted in severe deterioration of both the complex modulus (G*) and phase angle (δ). On the other hand, the utilization of 20% CRM resulted in major enhancement in both the G* and δ. Unfortuneately, these enhancements occur at the expense of the workability of the modified binder as a result of absorbtion of low molecular weight aromatics by the CRM and the decrease in the CRM interparticle distance [22]. The addition of 9% UMO with 20% CRM resulted in enhancement in both the G* and δ. It is expected that this combination will not significantly deteriorate the workability as the UMO will compensate for the light aromatics absorbed from asphalt in the CRM.

BTEX content of UMO

Table 2 shows results for the concentration of BTEX in all UMO samples collected and analyzed. Concentrations of each constituent of BTEX are shown in µg/g. It was found that within the time frame research was conducted; the BTEX content of sample 6 remained constant over time. Sample 6 was used for all interactions and testing in this research.

The fact that the BTEX content stayed constant over time in the UMO leads to concerns about if, when, and where BTEX could potentially exit UMO and enter the environment. This reaffirms that if UMO is used as an additive to asphalt binder, testing needs to be conducted to ensure BTEX does not enter the environment in a harmful manner.

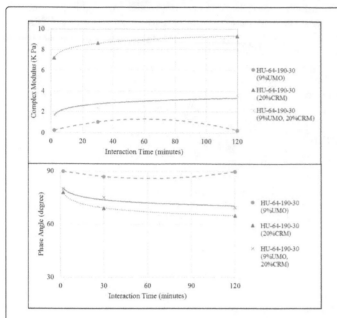

Figure 1: Dynamic mechanical analysis of asphalt binders modified with CRM and UMO.

Sample	Concentration of BTEX Components in UMO (μg/g)				
	Benzene	Toluene	Ethylbenzene	Xylene - p,m	Xylene -o
1	0	0	0	0	0
2	0	337.2	0	312.7	341.4
3	0	217	90.8	378.6	424.4
4	0	291.5	167.2	437.9	288.7
5	0	317.4	110.2	364.1	291.9
6	124	783	211.1	513.9	474.3
7	0	0	0	34.1	0
8	0	65.8	0	97.5	140.1
9	0	204.3	54.3	215.1	196.6
10	75.4	679.3	233.9	603.8	591.8

Table 2: BTEX Concentration in UMO.

Concerning sample source, it is noteworthy that UMO taken from semi-trucks (sample 1 and sample 7) contained almost no BTEX. This is likely due to the diesel fuel semi-trucks use and the fact that it burns cleaner than standard unleaded gasoline. The UMO sample with the next lowest BTEX content (sample 8) was known to have originally been fully synthetic oil. These results could indicate that synthetic oil has less of a tendency to absorb and retain BTEX while in use. In turn this could mean that used fully synthetic motor oil has a much lower BTEX content than most standard UMO.

Flash point results

Flash point tests were conducted by the NDDOT, and results are shown in Table 3. The addition of UMO does decrease the flash point of asphalt binders. Adding 15% UMO to PG 52-34 binder decreased the flash point from 278°C to an average of 245°C. While this decrease is substantial, it is still considered to be safe, and passes the Superpave specification of being greater than or equal to 230°C. Practical UMO percentages applied in pavement applications would likely be much less than 15%, so in general, the addition of UMO to asphalt binder should not be a safety concern with regards to flash point.

Binder Grade	UMO % by Weight	Flash Point (°C)
PG 52-34	0	278
PG 52-34	15	242
PG 52-34	15	248
PG 64-22	15	252

Table 3: Flash Point of Binders Containing UMO.

Batch leaching results

For reference and clarity, Table 4 and Table 5 displays all batch leaching values used in this report. It also displays BTEX content in leachate of control interactions that contain no UMO. This means that the unmodified asphalt binders tested do not readily leach BTEX. The time in Table 4 references the number of minutes from the start of the interaction to the time that the binder sample was taken. It is noteworthy that four control interactions were conducted and tested for PG 52-34 and PG 64-22 binders at temperatures of 160°C and 190°C. Results for all sampling times of all control interactions resulted in leachate BTEX concentrations of 0 mg/L. Results show that many leachate samples contain benzene at levels above the national drinking water standard of 0.005 ppm [23].

	BTEX Content of Leachate (mg/L)								
	Benzene			Toluene			Ethylbenzene		
Time (mins)	0	30	120	0	30	120	0	30	120
I	0	0	0	0	0	0	0	0	0
II	0	0	0	0	0	0	0	0	0
III	0	0	0	0	0	0	0	0	0
IV	0	0	0	0	0	0	0	0	0
V	0.05	0	0	0.156	0.104	0.032	0.028	0.015	0
VI	0.036	0	0	0.121	0.015	0.012	0	0	0
VIII	0.045	0	0	0.152	0.03	0.025	0.01	0	0
IX	0.014	0	0	0.062	0.038	0.024	0	0	0
X	0.012	0	0	0.074	0.025	0.014	0	0	0
XI	0.041	0	0	0.185	0.118	0.061	0.012	0	0
XII	0.023	0	0	0.13	0.075	0.063	0	0	0
XIII	0	0	0	0.067	0.039	0.012	0	0	0

XIV	0	0	0	0.073	0.051	0.038	0	0	0
XV	0	0	0	0.048	0.031	0.021	0	0	0
XVI	0	0	0	0.051	0.05	0.071	0	0	0
XVII	0	0	0	0.08	0.065	0.022	0	0	0

Table 4: Quantitative Results of BTEX Content in Leachate for benzene, toluene and ethylbenzene. (I. PG 64-22 at 160°C; II. PG 64-22 at 190°C; III. PG 52-34 at 160°C; IV. PG 52-34 at 190°C; V. PG 64-22 at 160°C (9% UMO); VI. PG 64-22 at 190°C (9% UMO); VII. PG 64-22 at 220°C (9% UMO); VII. PG 64-22 at 160°C (3% UMO); IX. PG 64-22 at 190°C (3% UMO); X. PG 52-34 at 160°C (9% UMO), XI. PG 52-34 at 190°C (9% UMO); XII. PG 52-34 at 160°C (3% UMO); XIII. PG 52-34 at 190°C (3% UMO); XV. PG 64-22 at 190°C (10% CRM, 3% UMO); XVI. PG 64-22 at 160°C (10% CRM, 3% UMO); XVII. PG 64-22 at 160°C (20% CRM, 9% UMO))

	BTEX Content of Leachate (mg/L)					
	Xylene - p,m			Xylene - o		
Time (mins)	0	30	120	0	30	120
I	0	0	0	0	0	0
II	0	0	0	0	0	0
III	0	0	0	0	0	0
IV	0	0	0	0	0	0
V	0.182	0	0	0.109	0.131	0
VI	0.072	0	0	0.005	0	0
VIII	0.033	0	0	0.032	0	0
IX	0.049	0	0	0.016	0	0
X	0.021	0	0	0.015	0	0
XI	0.029	0	0	0.05	0.026	0
XII	0.02	0	0	0.023	0	0
XIII	0.013	0	0	0	0	0
XIV	0.014	0	0	0	0	0
XV	0.01	0	0	0	0	0
XVI	0	0	0	0	0	0
XVII	0.079	0	0	0.03	0	0

Table 5: Quantitative Results of BTEX Content in Leachate for xylene-p,m and xylene-o. (I. PG 64-22 at 160°C; II. PG 64-22 at 190°C; III. PG 52-34 at 160°C; IV. PG 52-34 at 190°C; V. PG 64-22 at 160°C (9% UMO); VI. PG 64-22 at 190°C (9% UMO); VII. PG 64-22 at 220°C (9% UMO); VII. PG 64-22 at 160°C (3% UMO); IX. PG 64-22 at 190°C (3% UMO); X. PG 52-34 at 160°C (9% UMO), XI. PG 52-34 at 190°C (9% UMO); XII. PG 52-34 at 160°C (3% UMO); XIII. PG 52-34 at 190°C (3% UMO); XV. PG 64-22 at 190°C (10% CRM, 3% UMO); XVI. PG 64-22 at 160°C (10% CRM, 3% UMO); XVII. PG 64-22 at 160°C (20% CRM, 9% UMO))

Impact of interaction time on toluene content of leachate

While several components of BTEX did in fact leach from the modified binders tested, toluene is chosen as an indicator due to consistent results being found for it. Toluene was also consistently the most abundant component of BTEX found in leachate samples. Figure 2 shows the toluene content of leachate of binders containing UMO over interaction time. This data was taken from a wide variety of interaction conditions, and despite differences in UMO content, temperature, binder type, and CRM content, a negative trend is still present. In general, the total toluene content of these binders was between 0.05 and 0.15 mg/L for samples taken 2 minutes into the interaction. This concentration drops considerably for samples taken 30 minutes into the interaction, and drops even further for samples taken 120 minutes into the interaction. This clearly indicates that the longer the binder is heated in the interaction process, the less BTEX it contains.

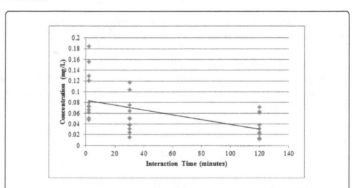

Figure 2: Toluene content of leachate of modified binders over interaction time for all samples investigated.

Impact of UMO content on BTEX content of leachate

As illustrated in Table 4, results clearly show that as the percentage of UMO added to asphalt binder increased, the concentration of BTEX in the leachate also increased. This is expected, as it is logical that if there is more BTEX present in a sample, then more BTEX will be leached from that sample.

Impact of interaction temperature on toluene content of leachate

Generally, as interaction temperature increased, toluene content in leachate decreased. Figure 3 shows this general trend by illustrating the drop in toluene content of leachate when increasing the interaction temperature from 160°C to 190°C. The data shown represents interactions containing 9% UMO and 0% CRM for both binder types.

The effect of interaction temperature on toluene content of leachate is perhaps more dramatically noticed when combined with changes in binder grade as discussed in the next section.

Impact of binder grade on toluene content of leachate

As illustrated in Figure 4, a marked difference in the toluene content of the leachate of the two binder types can be observed here. The PG 52-34 binder shows a much greater propensity to leach toluene than the PG 64-22 binder. This difference is especially noticeable over interaction time. This could be related to the higher amount of low molecular weight aromatics present in the lower grade asphalt, which might decrease the ability of the asphalt to integrate the soft UMO and thus lead to such behavior as compared to harder asphalt.

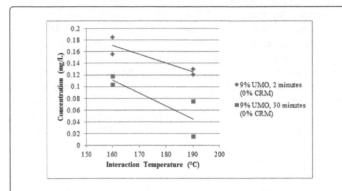

Figure 3: Toluene content of leachate of binders modified at different interaction temperatures.

Figure 4 also shows the relationship of decreasing toluene content of leachate over interaction time while also illustrating the difference between binder grades.

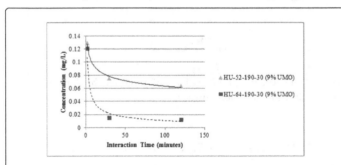

Figure 4: Toluene content of leachate of differing binder grades at 190°C.

Impact of CRM content on toluene content of leachate

It appears that it is possible that CRM affects the toluene content of leachate. Figure 5 shows leachate results comparing two interactions that are identical except for the CRM content. The most consistent and notable trend is observed in samples taken 2 minutes into the interaction process. For these 2 minute samples, the toluene content of leachate of binders containing both CRM and UMO is lower than that for those containing only UMO. This could mean that the CRM is absorbing and holding on to some of the UMO initially and causing a decrease in the toluene content of leachate. On the other hand, it could mean that the CRM is causing more BTEX to be released into the air. These possibilities can be further investigated through air testing.

BTEX content of air samples

Figure 6 illustrates the BTEX content comparison for the HU-64 asphalt samples interacted with 9% UMO with or without 20% CRM. It can be seen that after 30 minutes of interaction time the sample containing 20% CRM released higher amounts of benzene than the sample with no CRM.

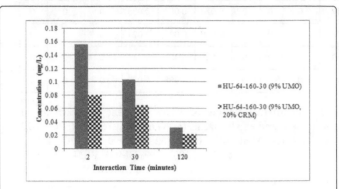

Figure 5: Impact of CRM on toluene content of leachate for binders containing 9% UMO at 160°C.

On the other hand, after 120 minutes of interaction time, almost no benzene was released from both samples investigated. For the toluene concentration in the air, the samples collected after 30 minutes of interaction time showed the same trend of higher concentration of toluene in the one having CRM over the one lacking it, where the toluene concentration was almost 250 ppm in the sample containing CRM, while it was about 190 ppm for the sample with no CRM. Opposite results can be seen for the samples collected after 120 minutes of interaction time, where the toluene concentration with slightly less for the sample containing CRM in comparison to the sample without CRM. A different trend can be seen for the ethylbenzene concentration in air for the samples with and without CRM after 30 minutes of interaction time. The concentration of ethylbenzene was about 80 ppm for the sample without CRM, while it was around 65 ppm for the samples with CRM. The same trend continues for the samples collected after 120 minutes of interaction time, but with lower concentrations. The same trend seen for ethylbenzene concentrations continues for the total xylene concentration in both the samples containing and lacking CRM, but with higher concentrations. It should be noted that xylene was the most released component of the four volatiles. After 30 minutes of interaction time, the concentration of the xylene for the sample without CRM was about 360 ppm, while it was about 240 ppm for the sample containing CRM. After 120 minutes of interaction time, the xylene concentration was almost identical for the samples with or without CRM, having a value of about 120 ppm.

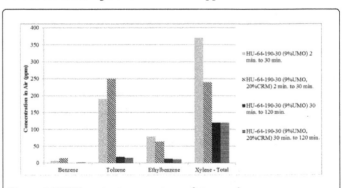

Figure 6: BTEX content comparison of air samples.

Relationship between leaching and air testing

Figure 7 shows that with increased interaction time, both the BTEX content of leachate and the BTEX content of air samples have decreasing trends.

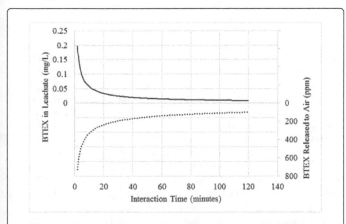

Figure 7: BTEX release for HU-64-190-30 (9% UMO) through leaching and air.

The amount of BTEX being released to the air is much greater than the amount of BTEX found in leachate. At high temperature over time, the BTEX is released into the air and becomes less present in the modified binder, therefore reducing the amount of BTEX available for leaching, and in turn reducing the BTEX content of leachate.

Conclusions

This research confirms that UMO content of modified asphalt binder clearly affects the BTEX content of leachate of that modified binder. As the UMO content of asphalt binder increases, the BTEX content of the leachate of that binder also increases. This is logical because as the BTEX that is available to be leached increases, the BTEX that is leached should tend to increase as well.

There is a marked difference in the leaching of BTEX over interaction time between 160°C and 190°C interactions. BTEX remains present in the binder (and in turn shows up in the leachate) further into the interaction at lower temperature. The interactions run at higher temperatures result in more BTEX exiting the binder and therefore not being present in the leachate.

Based on testing results for the interaction conditions in this research, binder type by itself affects BTEX content of leachate. In general, greater concentrations of BTEX are leached out of UMO modified PG 52-34 binder than are leached out of UMO modified PG 64-22 binder.

Binder type over time may affect BTEX content over time. This means temperature alone may not be the only prevailing force over time, and there may be interaction between binder grade and time that has an effect on the BTEX content of leachate. This could very well have something to do with the molecular structure and bonds of the two binder grades that were used in interactions. A trend has been observed, but additional investigation that is beyond the scope of this research may be required to determine the exact cause of this trend.

Binders containing both CRM and UMO appear to have a lower propensity to leach BTEX from samples taken at 2 minutes of interaction time than binders containing UMO alone at the same percentage. The CRM may initially be absorbing UMO and slowly releasing it over time at high interaction temperatures, but more testing is required to determine if this is the case.

There is a steep degradation of BTEX as indicated by leachate analysis over interaction time. Degradation of BTEX as indicated by leachate analysis over interaction time tends to occur at an increased rate for modified PG 64-22 binder when compared to modified PG 52-34 binder.

If UMO is employed in any pavement application, it is recommended that the UMO be fully tested for harmful contaminants first. Two of the ten UMO sources obtained in this research contained benzene. If UMO is employed in any pavement applications whatsoever, testing should be conducted to determine if it contains any benzene. The concentration of benzene in leachate of some binders containing 3% UMO was above national drinking water limits as well as North Dakota general water quality standards. The concentration of benzene in leachate of many binders tested containing 9% UMO was also above both national drinking water limits and North Dakota general water quality standards.

With regards to toluene aside from benzene, it is best to err on the side of caution and use a maximum UMO content of about 9% of the total binder weight. This conclusion is based on the UMO sampled in this study that contained the highest concentration of toluene. If a UMO source with an equivalently high concentration of toluene (783 μg/g) is employed at a level of 9% by weight of the total binder, leachate levels will remain below current limits for toluene content.

Generally, as interaction temperature increased, toluene content in leachate decreased. This could be related to the increase of release of toluene to the air with the increase of interaction temperature.

Acknowledgment

This material is based on the work supported by the National Science Foundation under Grant No. 0846861. Any opinions, findings, and conclusions or recommendations expressed in this material are those of the writer(s) and do not necessarily reflect the views of the National Science Foundation.

References

1. Thayumanavan, Pugazhendhi, Peter O. Nelson, Mohammad F. Azizian, Kenneth J. Williamson (2001) Environmental Impact of Construction and Repair Materials on Surface Water and Groundwater: Detailed Evaluation of Waste-Amended Highway Materials. Transportation Research Record 1743, Paper No. 01-3444.

2. DeDene, Christopher D., and Zhan-Ping You (2014) The Performance of Aged Asphalt Materials Rejuvenated with Waste Engine Oil. International Journal of Pavement Research and Technology 7: 145-152.

3. Villanueva, Aaron, Susanna Ho, LudoZanzotto (2008) Asphalt modification with used lubricating oil. Canadian Journal of Civil Engineering 35: 148-157.

4. Vazquez-Duhalt R1 (1989) Environmental impact of used motor oil. See comment in PubMed Commons below Sci Total Environ 79: 1-23.

5. Jia, Xiaoyang, Baoshan Huang, Benjamin F. Bowers,Sheng Zhao (2014) Infrared spectra and rheological properties of asphalt cement containing waste engine oil residues. Construction and Building Materials 50: 683-691.

6. Hidayah, Nurul, MohdRosli, Norhidayah, MohdEzree (2013) A short review of waste oil application in pavement materials.

7.   Simon AM Hesp , Herbert F. Shurvell (2010) X-ray fluorescence detection of waste engine oil residue in asphalt and its effect on cracking in service. International Journal of Pavement Engineering 11: 541-553.

8.   Hesp SAM, Shurvell HF (2012) Waste engine oil residue in asphalt cement. Proc., Seventh International Conference on Maintenance and Rehabilitation of Pavements and Technological Control.

9.   DeDene, Christopher Daniel (2011) Investigation of using waste engine oil blended with reclaimed asphalt materials to improve pavement recyclability. Dissertations, Master's Thesis and Master's Reports.

10.   Fact Sheet Attachment A (2005) Remediation General Permit Fact Sheet Excerpts. Accessed July 14, 2014.

11.   Leusch F, Bartkow M (2010) A short primer on benzene, toluene, ethylbenzene and xylenes (BTEX) in the environment and in hydraulic fracturing fluids.

12.   ATSDR (2007a) Toxicological profile for benzene. US Department of Health and Human Services, Agency for Toxic Substances and Disease Registry, USA.

13.   BTEX Fact Sheet - Maryland Department of the Environment. Baltimore, USA.

14.   Wang Z, Hollebone BP, Fingas M, Fieldhouse B, Sigouin L (2003) Characteristics of spilled oils, fuels, and petroleum products: 1. Composition and properties of selected oils. Environmental Protection Agency.

15.   Tayebali AA, Vyas B., Malpass GA (1997) Effect of Crumb Rubber Particle Size and Concentration on Performance Grading of Rubber Modified Asphalt Binders, Progress of Superpave (Superior Performing Asphalt Pavement): Evaluation and Implementation, ASTM STP 1322, R.N. Jester, Ed., American Society for Testing and Materials.

16.   Nelson PO, Huber WC, Eldin NN, Williamson KJ, Azizian MF, et al. (2000) Environmental Impact of Construction and Repair Materials on Surface and Ground Waters, Volume: IV. Laboratory Protocols. NCHRP Project 25-9. TRB, National Research Council, Washington DC, USA.

17.   Eldin NN, Huber WC, Nelson PO, Lundy JR, Williamson KJ, et al. (2000) Environmental Impact of Constrction and Repair Materials on Surface and Ground Waters, Final Report, Phases I and II, Volume II: Methodology, Laboratory Results, and Model Development. NCHRP Project 25-9. TRB, National Research Council, Washington DC, USA

18.   Nelson PO, Huber WC, Eldin NN, Williamson KJ, Azizian MF, et al. (2000) Environmental Impact of Construction and Repair Materials on Surface and Ground Waters, Volume I: Final Report. NCHRP Project 25-9. TRB, National Research Council, Washington DC, USA.

19.   Nelson PO, Huber WC, Eldin NN, Williamson KJ, Azizian MF, et al. (2000) Environmental Impact of Construction and Repair Materials on Surface and Ground Waters, Volume III: Phase III Methodology, Laboratory Results, and Model Development. NCHRP Project 25-9. TRB, National Research Council, Washington DC, USA.

20.   Nelson PO, Huber WC, Eldin NN, Williamson KJ, Azizian MF, et al. (2000) Environmental Impact of Construction and Repair Materials on Surface and Ground Waters, Volume IV: Laboratory Protocols. NCHRP Project 25-9. TRB, National Research Council, Washington DC, USA.

21.   NIOSH method 1501 hydrocarbons, aromatic. In: Eller, P.M. (Ed.), NIOSH Manual of Analytical Methods (1994) NIOSH (National Institute for Occupational Safety and Health). DHHS (NOSH), Cincinnati, USA.

22.   Gawel I, Stepkowski R, Czechowski F (2006)Molecular interactions between rubber and asphalt. Industrial & engineering chemistry research 45: 3044-3049.

23.   Basic Information about Benzene in Drinking Water.

11

The Reuse of Waste Electrical and Electronic Equipment (WEEE). A Bibliometric Analysis

Dolores Queiruga[1] and Araceli Queiruga-Dios[2]

[1]Department of Economics and Business Administration, University of La Rioja, Edificio Quintiliano. C/ La Cigüeña 60. 26006 Logroño, Spain, E-mail: dolores.queiruga@unirioja.es

[2]Department of Applied Mathematics, ETSII. University of Salamanca, Avda. Fernandez Ballesteros 2, 37700-Bejar, Salamanca, Spain, E-mail: queirugadios@usal.es

*Corresponding author: Dolores Queiruga, Department of Economics and Business Administration, University of La Rioja, Edificio Quintiliano. C/ La Cigüeña 60. 26006 Logroño, Spain
E-mail: dolores.queiruga@unirioja.es

Abstract

The management of Waste Electrical and Electronic Equipment (WEEE) has much scope for improvement in almost all countries. Some legislation, such as the European Directive, aims to increase reuse, to prevent such equipment from being recycled without being reused. In this sense, the reuse of WEEE has a huge interest from an economic, environmental, legal, and social viewpoint. However, this has not been studied much so far. This paper adopts a bibliometric analysis of the scholarly literature that has addressed the study of reuse of WEEE. For the purpose, we have examined 32 papers on this topic, indexed up to 2014 in the Scopus database, and have identified trends and opportunities related to the what, how, and where of the research in this matter.

The results show the need for further investigations on certain aspects of reuse in order to overcome the barriers that exist in different countries. In particular, further case studies are needed in countries that practice greater reuse. Further research is also necessary in logistics and the reasons why different economic agents do not reuse WEEE. Besides, it is also important to understand the aspects of strategic human resource management and operation management of reuse centers.

Keywords: Waste; Electronic equipment; Reuse; Bibliometric

Introduction

In recent years, there has been a growing concern about waste electrical and electronic equipment (WEEE) management and this is reflected in many academic case studies that elucidate how their management is or could be in different countries, such as the United States [1], China [2], Japan [3], Switzerland [4,5], United Kingdom [6], Germany [7], or Spain [8].

One of the possibilities for managing electrical and electronic equipment at the end of their lifecycle is their reuse. The European WEEE Directive 2002/96/EC presented a hierarchy of actions for the management of WEEE that reuse should assume priority over recycling. According to this legislation, reuse, recycling, and recovery targets have to be met for 10 different categories of electrical and electronic equipment by EU member countries. However, the Directive did not specify separately the percentage of recycling and reuse. In fact, the need for reuse was hardly mentioned. The minimum requirements included in the legislation suggested only recycling, and rarely, reuse [9]. Therefore, the result of the application of this Directive was that member countries installed only the necessary infrastructure for recycling (e.g. in Germany [9] or in Spain [10]). In practice, the WEEE management system operates in such a way that the countries mainly recycle; and the reuse is rare and only by just a few social stakeholders [11-13].

In some countries, such as Spain, much of the WEEE collected (approximately 70%) goes to unauthorized and uncontrolled players to be treated outside the official channels provided for this purpose, i.e. [8,14], the waste does not reach its rightful destination, which is an authorized WEEE treatment plant, but are irregularly or illegally "lost." It is unknown where exactly this waste ends.

Paradoxically, this situation is not inconsistent with the fulfillment of the 2003 Directive. This means that this situation could be maintained because the law required only recycling of a certain amount (4 kg per inhabitant per year). Nevertheless, it is expected that not so many equipment would get lost with the new legislation in place, because the required amount to be reused and recycled depends on the number of the equipment placed on the market.

In 2012, the new Directive 2012/19/EU came into force and it was incorporated into the national legislation in 2014. Until December 31, 2015, the collection of 4 kg per inhabitant per year of WEEE from private households will still be required [15]. However, from 2016, the minimum collection rate that is to be achieved is 45% of the average weight of electrical and electronic equipment (EEE) placed on the market by each member state in the three preceding years. From 2019, a choice between two options is possible: either 65% of the average weight of EEE placed on the market, or 85% of WEEE generated by the member state. This implies that the requirements for recycling and reuse are no longer measured as a fixed amount depending on the number of inhabitants, but are related to the amount of products placed on the market. This new Directive emphasizes the preparation of WEEE for reuse [9]. As suggested by the new legislation, the amount of WEEE to be treated is increasing, so more equipment must enter the waste management system. In Spain, at present, there are some reuse plants near shutdown because they do not receive enough equipment. For example, the company Ekorrepara is near bankrupt because it is receiving insufficient equipment passing through its system. In the past, the Integrated Management Systems (IMS) used to provide the equipment, but now, they have opted for a recycling plant [16]. Paradoxically, the demand for second-hand EEE is exceeding the supply [17].

There is an ongoing debate about the benefits of reuse. Some authors argue that even intensive product reuse of EEE reduces total resource consumption of a highly developed industrial economy by less than 1% [18]. Some authors argue that recycling is not always profitable [19]. But for many others, reuse may involve several environmental and socioeconomic benefits [20], such as extending the useful life of equipment, protecting the environment, or facilitating access of the unemployed to jobs [18], and also facilitating the acquisition of equipment by people with few resources. However, the technological, economic, social, and environmental structure of the reuse chain is largely unknown [21]. Furthermore, reuse is necessary to facilitate compliance with the European Directive.

In the scientific literature, the management of WEEE for recycling has been widely studied from both a logistical and a technological point of view. However, there are considerably less studies on reuse. It is important to know what has been done and what remains to be done in a field with different opinions. It is still unknown how reuse may be cost-effective and beneficial to the environment and society in a way in which recycling can be cost-effective and beneficial to the environment and society. In addition reuse plants need to improve their management and relations with other economic and social actors in order to get more equipment to repair, and thereby, improve their results. Again, from a legislative point of view, reuse should be increased. Due to this research need, the objective that had been suggested was to make a review of the literature addressing what, how, and where reuse has been researched so far, allowing us to discover what remains to be learnt in the field of WEEE reuse. To achieve this, we will address three key issues: (1) what is being researched, i.e., what specific waste management activities have been studies from what perspectives and what products category; (2) how to research, i.e., what research methodologies are used; and (3) where has it been investigated, i.e., which geographical areas has the work been carried out in and in which journals are the articles disseminated? To obtain the necessary information, we adopted a bibliometric standpoint to analyze the scholarly literature that addresses the study of WEEE reuse.

To do this, we analyzed the academic papers included in the Scopus research database. The main interest of this article, as with the majority of bibliometric studies, is that it allows the research conducted so far to be assessed, while, at the same time, helping to identify current weaknesses and future opportunities. Moreover, reviews and studies of this nature help prospective researchers to situate and contextualize their contributions to the field of study in question [22-25].

The paper is organized into 6 sections. Section 2 describes the methodology used in the study. Section 3, 4 and 5 focuses on elucidating the what, how, and where, respectively, of the research into WEEE reuse. Section 6 summarizes the main conclusions, while singling out some of the study limitations, which, in turn, pose certain challenges for future research.

Methodology

To conduct a systematic review of the literature, we followed following steps: identify the field of study and period of analysis; select information sources; and search, manage, and debug the results and analyze them [26]. This study has centered on the scholarly contributions published up to 2014 and indexed in the Scopus database. The Scopus interdisciplinary database indexes more than 14,000 peer-reviewed publications. The use of a multidisciplinary database was positive for our review because the reuse topic can be explained from a number of different perspectives [27], such as business, technical, environmental, or social. We noted beforehand that some of the more representative journals in the field of management and environment (e.g., Resources, Conservation & Recycling, Waste Management, and Journal of Cleaner Production) were indexed in the database. We, therefore, concluded that the Scopus database would provide a reasonably accurate view of the research carried out.

A search was, therefore, conducted using two keywords: "reuse" and "WEEE," whereby we called up all those contributions that mentioned both the terms in any one of the registered fields in the database (title, keywords, abstract, authors' details, etc.). The only other restriction was that the articles must have been published in scholarly journals through a peer-review process. In June 2014, these criteria resulted in overall 66 papers. Regarding the subsequent work of classifying and ordering, each of these papers was read and analyzed by at least two researchers independently, with the intervention of a third in case of difference in opinions.

Largely owing to the use of such open search criteria, the process of analysis and classification also involved a screening of the initial sample—with 34 papers being discarded because although they contained the keywords, the subject area was out of purview of the study. This is explained by the fact that, for example, some of the words appeared only in the authors' registration details; they received only a cursory mention in the text speaking about the legislation targets. Some other papers used the WEEE legislation as an example pertaining to another sector, as in the case of cars or packaging reuse. There are some published studies related to reuse as secondary raw materials. The final number of papers considered in this research, therefore, was 32 (see annex 1). It is a smaller number of papers than in other bibliometric reviews because it is a current topic with not much available literature.

What is Being Researched?

Scope of research

According to the European Directive, reuse is defined as "any operation by which WEEE or components thereof are used for the same purpose for which they were conceived, including the continued use of the equipment or components thereof which are returned to collection points, distributors, recyclers or manufacturers" (Art. 3, d, WEEE Directive). Reuse is one of the possibilities of WEEE management. Therefore, for this bibliometric review, we describe the context of the reuse as part of WEEE management. Considering the whole system from production to waste management, we have classified the papers according to the chosen approach, i.e., according to the volume of items and activities considered during the analysis.

Thus, we can distinguish from the papers that take a broad approach and consider reuse as one option among others to those that focus exclusively on reuse. To group the papers, depending on the scope, we suggest four possibilities: (1) Direct and reverse logistics; (2)

Collection, reuse, recycle, disposal; (3) Reuse, recycle, disposal; and (4) Reuse.

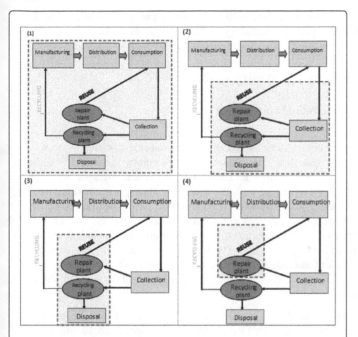

Figure 1: Four possibilities to study the management of WEEE, if reuse is included.

Some of the papers involved in this review analyze the management of WEEE in general; others focus on specific activities such as collection, reuse, recycling, and disposal. However, most of the documents focus on two activities: reuse and recycling. These are the two WEEE management alternatives after use; therefore, most of the studies are centered on these two key options. These papers are too general to take into account the whole supply chain, and are not so specific as to consider only reuse; and they focus on what the possibilities are with the devices.

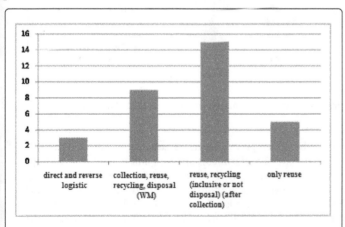

Figure 2: Distribution of papers by scope of research.

Figure 2 shows the number of items for each possibility of research. When interpreting this graph, it must be remembered that these are not articles on the management of WEEE in general, but only those that include reuse.

This means that there are many papers not listed here that include both, the direct and reverse logistics. For this reason, the fact that there are few items in the first category, "direct and reverse logistics" does not mean it is a topic that needs further research, but that articles on the topic of reuse are realized in the context of reverse logistics.

It is remarkable that the first research on this topic, including reuse, appears in 2002 [28]; however, if we look for the words "WEEE management" in the same database (Scopus), we find papers dating back to 1998. It has also been drawn our attention that only four papers refer exclusively to reuse and these are concentrated mainly between 2012 and 2014. This indicates that reuse, as an option to reduce waste, has not received much attention on its own, but does so only as an option linked to recycling. With the whole research activity centered around logistics and case studies, solutions can be given in terms of economic, social, and environmental terms. At the same time, countries can learn from the case studies from other countries and could provide recommendations to others. It would, therefore, be useful to lay more emphasis on research that is exclusively on reuse.

Organizational focus

It is also important to know if the papers take an inter- or intra-organizational focus, that is, if they focus on issues related to logistics, transportation, or products storage, or, if instead, they address technological topics, operations, or decisions in plants. Figure 3 groups the papers under each approach.

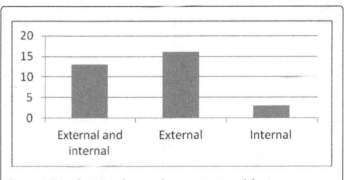

Figure 3: Distribution of papers by organizational focus.

The results show that there is little research exclusively carried out inside the plants. This disparity could be understood by the fact that reparation has always been carried out in most sectors, and therefore, does not require such an immediate technology development. In this sense, it would not be necessary to increase research on reuse plants technology.

However, it is important to consider that other aspects and decisions of reuse plants, such as strategic management, marketing, or human resources management, are not addressed in any paper. These topics have been extensively studied in factories, but not in recycling or reuse plants, which have some issues that differ from factories (these were analyzed, for example, by Rogers et al. [29]).

As Kissling et al. [30,31] state, the lack of products to repair and to put back on the market is a barrier for reuse. In addition, different stakeholders involved in the management, such as manufacturers or recyclers, also pose barriers.

Therefore, more knowledge and research in these areas would be important to overcome the barriers, for example, about the proper use of resources and capacities of reuse plants. Another possible area of research is the management of importing used products. Yet another possibility would be analyzing the implications of establishing new agreements with the stakeholders involved. However, reuse plants have traditionally acted within local markets and such decisions may be difficult for them.

Product category

To define the product categories, we used the 10 categories proposed by the European Directive as the most known and used in scientific papers. At the same time, this division includes all types of devices: (1) Large household appliances; (2) Small household appliances; (3) Information and Communication Technologies; (ICT); (4) Consumer equipment; (5) Lighting equipment; (6) Electrical and electronic tools; (7) Toys, leisure, and sports equipment; (8) Medical devices; (9) Monitoring and control instruments; and (10) Automatic dispensers.

The results show that most of the articles address the management of WEEE from a general perspective (see Figure 4). This is because the law is the same for all devices, and therefore, the design of logistics systems that might be used by everyone should be proposed. However, some devices have specific characteristics, such as PCs or mobile phones, which are replaced more than others. Large appliances also have distinctive characteristics because of their size, and so also refrigerators, due to the special treatment of the toxic substances that they contain.

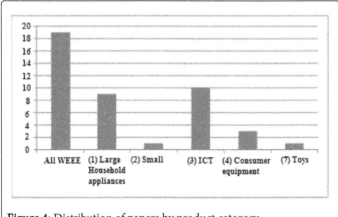

Figure 4: Distribution of papers by product category.

There is certainly very little research in categories such as toys— only one paper. Toys, like other household appliances, are sometimes unusable, but in other cases, they are discarded while they are still in working conditions. When this happens, it would be important to conduct further research to improve the resale market within our own country or in foreign countries.

The other products and categories, such as lighting equipment, electrical and electronic tools, medical devices, monitoring and control instruments, automatic dispensers, for which there is no article, are basically very specific non-domestic use equipment pertaining to a specific profession. It can be expected that they will be replaced by modern technology when they stop working properly with the new ones. It would make sense to analyze the suitability of the export of these products to countries lagging behind in the technology. On the

other hand, from the social point of view, investment in reuse plants would be beneficial to developing countries.

As for papers dealing exclusively with reuse, they are related to all WEEE in general, including ICT and large household appliances. Therefore, there is no difference from other papers with a wider scope.

How Reuse is Investigated?

Methodological approaches

As reuse can be studied from any branch of study, such as management, engineering, production management, sociology, or law, we have classified the papers into two major generic methods: theoretical and empirical. Under the theoretical method, we have included those dealing with the legislation status, waste management in a specific country, or a management model. The papers, which incorporate the development of mathematical models or conduct interviews with consumers or stakeholders, are included in the empirical category. Papers with a cost-benefit analysis with the amounts of discarded products and others with software development for information management are also included in the empirical section.

The results show that the number of items in each method is comparable (Figure 5): 13 theoretical work papers and 19 empirical. This highlights that it is not only necessary to develop mathematical models and calculate the most efficient logistics systems to properly manage waste, it is also important to be familiar with case studies about how other countries apply the law, the role of each stakeholder, and the relationships between them.

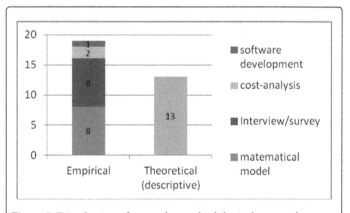

Figure 5: Distribution of papers by methodological approach.

Among the 19 empirical papers, 8 use mathematical models, 8 include interviews, 2 perform cost analysis, and the last one develops software. Regarding the overall waste management or WEEE recycling topics, some software applications have been developed to know the waste stream (e.g., Ofiraee is a Spanish virtual office that receives the request to collect the WEEE from collection points and notifies the Integrated Management Systems). It would be interesting to develop software tools to centralize information about the status of obsolete equipment, connecting the collection points with the reuse plants.

Disciplinary perspective

The subject of WEEE management, in general, and reuse, in particular, can be studied from different disciplines and different points of views. On the one hand, if the equipment is reused, it takes longer to be a waste product, which affects the environment. On the other hand, it is a requirement of the legislation. Finally, it is necessary to know the role of the agents involved and to calculate the most suitable logistics and their costs. Due to the need to consider all these perspectives of reuse, we have distinguished four points of view: (1) economic, (2) social, (3) environmental, and (4) legislative.

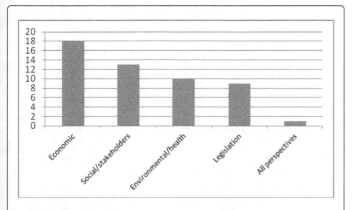

Figure 6: Number of papers for each perspective.

Even though the legislative perspective appears less frequently (Figure 6), we should always be aware that compliance with the legislation is an underlying factor in most of the literature on this topic. This is evidenced by the significant increase of scientific papers directly or indirectly arising from the entry into force of new legislations such as the European Directive.

Whatever the focus of the papers, the underlying issue is how best to manage waste. However, since the producer is financially responsible for waste management in many countries, it would be interesting to also investigate the management of waste from the point of view of strategic management or business management, and not just from the point of view of waste management.

Where has it been Investigated?

Geographic scope of research

The results obtained according to the geographical scope of research are presented in Figure 7.

Some papers analyze waste management in several countries. In such cases, we have added each country involved in the graph. What we wanted to know was which countries participate and which do not. Most of the papers under examination belong to Europe and Asia. Some of them are centered in specific cities, but others are about the whole country. There are some other papers from major geographical areas such as Latin America or North America. The reason why there is a lot of research in Europe may be because the European Directive requires some changes and the researchers want to know how they could bring about these changes. In the case of Asia, there is a problem with importing products and informal waste management systems. Researchers try to raise these relevant issues.

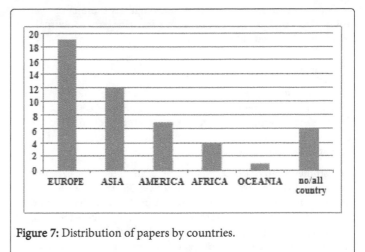

Figure 7: Distribution of papers by countries.

Of particular interest is the low number of papers dealing with North America and Africa, as in both continents, research about the increase of reuse would be extremely important: In the case of America, it is a country that traditionally exports WEEE to other countries. Instead of exporting, the environmental, social, and economic benefits could be analyzed, repairing WEEE and promoting the market for second-hand goods or even repairing and exporting. Africa could also consider equipment reuse.

Journals where the articles are published

Figure 8 shows the graph with a list of all the journals involved in this review.

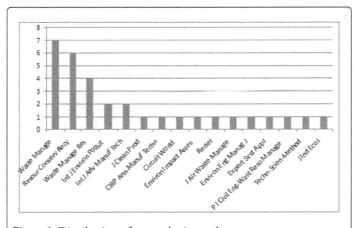

Figure 8: Distribution of papers by journal.

The reuse topic is researched from an environmental, engineering, or business point of view. This is reflected in the fact that most journals that publish articles on this subject are concerned about waste management and resource conservation: Waste Management (7 papers); Resources, Conservation and Recycling (6 papers); Waste Management and Research (4 papers). The remaining items are divided into one or two papers per journal (environment, technology, or industry journals).

Although it is true that reuse may favor the environment, and that logistics management and process improvements are needed, it is also true that other journals could research reuse in a multidisciplinary way, such as with human resources or sociology, as there are social enterprises dedicated to reuse too. Strategic management journals could also contribute much to this topic.

Conclusions

After answering the questions what, how, and where, the work has highlighted, on the one hand, existing reuse barriers, and on the other, the need for further research. Therefore, for each barrier to reuse, we propose more intense research in the area. Some specific issues that we could not answer after our analysis are:

- Some studies analyze the barriers of reuse in different geographic areas (e.g., [31] in Africa, Latin America, North America, and Europe), but it is important to analyze it in each country through case studies. As numerous case studies on WEEE recycling have been published, reuse could also be likewise analyzed. Thus, some countries could learn from the experience of others to overcome their barriers. It is important to conduct these case studies in developing countries too, since they are the recipients of numerous obsolete equipment from other countries.

- The second research topic could help overcome the most important reuse barriers, i.e., the lack of access to sufficient volumes of equipment used for different reasons [30,31]. It is necessary to study the reasons from the point of view of consumers, retailers, and manufacturers. It is possible that new contracts between companies or government aid could improve this situation.

- According to the results of our analysis, the issue of reuse is treated mainly from a technical and environmental point of view. However, to enhance the number of equipment obtained from reuse plants, there are certain aspects of the plant's decision-making process that could be improved. Therefore, we conclude that further research is necessary from the point of view of business management, strategic management, and human resources.

- For WEEE recycling, logistics systems have been developed in all countries and there are new operators, such as virtual office registration to register the number of equipment or Integrated Management Systems. In the same way, it is important to research the same as related to reuse. Logistics costs are also a barrier to reuse [30,31].

We can conclude that there are few studies on reuse and increasing the numbers could contribute to understanding how reuse is dealt in different countries and how to improve this activity. As a conclusion of the whole analysis, we can say that if reuse is not studied exclusively, there is a risk of considering it as a marginal activity.

This work is not without its limitations—which can lead to new research opportunities. We have used only one "database." It would be interesting to use other databases, which may slightly increase the number of items that can be considered.

Acknowledgements

This research was partially funded by the Spanish Government and FEDER funds through research project ECO2013-47280-R. Aid was also received from the research project SA083A12-1 financed by the Consejeria de Educacion de la Junta de Castilla y Leon (Regional Ministry of Education of Castile and Leon).

References

1. Kang HY, Schoenung JM (2005) Electronic waste recycling: A review of U.S. infrastructural and technology options. Resources, Conservation and Recycling 45: 368-400.

2. Veenstra A, Wang C, Fan WJ, Ru YH (2010) An analysis of E-waste flows in China. International Journal of Advanced Manufacturing Technology 47: 449-459.

3. Aizawa H, Yoshida H, Sakai S (2008) Current results and future perspectives for Japanese recycling of home electrical appliances. Resources, Conservation & Recycling 52: 1399-1410.

4. Streicher-Porte M, Widmer R, Jain A, Bader HP, Scheidegger R, et al. (2005) Key drivers of the e-waste recycling system: Assessing and modelling e-waste processing in the informal sector in Delhi Environmental Impact Assessment Review 25: 472-491.

5. Hischier R, Wagner P, Gauglhofer J (2005) Does WEEE recycling make sense from an environmental perspective? The environmental impacts of Swiss take-back and recycling systems for waste electrical and electronic equipment (WEEE). Environmental Impact Assess Review 25: 525-539.

6. Turner M, Callaghan D (2007) UK to finally implement the WEEE Directive. Computer Law & Security Report 23: 73-76.

7. Spengler T, Luger T, Herrmann C (2010) Implementation of the WEEE-directive. Economic effects and improvement potentials for reuse and recycling in Germany. International Journal of Advanced Manufacturing Technology 47: 461-474.

8. Queiruga D, González-Benito J, Lannelongue G (2012) Evolution of the electronic waste management system in Spain. Journal of Cleaner Production 24: 56-65.

9. European Parliament and Council (2012) Directive 2012/19/UE of the European Parliament and of the Council of 4 July 2012 on Waste Electrical and Electronic Equipment (WEEE), Brussels, Belgium.

10. Roller G, Führ M (2008) Individual Producer Responsibility: A Remaining Challenge under the WEEE Directive. Reciel 17: 277-283.

11. Huisman J, Magalini F, Kuehr R, Maurer C, Ogilvie S, et al. (2007) 2008 Review of Directive 2002/96 on Waste Electrical and Electronic Equipment (WEEE). Final report. United Nations University, Japan.

12. O'Connell M, Hickey S, Fitzpatrick C (2010) Investigating Reuse of B2C WEEE in Ireland. Proceedings of the IEEE International Symposium on Sustainable Systems & Technology (ISSST), Washington, USA.

13. Gamberini R, Gebennini E, Rimini B (2009) An innovative container for WEEE collection and transport: Details and effects following the adoption. Waste Management 29: 2846-2858.

14. OCU-Compra Maestra (2012) Reparación ordenadores portátiles 371.

15. European Parliament and Council (2003) Directive 2002/96/EC of the European Parliament and of the Council of 27 January 2003 on Waste Electrical and Electronic Equipment (WEEE), Brussels, Belgium.

16. Alcántara M (2012) EkorreparaS.Coop. Centro para la Reutilización Red Social Koopera. Primeras Jornadas RELEC de preparación para la reutilización. Sevilla.

17. Rubio García L, Ramos Álvarez A, Atienza de Andrés M (2012) Preparación para la reutilización: experiencia en el tratamiento de RAEEs de las entidades recuperadoras de AERESS en el contexto de aprobación de la nueva Directiva 2012/19/EU. Congreso Nacional del medioambiente.

18. Truttmann N, Rechberger H (2006) Contribution to resource conservation by reuse of electrical and electronic household appliances. Resources, Conservation and Recycling 48: 249–262.

19. Barba-Gutierrez Y, Adenso-Díaz B, Hopp M (2008) An analysis of some environmental consequences of European electrical and electronic waste regulation. Resources, Conservation and Recycling 52: 481-495.

20. Rreuse (2012) Challenges to boosting reuse rates in Europe.

21. Baker S, King A (2007) Organising Reuse: Managing the Process of Design For Remanufacture (DFR). POMS 18th Annual Conference Dallas, Texas, USA.

22. Qureshi AS, McCornick PG, Qadir M, Aslam Z (2008) Managing salinity and waterlogging in the Indus Basin of Pakistan. Agricultural Water Management 95: 1-10.

23. González-Benito J, Lannelongue G, Alfaro-Tanco JA (2013) Study of supply-chain management in the automotive industry: a bibliometric analysis. International Journal of Production Research 51: 3849-3863.

24. Li Si Y, Hao Chun B, Feng Chuan P, Wang Li H, Liu Y (2014) A bibliometric analysis on acidophilic microorganism in recent 30 years. International Journal of Waste Resources 4: 1-7.

25. Pérez-Belis V, Bovea MD, Ibáñez-Forés V (2015) An in-depth literatures review of the waste electrical and electronic equipment context: Trends and evolution. Waste Management & Research 3: 3-29.

26. Medina-López C, Marín-García J, Alfalla-Luque R (2010) Una propuesta metodológica para la realización de búsquedas sistemáticas de bibliografía. Working Papers on Operations Management 1: 13-30.

27. Kissling R, Coughlan D, Fitzpatrick C, Boeni H, Luepschen C, et al. (2013) Success factors and barriers in re-use of electrical and electronic equipment. Resources, Conservation & Recycling 80: 21-31.

28. Hume A, Grimes S, Boyce J (2002) Environmental product attributes in end-of-life management in the UK. Problems encountered with informational systems for the management of end-of-life IT and office equipment. International Journal of Environment and Pollution 18: 126-137.

29. Rogers DS, Tibben-Lembke RS (1998) Going Backwards: Reverse Logistics Trends and Practices. University of Nevada, Reno. Center for Logistics Management.

30. Kissling R (2011) Best Practices in Re-Use. Success Factors and Barriers for Re-use Operating Models. StEP (Solving the e-waste problem).

31. Kissling R, Coughlan D, Fitzpatrick C, Boeni H, Luepschen C, et al. (2013) Success factors and barriers in re-use of electrical and electronic equipment. Resources, Conservation & Recycling 80: 21-31.

Annex 1

1. Manomaivibool P, Hong, JH (2014) Two decades, three WEEE systems: How far did EPR evolve in Korea's resource circulation policy? Resources, Conservation & Recycling 83: 202-212.

2. Wang L, Vincent Wang X, Liang G, Váncz J (2014) A cloud-based approach for WEEE remanufacturing. CIRP Annals-Manufacturing Technology 63: 409-412.

3. Ongondo FO, Williams ID, Dietrich J, Carroll C (2013) ICT reuse in socio-economic enterprises. Waste Management 33: 2600-2606.

4. Kissling R, Coughlan D, Fitzpatrick C, Boeni H, Luepschen C, et al. (2013) Success factors and barriers in re-use of electrical and electronic equipment. Resources, Conservation & Recycling 80: 21-31.

5. AchillasCh, Aidonis D, VlachokostasCh, Karagiannidis A, Moussiopoulos N, et al. (2013) Depth of manual dismantling analysis: A cost-benefit approach. Waste Management 33: 948-956.

6. Gui L, atasu A, Ergun O, Toktay LB (2013) Implementing Extended Producer Responsibility Legislation: A Multi-stakeholder case analysis. Journal of Industrial Ecology 17: 262-276.

7. Dwivedy M, Mittal RK (2012) An investigation into e-waste flows in India. Journal of Cleaner Production 37: 229-242.

8. Solé M, Watson J, Puig R (2012) Proposal of a new model to improve the collection of small WEEE: A pilot project for the recovery and recycling of toys. Waste management & research 30: 1208-1212.

9. Herat S, Agamuthu P (2012) E-waste: A problem or an opportunity? Review of issues, challenges and solutions in Asian countries. Waste management & research 30: 1113-1129.

10. Ongondo FO, Williams ID (2011) Mobile phone collection, reuse and recycling in the UK. Waste Management 31: 1307-1315.

11. Townsend TG (2011) Environmental Issues and Management Strategies for Waste Electronic and Electrical Equipment. Journal of the Air & Waste Management Association 61: 587-610.

12. Rousis K, Moustakas K, Stylianou M, Papadopoulos A, Loizidou M (2011) Management of waste from electrical and electronic equipment in Cyprus- a cases study. Environmental Engineering and Management Journal 10: 703-709.

13. Ongondo FO, Williams ID, Cherrett TJ (2011) How are WEEE doing? A global review of the management of electrical and electronic wastes. Waste Management 31: 714-730.

14. Bereketli I, ErolGenevois M, EsraAlbayrak Y, Ozyol M (2011) WEEE treatment strategies' evaluation using fuzzy LINMAP method. Expert Systems with Applications 38: 71-79.

15. Achillas C, Moussiopoulos N, Karagiannidis A, Vlachokostas C, Banias G (2010) Promoting reuse strategies for electrical/electronic equipment. Proceedings of Institution of Civil Engineers. Waste and Resource Management 163: 173-182.

16. Fortin J (2010) The rise of the Ressourceries' network. Techniques - Sciences – Methods 9: 40-47.

17. Yu J, Williams E, Ju M, Shao Ch (2010) Managing e-waste in China: Policies, pilot projects and alternative approaches. Resources, Conservation & Recycling 54: 991-999.

18. Walther G, Steinborn J, Spengler Th, Luger T, Herrmann Ch (2010) Implementation of the WEEE-Directive – economic effects and improvement potentials for reuse and recycling in Germany. International Journal of Advanced Manufacturing Technology 47: 461-474.

19. Geyer R, Blass VD (2010) The economics of cell phone reuse and recycling. International Journal of Advanced Manufacturing Technology 47: 515-525.

20. Papaoikonomou K, Kipouros S, Kungolos A, Somakos L, Aravossis K (2009) Marginalised social groups in contemporary weee management within social enterprises investments: A study in Greece. Waste Management 29: 1754-1759.

21. Roller G, Führ M (2008) Individual Producer Responsibility: A remaining Challenge under the WEEE Directive. Reciel 17:277-283.

22. Aizawa H, Yoshida H, Sakai S-I (2008) Current results and future perspectives for Japanese recycling of home electrical appliances. Resources, Conservation & Recycling 52: 1399-1410.

23. Yang J, Lu B, Xu Ch (2008) WEEE flow and mitigating measures in China. Waste Management 28: 1589-1597.

24. Rousis K, Moustakas K, Malamis S, Papadopoulos A, Loizidou M (2008) Multi-criteria analysis for the determination of the best WEEE management scenario in Cyprus. Waste Management 28: 1941-1954.

25. Babu BR, Parande AK, Basha CA (2007) Electrical and electronic waste: A global environmental problem. Waste management & research 25: 307-318.

26. Nicol S, Thompson S (2007) Policy options to reduce consumer waste to zero: comparing product stewardship and extended producer responsibility for refrigerator waste. Waste management & research 25: 227-233.

27. Truttmann N, Rechberger H (2006) Contribution to resource conservation by reuse of electrical and electronic household appliances. Resources, Conservation and Recycling 48: 249-262.

28. Streicher-Porte M, Widmer R, Jain A, Scheidegger R, Kytzia S (2005) Key drivers of the e-waste recycling system: Assessing and modelling e-waste processing in the informal sector in Delhi. Environmental Impact Assessment Review 25: 472-491.

29. Darby L, Obara L (2005) Household recycling behavior and attitudes towards the disposal of small electrical and electronic equipment. Resources, Conservation and Recycling 44: 17-35.

30. Goosey M (2004) End-of-life electronics legislation – an industry perspective. Circuit World 30: 41-45.

31. Hume A, Grimes S, Boyce J (2002) Environmental product attributes in end-of-life management in the UK. Part I: An end-of-life eco-declaration for waste IT and office equipment. International Journal of Environment and Pollution 18: 109-125.

32. Hume A, Grimes S, Boyce J (2002) Environmental product attributes in end-of-life management in the UK. Part II: Problems encountered with informational systems for the management of en-of-life IT and office equipment. International Journal of Environment and Pollution 18: 126-137.

12

Towards the Circular Economy: Waste Contracting

Henning Wilts* and Alexandra Palzkill

Wuppertal Institute for Climate, Environment and Energy, Germany

*Corresponding author: Henning Wilts, Wuppertal Institute for Climate, Environment and Energy, Germany
E-mail: henning.wilts@wupperinst.org

Abstract

The transition towards a circular economy is high on the political agenda and support for innovative business models can be seen as one of the key strategies for its implementation. Nevertheless most of these business models rely on an increasing generation of waste and thus undermine the prevention of waste as top of the waste hierarchy. The paper aims to link this debate to more systemic eco-innovations that offer economic market potentials by reduced material inputs and waste generation. This directs the attention to sufficiency strategies that surpass the level of individual consumer choices and regards the potentials of entrepreneurial sufficiency strategies. It takes the example of waste contracting modelsin Germany as a possible approach of resource-light business models that provide existing utility aspects with altered consumption patterns and decreased resource consumption. It describes environmental and economic benefits and draws conclusions on necessary policy framework conditions.

Keywords: Circular economy; Waste prevention; Contracting; Innovative business models

Introduction

Strategies for the decoupling of consumption patterns, waste generation and related environmental burdens predominantly focus on technical efficiency and consistency. The overall aim is to reduce the use of resources with the help of technological progress and closed material cycles. Besides, so-called sufficiency strategies are described in the literature, which not only aim at a relative decrease of resource consumption, but by means of strategies of "less, slower" or "more regional" also aspire to achieve an absolute reduction in resource use and consequently waste generation. However, sufficiency strategies, if even discussed, mostly consider individual consumer choices [1-4]. Especially in the face of the further increasing global resource consumption [5], this is an astonishing fact and directs the attention to sufficiency strategies that surpass the level of individual consumer choices and regards the potentials of entrepreneurial sufficiency strategies.

A common argument for the inadequacy of simple, usually technically dominated strategies is the so-called 'rebound effect' [6-8], which implies that successfully improved efficiency can lead to increased consumption, which again exhausts a part of the achieved savings – in unfavourable cases even overcompensates them. In terms of the rebound effect, very diverse causes can be found. Regarding direct cost savings in households, magnitudes of 10-30 % are specified, while for indirect, macroeconomic effects only few, strongly varying sources are available, which, however, identify rebound effects of up to 100% [9]. On the other hand, the complete and equivalent substitution of all current goods and services in a "consistent" manner in terms of an environmentally sound resource use (e.g. through cradle to cradle approaches) is not foreseeable at this point in time – especially not under consideration of widely differing spatial and temporal problem shifts [5]. Therefore, hoping for the necessary technological leaps through market incentives is highly risky: "For all these reasons, consistency is indispensable – but alone not sufficient for the actuation of a sustainable development" [10-11].

Against this background this paper analyses innovative contracting business models in the field of waste management based on waste fee savings from waste prevention and improved waste separation in Germany. It aims to give new insights regarding how the overall concept of sufficiency could enrich discussions about a circular economy, how contracting can create new business opportunities in the waste sector and which policy instruments might increase the market uptake of such promising innovations. The paper is structured as follows: Chapter 2 describes the analytical framework and the methodology of the analysis, chapter 3 introduces the empirical foundation of the paper. Based on this theoretical and empirical basis, the final chapter draws conclusions on key success factors, the necessary political framework and further research.

Theoretical Background

Key terms and definitions

The term of sufficiency is defined in different ways in the literature. The following takes on a definition of Fischer et al.: "The term ,sufficiency' refers to changes in consumption patterns which facilitate operation within the ecological bearing capacity of the earth, whereby utility aspects of consumption are changing". Hence, it is clearly distinguished from efficiency and consistency: "for these two strategies assume that with a lower environmental consumption, benefits do not change: Efficiency quantitatively reduces resource inputs or emission outputs in proportion to the generation of the same benefits; while consistency achieves the same through another technology which is equally environmentally sound on a large scale. Sufficiency, however, is accompanied by modifications of the benefit package" [9]. Assessment thereby in each case takes place based on individual preferences: On the one hand, sufficiency is therefore often outlined as "renunciation" or "lower welfare" [12]. On the other hand, terms like the "succeeding life" or the "right measure" [10,13] suggest that sufficiency could also bring forth an individual benefit – for example "time prosperity" [13], "freedom from excess" [14] or "the good life" [15]. This perception also addresses fundamental criticism brought for e.g. by Huber [16] who points out the often strongly paternalistic underlying assumptions of sufficiency policies.

For this paper we applied an analytical framework developed by Sachs [17] that distinguishes four fundamental sufficiency strategies (the so-called 4 D's) that oftentimes correspond with modified satisfaction of such benefit packages:

1. Decluttering (in terms of absolute reduction of the number and diversity of consumed/acquired products),
2. Deceleration (in terms of a reduction of the frequency of consumption),
3. Decommercialisation (in terms of the subsistence economy of DIY and production instead of commodification) and
4. Deconcentration (in terms of a simplification and regionalization of value chains).

Sufficiency as Consequence of Innovative Business Strategies

Despite its potential for resource conservation and the now recognized necessity of subsistence strategies for the reduction of global resource consumption, aside from individual consumer choices only few scientific approaches deal with the question and promotion of sufficiency strategies (on a political and/or entrepreneurial level) (first approaches to sufficiency policy see: [9,15]; on entrepreneurial sufficiency strategies [18-21]. Especially for business strategies targeted on sufficiency, which provide existing utility aspects with altered consumption patterns and decreased resource consumption, there is a considerable conceptional and instrumental deficit [22]. A deepening of implications specific to the company is therefore also hard to find, due to missing alternative conceptional landmarks. According to Paech, "new management concepts" [18] are in demand, which allow for a substitution or modification of existing consumer demands and develop new business models accordingly.

Along the abovementioned heuristics of the 4 D's, business models are indeed identifiable within which sufficiency becomes driver for a business case, or which take on trends towards sufficient behaviour [9,22] including the provision of services instead of resource intensive goods (e.g. Carsharing); satisfaction of needs in the sense of "less is more" (simplify your life) (decluttering) or the establishment of offers that enable a longer use phase (e.g. cheap repair and replaceable batteries in laptops) (deceleration).

The following case study will apply this analytical framework to a specific case study – waste contracting in different German metropolitan regions. For the research several expert interviews have been conducted and public available as well as internal planning documents have been analyzed.

Contracting

Contracting as contribution to resource efficiency

A possible approach of such resource-light business models that provide existing utility aspects with altered consumption patterns and decreased resource consumption are contracting models, which have predominantly been developed in the energy sector up to now: They belong to the few business models which turn a "less" into an established business case. Performance contracting in the energy sector is a contractually agreed service between the owners of buildings and tenants on the one side and specialised energy service enterprises, the contractors, on the other side. The contractors optimize energy supply, mostly through a mixture of classical and technological efficiency measures, but also energy use by means of sufficiency measures and profits from saved energy costs over an extended contract term.

The benefits of contracting can include, next to cost savings, the involvement of external experts in the modernisation of infrastructures, the avoidance of short-term high investment costs for optimized infrastructures for the owner as well as the transfer of investment risks to the contractor and the reduction of liability risks, among others: "The actual advantage of contracting is, however, that the technical expertise of the contractor can be used for planning, construction and operation of structural and supply and control technical measures and the risk and possibly the financing can be transferred to him". So far, contracting seems to be a classical financing strategy of efficiency measures for energy conservation. However, it can also be held as sufficiency strategy due to the fact that on both sides, that of the owner of the building and that of the tenant, the structures of need satisfaction change through energy supply and consumption.

Case study waste contracting

Based on the experiences outlined above, further significant resource efficiency potentials on the basis of contracting models can be assumed, for these can offer efficiency strategies combined with sufficiency strategies. For example, it might be considered to transfer the principle of performance contracting from the energy sector to the waste sector, as the Kiel enterprise Innotec that was founded in 1996 tries to realize. The innovation targets at a „financial incentive model" for improved waste separation by tenants and therefore sustainable cost reductions in the operation of residential properties [23]. It aims at key similarities of waste and energy: technical grid-based infrastructures with high shares of sunken investments, relevant negative externalities and public regulation for consumption-based fees: Thus the business model is based on two different pillars:

The area "conventional waste management" consists of an intensive personal counselling of tenants regarding correct waste separation and prevention of waste in different languages with campaigns specific to target groups such as "There is space in the smallest kitchen". The focus of these activities, however, is on the visual inspection of residual waste bins which is conducted by personnel of Innotec a few times per week, the so-called property supervisors, and sorts out bulky and large volume waste such as cardboard boxes or other large packaging in order to achieve a reduction in residual waste amounts [23]. Even after intensive counselling, this task has to be performed repeatedly and regularly, as counselling alone does not lead to lasting changes in waste separation behaviour. Innotec hereby emphasizes that from the point of view of waste legislation this doesn't involve the sorting of waste in technical terms but simply a correction of filling mistakes. In doing so, neither a withdrawal of recyclable materials – all waste liable to collection for disposal are in fact left to local authority waste management services – nor waste compaction takes place. Besides this, the enterprise also grants a regular cleaning of the waste bin locations.

Moreover, Innotec offers technical systems for a "consumer-related waste management" as an extension of its offer, which enable an individual registration and accounting of residual waste amounts by means of "waste identification" systems. Waste locks are hereto installed in residual areas which outwardly hardly differ from normal standard washed concrete boxes but where waste can only be disposed with a transponder chip, while being weighed at the same time Alternatively, waste containers are installed which can be filled with either 5l or 20l waste sacks and only individually register the number of fillings. The individual attribution achieves that residents only have to pay for their individual amount of residual waste and therefore have a real incentive to save money on fees by optimized sorting and waste-conscious consumption.

Even if the approach of individual transponder chips is based on a technical innovation, it particularly bets on a behavioural change of tenants and therefore combines efficient technology with altered consumption patterns for the satisfaction of needs. Though the focal point of Innotec's services is the correction of filling mistakes directly at the waste location, an essential component of the concept is, according to Innotec, also the information of tenants and the development of a fundamental awareness of costs and saving potentials in the area of waste [24].This necessary knowledge transfer has to be adapted to the gradual alteration of collection infrastructure very carefully; among other things, tenants regularly have to be informed about purpose and success of individual measures (more "yellow bins", introduction of an organic waste bin, introduction of waste locks, restriction of filling volumes etc.) [24]. The involvement of tenants in the concept is of central significance, also in order to prevent economic incentives that lead to waste being disposed in the environment in order to save costs. Investigations of the Witzenhausen institute after the introduction of waste locks showed however, that through intensive counselling no increase in wild waste deposits was noticed and the cleanliness at waste locations even increased considerably. The institute states that "after the introduction of the locks, no increase in environmental pollution and no wild waste deposits or illegal disposal of waste occurred in the investigated area and its surroundings" [25]. Meanwhile the enterprise supervises about 250 housing companies with 630,000 housing units and 1.5 Million tenants throughout Germany [26]. Innotec concentrates its activities as far as possible on social or publicly promoted residential construction, where tenants often have no distinct awareness for waste separation or prevention due to language barriers, education level or social environments. In high-grade apartments, however, the situation is often met that inhabitants are quite willing to spend 300 Euros or more per year on waste fees if it grants them sufficiently high volumes of residual waste at any time and the level of incidental costs does not form the relevant criterion in the decision to rent [27].

Contribution to sustainable resource management and reduction of resource consumption

Scientifically accompanied evaluations of Innotec projects in Hamburg and Erfurt revealed that the introduction of a consumer-oriented accounting of fees especially increased the amount of separately collected materials and therefore more materials could be directed to material recycling. This leads to a decrease in waste-related CO2 burdens of up to 60 % [28].

In this way, the amount of collected packaging from the dual system in e.g. Erfurt could be increased by 60%, while separately collected organic waste even increased by 400%. Residual waste generation specific to inhabitants could hereby be reduced from 3.85 kg per week to 0.8 kg per week, see Figure 1[29].

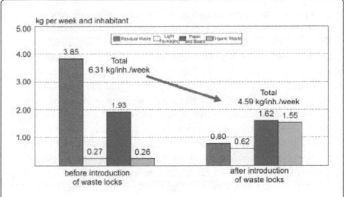

Figure 1: Changes in waste streams through the introduction of waste locks [29].

Taking a broader view on all recycling paths such as e.g. recycling depots, a residual amount of 800g per inhabitant and weak results, which can either be effectively prevented or shifted, e.g. to waste disposal at work. The accompanying scientific study comes to the following conclusion: "The introduction of waste locks encouraged the affiliated inhabitants to realize the fundamental waste economic ideas of prevention and recycling in their disposal behaviour" [29].

Triggers and required framework conditions of waste contracting

Starting point and occasion for the introduction of a sufficiency strategy in the waste sector were particularly the dramatically increased waste fees in the 1990s that turned waste fees into an effectively relevant cost block of residential incidental costs. The monthly cost of waste disposal and street cleaning in Germany amount to approximately 0.17 Euros per square meter on average – thus highlighting the importance of framework conditions like landfill bans and specific technical treatment obligations in Germany, inter alia aiming at economic incentives for high quality waste recycling or waste prevention. Another important driver for innovation is the intensified competition on the housing market. In the face of the demographic development in Germany, homeowners and especially big housing associations can no longer expect to find tenants for their housing spaces at arbitrary prices. Therefore property owners try to secure the competitiveness of their renting objects by means of efficient facility management that aims at low incidental costs regarding, among others waste.

Sorting analyses have revealed that particularly in the area of social housing waste disposal is "especially inefficient" because of the widely anonymous collection in large bins, meaning that on the one hand waste is not correctly separated and on the other hand significant amounts of light packaging can still be found in residual waste subject to charges. Added to this is the specific situation that rent in these places is often entirely paid by the social security office, so that the level of waste fees offers no economic incentive to inhabitants. This also signifies that waste volumes, which usually represent a benchmark of cost calculation, are not considered sufficiently. A classical example for this isfor example the disposal of high-volume cardboard boxes or bulky waste through residual waste bins. „Anonymity effects and convenience" [27] lead to classical cost externalisations, caused by the individual conviction of every inhabitant that for example his disposal of bulky waste through the residual waste bin saves him the trip to the recycling depot or the commissioning of the large refuse disposal unit (which is often even free of charge) and therefore reduces his costs. Especially the large 1,100l containers common in residential construction even enable e.g. the disposal of whole cupboards etc. through the residual waste bin. In these housing complexes it could be empirically observed that every offered waste volume has been used in the end; the amount of generated waste followed the available volume (up to a specific maximum limits of approximately 180l per person and week - however, the differences would be clearly lower if real weight would be registered instead of volume). These different effects lead to the fact that residual waste amounts and corresponding charges in social housing were significantly higher, on average about 3-4 times, than the comparative figures for e.g. one-family-houses, which dispose about 30l of waste per week and person on average, whereby in this case the correlation between the level of charges and residual waste generation is clearly recognizable. In the face of these tremendous cost savings potentials, Innotec was able to grow very fast especially in congested areas with large numbers of social housing complexes.

Business case and involved actors

Waste Contracting has its business case through the cost savings with regard to residual waste disposal. For landlords the total costs of waste disposal in past projects, depending on waste economic framework conditions, decreased by 20-50%, in individual cases by up to 70%, due to subsequent sorting [26] – hence properties can be offered at competitive conditions on the market. Innotec is exclusively funded by a profit sharing of these saved costs. Usually a pilot phase of five years is arranged in projects, during which Innotec obtains 75% of the saved residual waste fees in order to refinance installation engineering and the initially personnel-intensive counselling of tenants. After this pilot phase, Innotec still receives 50% of the savings in case the project is continued. A conducive factor has proven to be the fact that the owner's expenditures for waste management services can be applied to the income tax as household-related services.

Advantages result also for tenants, as the emerging costs of the residential area are no longer calculated on a flat-rate basis per head or square meter, but according to the individual waste behaviour. According to surveys of the German Tenant's Association, 65% of all tenants wish for a more consumer-related calculations of running costs [28].Moreover, the optimized separated collection and the correction of filling mistakes altogether leads to a clear reduction in residual waste fees. Innotec argues that cities also benefit from optimized waste management: On average, 20% of the inhabitants of large residential complexes are welfare recipients, whose decreased rental charges lead to lower rental subsidies. Taking Duisburg as an example, savings in rental charges of 1.7 Million Euros would lead to municipal savings of 180,000 Euro [30].

Conclusions

The case study and the triggered political discussions show that on the one hand, sufficiency strategies are no longer a simply private question of individual consumer choices. Private consumer choices, sufficient or not sufficient, are always dependent on several impact factors and enabled or restricted by economic, political and infrastructural framework conditions. Due to existing incentives and offers it can be assumed that the increased implementation of more sufficient lifestyles can hardly be simply promoted by the demand side of the market. Due to the fact that the sole addressing of the consumer

in order to implement more sufficient lifestyles constitutes an excessive demand [31], sufficient business strategies and offers are needed, just as directional political framework conditions promoting sufficiency.

On the other hand, initiatives like Innotec, which target at the prevention of waste as a component of their business model, show that sufficiency strategies as business cases can be successfully implementable. Innotec has by no means been initiated out of the normative idea of a low-waste economy, but because a lucrative business field was expected. Even the selection of projects doesn't occur from a perspective of maximum contribution to resource efficiency, but from the angle of maximum return. Nevertheless, the business model clearly contributes to more sufficient lifestyles which by now significantly influence the business interests of established waste management actors that have invested in waste management infrastructures – partly financed by fee payments.

Despite of relevant resource efficiency potentials, demands for national sufficiency politics have moral concerns, as they would intervene in fundamental rights and freedom of the individual in a particular manner and quality based on the (alleged) claim of renunciation. This, however, relates the concept of freedom solely to the freedom of the consumer and therefore strongly curtails it. In contrast, politics as facilitator of "positive freedom" (Thomas Green Hill) creates liberties for the development of individual ways of life without restricting the freedom of others [15].

In this way, every form of political framework setting (even the omission of such) always constitutes an influence on the possibilities and liberties of the individual. Therefore the reference to a restriction of the market doesn't release politics from the liability of creating suitable framework conditions that contribute to waste prevention and other sufficient lifestyles: "As little as politics alone can create sufficient lifestyles, as hard it is without them, according to our conviction" [9]. The German Advisory Council on Global Change (WBGU) justly refers to the necessity of a "shaping state with extended participation", that initiates "search processes" and directs them by means of frameworks and agenda setting [32].

Instruments for the promotion of sufficient contracting models

The basic approach of contracting raises the question to what extent the application of public resource political instruments is necessary at all: "Contracting as operation and financing model generally doesn't require separate promotion: In order to stay successful in the long run, contracting has to be economic per se" [33]. Especially with the herein chosen focus on the promotion of sufficient lifestyles, different starting points result for the promotion of contracting models and the overcoming of information deficits, high transaction costs due to lacking experience as well as path dependencies and bureaucratic procedures.

In Baden Wuerttemberg, the issues most relevant for the representatives of suppliers, consumers and financial management have been discussed and central recommendations were expressed in the framework of a contracting offensive. The focus was on contracting models in the area of energy; however, these approaches are also widely transferable to the area of resource efficiency:

Communications-initiative contracting: Especially in the area of resource efficiency, contracting possibilities are not yet sufficiently known. This was addressed by a communications initiative that aimed to predominantly target at political decision-makers, to whom the potentials of such models are often not very familiar. As a further circle of addressees, the supporting partners for enterprises in the area of finances, especially accountants, tax consultants and banks, were addressed as the parties trusted by decision-makers in enterprises that could take on an important role as "contracting multipliers". For the topic of resource contracting, a lack of project developers and consultants exists which could support and counsel contracting customers as experienced, competent and neutral partners. The aim should be the qualification of engineers, planners, architects and actors of existing resource efficiency counselling institutions as project developers. A comparable lack exists on the supply side.

Extending financing options for contracting: If a contractor can offer financing to the client at the same time, his market position improves. Adapted financing models are particularly missing for small projects in conjunction with new and smaller contractors. Therefore the framework of the Baden-Wuerttemberg project recommended the development, testing and distribution of new financing models together with principal banks and the federation of cooperatives which are not directly oriented towards the credit standing of the contractor or its counterpart (e.g. project-based financing). Furthermore, successful contracting models require hedging instruments such as contingency insurance or indemnity bonds in order to minimise the risk for both the contracting client and the contractor.

Possibilities of transmission

With a view on the transmission of the contracting approach to the topic of waste prevention, the question arises for which further application fields contracting is feasible, or on which areas the support of sufficiency business models should focus. Figure 2 shows the distribution of the total resource consumption in Europe to different consumption areas. Against this background, an example would be to focus on the prevention of food waste, as significant economic and ecologic potentials could be generated here – food waste represents a significant share of residual waste, shows large prevention opportunities and at the same time massive resource conservation potentials [34,35].

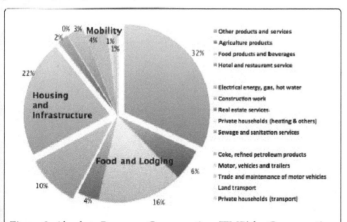

Figure 2: Absolute Resource Consumption (TMR) by Consumption Areas.

Fischer et al. [36], following Bilharz [37], refer to the criteria of ‚radiating effect' and the potential for structural change in the selection of topics for sufficiency and contracting strategies. 'Radiating effect'

means that a measure is so convincing that it is suitable to find imitators. The 'potential for structural change' describes a measure who changes the individual situation or the social practice in a way that the achieved environmental disburdening is most likely permanent [36]. Against this background, they recommend the use of cars, electricity saving, dietary shifts to Mediterranean food and living space reductions as interesting fields of action for sufficiency – though not all cases reveal how business models could be built upon these topics. Especially the topic of preventing food waste would be an important field of action for structural change due to its emotional charge and high radiating effect as well as its economic and ecological potentials.

Further Research Questions

The example of waste contracting shows that sufficiency can successfully function as business case and opens up business opportunities. However, existing concepts for sufficiency business strategies in business administration are only at the beginning stage: For an extended understanding of sufficiency-promoting business models, considerably extended knowledge about existing and prospectively possible sufficiency-based business models is required:

- What are success factors of existing business models, where have social trends for sufficiency successfully been adapted in the framework of innovative business ideas? How can these business models and their success factors be systematized? [21]. In which way is the specific extended producer responsibility scheme for waste packaging a necessary framework condition for such approaches?

- Who are relevant veto players whose traditional business models might be threatened by concepts of sufficiency and waste prevention, e.g. companies with large sunk costs in waste treatment facilities?

- Which sectors hold particular development potentials? How can traditional business models in these sectors be restructured in order to contribute to resource efficiency? [38].

References

1. Stengel O (2011) Suffizienz: Die Konsumgesellschaft in der ökologischen Krise. München

2. Linz M, Bartelmus P, Hennicke P, Jungkeit R, Sachs W, et al. (2002) Von nichts zuviel. Suffizienz gehört zur Zukunftsfähigkeit. Über ein Arbeitsvorhaben des Wuppertal Instituts. Wuppertal Institute for Climate, Environment, Energy, Germany .

3. Linz M (2012) Weder Mangel noch Übermaß: Warum Suffizienz unentbehrlich ist. München, Germany.

4. Paech N (2012) Befreiung vom Überfluss. Auf dem Weg in die Postwachstumsökonomie. München, Germany.

5. Bringezu S, Bleischwitz R (2009) Sustainable Resource Management. Trends, Visions and Policies for Europe and the World, Greenleaf Publisher, Sheffield, UK.

6. Madlener R, Alcott B (2011) Herausforderungen für eine technisch-ökonomische Entkoppelung von Naturverbrauch und Wirtschaftswachstum unter besonderer Berücksichtigung der Systematisierung von Rebound-Effekten und Problemverschiebungen. Enquete-Commission 'Wachstum, Wohlstand, Lebensqualität' of the German Bundestag, Berlin, Geermany.

7. Hertwich EG (2005) Consumption and the Rebound Effect: An Industrial Ecology Perspective. Journal of Industrial Ecology 9: 85-98.

8. Jenkins J, Nordhaus T, Shellenberger M (2011) Energy Emergence. Rebound and Backfire as Emergent Phenomena. The Breakthrough Institute, CA, USA.

9. Heyen DA, Fischer C, Barth R, Brunn C, Grießhammer R, et al. (2013) Suffizienz: Notwendigkeit und Optionen politischer Gestaltung. Öko-Institut Working Paper, Germany.

10. Linz M (2002) Warum Suffizienz unentbehrlich ist. In: Von nichts zuviel. Suffizienz gehört zur Zukunftsfä- higkeit. Über ein Arbeitsvorhaben des Wuppertal Instituts; M. Linz et al. (Ed.) Wuppertal Institutefor Climate, Environment, Energy, Germany.

11. Linz M (2004) Weder Mangel noch Übermaß. Über Suffizienz und Suffizienzforschung. Wuppertal Institute for Climate, Environment, Energy, Germany.

12. Alcott B (2007) The sufficiency strategy: Would rich-world frugality lower environmental impact? Ecological Economics 64: 770-786.

13. Linz M (2006) Was wird dann aus der Wirtschaft? Über Suffizienz, Wirtschaftswachstum und Arbeitslosig- keit, Germany.

14. Paech N (2011) Vom grünen Wachstumsmythos zur Postwachstumsökonomie. In: Welzer, H./Wiegandt, K. (ed.)Perspektiven einer nachhaltigen Entwicklung, Frankfurt, Germany.

15. Schneidewind U, Zahrnt A (2013) Damit gutes Leben einfacher wird - Perspektiven einer Suffizienzpolitik. Oekom Verlag, München, Germany.

16. Huber J (2000) Industrielle Ökologie. Konsistenz, Effizienz und Suffizienz in zyklusanalytischer Betrachtung. In: Simonis, Udo Ernst (ed): Global Change. Baden-Baden.

17. Sachs W (1993) Die vier E's: Merkposten für einen maß-vollen Wirtschaftsstil. Politische Ökologie 11: 33.

18. Paech N (2005) Nachhaltiges Wirtschaften jenseits von Innovationsorientierung und Wachstum⊠: eineunternehmensbezogene Transformationstheorie. Marburg: Metropolis.

19. Reichel A, O'Neil D, Bastin C (2010) 'Enough Excess Profits: Rethinking Business'. In: O'Neill, D., Dietz, R., Jones, N. (Eds.) Enough is enough. Ideas for a sustainable economy in a world of finite resources: 87–94. Leeds: Center for the Advancement of the Steady State Economy (Arlington, Virginia, USA), 2010; Economic Justice for All (Leeds, UK).

20. Reichel A, Seeberg B (2011) The Ecological Allowance of Enterprise: An Absolute Measure of Corporate. Journal of Environmental Sustainability 1: 1-14.

21. Sommer A (2012) Managing Green Business Model Transformations. Dissertation. Springer Verlag.

22. Schneidewind U, Palzkill A (2011) Nachhaltiges Ressourcenmanagement als Gegenstand einer transdisziplinären Betriebswirtschaftslehre – Suffizienz als Business Case.In: Corsten, H., Roth, S. Nachhaltigkeit – Unternehmerisches Handeln in globaler Verantwortung, Wiesbaden.

23. Innotec Abfallmanagement GmbH (n. d.) Referenzen. Ausgewählte Kunden im Konventionellen Abfallmanagement.

24. Innotec Abfallmanagement GmbH (2004)GWH setzt auf Abfallmanagement. Pilotprojekt in Bad Vilbe, Germany.

25. Witzenhausen-Institut (2007) Wissenschaftliche Begleitung der Einführung von Müllschleusen in der Stadt Erfurt. Final Report. Witzenhausen, Germany.

26. Innotec Abfallmanagement GmbH (2009) Der Innotec-Quotient. Modernes Abfallmanagement. Brochure.

27. Wilts H (2014) Nachhaltige Innovationsprozesse in der kommunalen Abfallwirtschaftspolitik – eine vergleichende Analyse zum Transition Management städtischer Infrastrukturen in deutschen Metropolregionen. Dissertation at the TU Darmstadt, Research group, Spatial and Infrastructure Planning, Darmstadt, Germany.

28. Hunklinger R (2011)Die intelligente Tonne. Innotec bietet modernes Abfallmanagement: Mülltonnen, deren Chip die Einwürfe zählt. Cleantech Magazin, Ausgabe.

29. Kern M (2007) Ergebnispräsentation von verursachergerechten Abfallmanagementsystemen im Wohnungsbau am Beispiel Müllschleusen in Erfurt. Practical forum. Witzenhausen, Germany.

30. Innotec Abfallmanagement GmbH (2011) Was leisten private Anbieter im Entsorgungs-Standortservice? Recycling Portal, Europe.

31. Grunwald A (2012) Against Privatisation of Sustainability - Why Consuming Ecologically Correct Products Will Not Save the Environment. GAIA - Ecological Perspectives for Science and Society 19: 178-182.

32. WBGU - Wissenschaftlicher Beirat der Bundesregierung Globale Umweltveränderungen (2011)Welt im Wandel – Gesellschaftsvertrag für eine Große Transformation; Berlin, Germany.

33. EnergieAgentur.NRW (2014) Förderprogramme & Contracting.

34. FAO (2012) Global Food Losses and Food Waste. Interpack2011 Düsseldorf, Germany.

35. UNEP (2013) Food Waste Facts.

36. Fischer C, Grießhammer R (2013) Suffizienz: Begriff, Begründung und Potenziale. Working Paper, Öko-Institut, Germany.

37. Bilharz M (2008) "Key Points" nachhaltigen Konsums; Metropolis, Marburg, Germany.

38. Palzkill A (2012) Business model resilience in the context of corporate sustainability transformation. Conference Paper. The 18th Greening of Industry Network Conference, 2012: 22–24.

13

Effect of Fertilization and Irrigation on Plant Mass Accumulation and Maize Production (*Zea mays*)

Paschalidis X[1], Ioannou Z[2]*, Mouroutoglou X[1], Koriki A[1], Kavvadias V[2], Baruchas P[3], Chouliaras I[4] and Sotiropoulos S[5]

[1]*Technological Educational Institute of Kalamata, Antikalamos, 24100, Kalamata, Greece*

[2]*Hellenic Agricultural Organization- DEMETER, Soil Science Institute of Athens, Greece*

[3]*Technological Educational Institute of Western Greece, Koukouli, Patra, Greece*

[4]*Mediterranean Agronomic Institute of Chania, AlsylioAgrokepio, Chania, Crete, Greece*

[5]*Greek Agricultural Insurance Organization, Nafpliou& Al. Soutsou, 22100, Tripoli, Greece*

*Corresponding author: Ioannou Z, Hellenic Agricultural Organization - DEMETER, Soil Science Institute of Athens, 1 Sof. Venizelou Str., Lykovrissi, 14123, Attiki
E-mail: zioan@teemail.gr

Abstract

The efficiency levels of nitrogen fertilization on growth and yield of maize (*Zea mays*) at two different levels of irrigation were examined. The experimental work was carried out at the farm of the Technological Educational Institute of Kalamata. The soil characteristics include: sandy clay soil texture, 11.07% $CaCO_3$, slightly acidic to neutral pH, non- saline, sufficient organic matter, adequate nitrogen, phosphorus, potassium and magnesium concentrations. Plot dimensions were equal to 3.0x4.0 m with four rows of plants per 0.75 cm in each plot of which the two inner rows represent the experimental surface. The experimental design was a Randomized Block Design. The experiment consisted of six treatments in three replications, with two levels of soil water capacity (70 and 40% respectively). N levels were 0, 160, 240 kg/ha, while the P and K levels were kept constant at 100 kg/ha. The amounts of P, K and 30% of N were added to the basic fertilization before sowing. The remaining N amount separated to two doses, at different growth stages of maize and incorporated through the irrigation system. The type of fertilizers was ammonium sulfate, ammonium nitrate, superphosphate and potassium sulfate. Based on the experimental data, it was found that the fed conditions greatly influence the nature and direction of the processes involved in developing plant, the addition of nutrients, regardless of treatment combination and their dosages, affected positively the plant growth, the fresh plant mass accumulation and the weight of 1000 grains compared to plants cultivated in soil without nitrogen. The total plant weight increase was equal to 59.13% compared to plants cultivated in soil with lack of nitrogen maintaining N, P, K levels to 240, 100, 100 kg/ha and soil water capacity to 70%. High seed yields were observed with the addition of 160 and 240 kg/ha N, constant levels of P and K, and 70% of soil water capacity. Low seed yields were observed when the level of irrigation was 40% of soil water capacity regardless the added amount of N, P and K. This can be interpreted that fertilizers have a high impact on crop yield, when combined with the appropriate level of irrigation.

Keywords: *Zea mays*; Irrigation levels; Sandy clay soils; Fertilizers

Introduction

In Greece, maize is cultivated mainly in Macedonia, Thrace, Central Greece and Peloponnese. The annual output reaches around 1.5 million tons. Historically maize is one of the major Greek crops. It was grown mainly for grain and biomass production due to its higher yield compared to other cereals. Maize seed is used in animal husbandry, in the human diet and in the production of by-products for the industrial sector. The biomass intended for silage, fresh or dried fodder, and for paper, ethanol and biofuel production. Among all the nutrients, nitrogen has the greatest effect on maize yield, particularly in light textured soils [1]. According to literature [2], the production of 1000 kg grain/ha require 19.4 kg of nitrogen, 2.7 kg of phosphorus, 13.8 kg of potassium, 1.4 kg of magnesium, 2.7 kg of calcium and small amounts of trace elements. In Messinia region (South Peloponnese), maize cultivation realized from light to medium texture soils, slightly acidic environment, where nitrogen leached from the surface layer, mainly due to the high rainfall and large annual irrigation doses. The excessive use of nitrogen fertilizers resulted to high nitrate concentrations in groundwater especially during the period of major nutrient requirements. Maize is very demanding in water, lack of which decrease its yield. The composition of 1 kg of dry maize matter requires the adsorption of 350-400 kg of water [3,4].

Studies have shown the efficiency of soil amendments such as zeolite, bentonite and zeolite – bentonite regarding the retention of nitrate ions, from maize (*Zea mays*). Two doses of nitrogen were used (400 and 800 kg N ha[-1]) in the form of NH_4NO_3. According to the statistical analysis of the greenhouse experimental data, nitrogen fertilizers influenced strongly the development of maize, bentonite and zeolite – bentonite as soil amendments increased the height of the plants in the dose of 800 kg N ha[-1]. Moreover, all the used soil amendments reduced the concentration of nitrate nitrogen in soil and plants [5].

The aim of the present study is the examination of the efficiency levels of nitrogen fertilization on growth and yield of maize (*Zea mays*) at two different levels of irrigation (40 and 70%, respectively). Moreover, the accumulation of maize organic dry biomass was also examined. The experimental design was a Randomized Block Design which consisted of six treatments in three replications, with two levels of soil water capacity. Fertilizers that were used, were ammonium sulfate, ammonium nitrate, superphosphate and potassium sulfate.

Materials and Methods

Maize was cultivated in the farm of the Technological Educational Institute (TEI) of Kalamata. The soil of the farm is a typical light textured sandy clay soil of the Messinia region in South Peloponnese. Other soil characteristics include: 11.07% of $CaCO_3$, pH equal to 6.39, non- saline soil. Plot dimensions were equal to 3.0x4.0 m with four rows of plants per 0.75 cm in each plot of which the two inner rows represent the experimental surface. The experimental design was a Randomized Block Design. The completely randomized experimental design consisted of six treatments in three replications, with two levels of soil water capacity (70 and 40% respectively). N levels were 0, 160, 240 kg/ha, while the P and K levels were kept constant at 100 kg/ha. The whole amounts of P, K and 30% of N were added to the basic fertilization before sowing. The remaining N amount separated to two doses, at different growth stages of maize and incorporated through the irrigation system. The type of fertilizers was ammonium sulfate (21-0-0), ammonium nitrate (34.5-0-0), superphosphate (0-20-0) and potassium sulfate (0-0-45). The simple hybrid maize used was "Aris". Seeding took place by hand at April 23rd, 2008. Crop irrigation was through the artificial rain and soil moisture was maintained at the levels of 70 % and 40 % of soil water capacity during cultivation. The appropriate cultivation treatments were applied to all experimental plots uniformly.

Measurements were taken for plant growth in two stages, the first is the development of corn two months after sowing and the second is at the end of its biological cycle. The measurements were made on each block separately and then the average plant growth per treatment was also determined. Plant tissue samples obtained at ripeness corn. The processing of the experimental data was performed by ANOVA analysis. The comparison of the averages by Dunkan test at a significance level of 0.05 was also examined.

Results and Discussion

During the experiment, morphological characteristics concerning the color appearance of leaves during the life-cycle of maize were observed. Comparing maize crops, where fertilizers were applied, with crops, which were grown in fertilized plots with no nitrogen (N0-P10-K10), color differences appeared in the 5th-6th normal leaves. The plants, which fertilized with nitrogen, have intensely green color in leaves compared to those which developed to soil with lack of nitrogen.

Plant growth measurements were taken in two stages, the first one represents the development of maize two months after sowing (June 26th) and the second one the development of maize at the end of its biological cycle (September 11th). The measurements were made on each block separately and then the average plant height per treatment was determined.

According to Table 1, the addition of phosphorus and potassium (N0-P10-K10) has a positive effect on plant height. All treatments had a constant level of potassium and phosphorus equal to 100 kg/ha. The average plant height with 70% soil water capacity ranged to 157.0 cm without the addition of nitrogen, to 247.33 cm with the addition of 160 kg N/ha and to 257.17 cm with the addition of 240 kg N/ha.

Treatments with 40% soil water capacity had an average plant height close to 227.0 cm without the addition of nitrogen, to 224.0 cm with the dose of 160 kg N/ha and to 218.50 cm with the dose of 240 kg N/ha.

Treatments			Plant height (cm)	
a/a	Fertilizers	Soil water capacity	June 24th	September 11th
A1	N0-P10-K10	70%	148.33	157
A2	N16-P10-K10	70%	188.83	247.33
A3	N24-P10-K10	70%	200.67	257.17
B1	N0-P10-K10	40%	162.83	227
B2	N16-P10-K10	40%	152.83	224
B3	N24-P10-K10	40%	159.83	226.5

Table 1: Effect of nitrogen levels and soil water capacity on the plant height of maize.

The addition of nitrogen (N16-P10-K10 & N24-P10-K10) influences significantly the plant heights when it was combined with 70% soil water capacity. Low soil water capacity around 40% did not increase the plant height regardless the addition amount of nitrogen in plants.

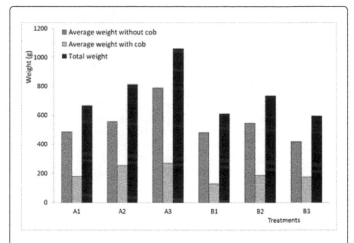

Figure 1: Effect of nitrogen levels and soil water capacity on the total maize weight (g/plant).

The effect of fertilizers and soil water capacity on seed fresh weight and on the fresh weight of 1000 grains is presented in Table 2. There was an increasing trend of weight in treatments with high soil moisture (70%) in contrary to the treatments with low soil moisture (40%). The increase of nitrogen level, e.g. N16-P10-K10 and N24-P10-K10 with 40% of soil water capacity maintained constant the seed fresh weight and the fresh weight of 1000 grains.

Maintaining the level of potassium and phosphorus constant and increasing the amount of nitrogen and soil water capacity, the seed fresh weight and the fresh weight of 1000 grains increased. According to literature the application of 100 kg N/ha increased biomass by 25-42% while irrigation increased grain yield by 59% [6].

Treatments	Average weight of fresh seeds (g/plant)	Fresh weight of 1000 grains (g)
A1	147.52	240.32
A2	211.98	293.85
A3	227.25	322.73
B1	101.81	234.68
B2	157.28	260.12
B3	141.68	258

Table 2: Effect of nitrogen levels and soil water capacity on seed fresh weight (g/plant) and on the fresh weight of 1000 grains (g).

Table 3 presents the data regarding the average yield of maize fertilization and irrigation levels. The lowest average maize yield (8110.0 kg/ha) was presented in treatment without nitrogen and irrigation at 40% of the soil water capacity. The highest average maize yield (19630.0 kg/ha), was presented in treatment with nitrogen level around 240 kg/ha and irrigation at 70% of the soil water capacity. Treatments had a constant level of potassium and phosphorus equal to 100 kg/ha.

Treatments	Maize Yield Replicates (kg/ha)			Average yield(kg/ha)
	1st	2nd	3rd	
A1	11730	8860.6	10460.6	10350cd*
A2	15330.3	22330.3	14660.6	17440ab*
A3	17290.3	19860.6	21730.3	19630a*
B1	9930.3	7330.3	7060.6	8110d*
B2	17130.3	14400	12730.3	14760bc*
B3	12200	14660.6	12530.3	13130bc*

Table 3: Effect of nitrogen levels and soil water capacity on average maize yield.*Means in the same column followed by the same letter(s) are not significantly different according to Duncan Multiple Range Test at 0.05 level of significance.

Treatments with nitrogen levels equal to 160 and 240 kg/ha, respectively, and 70 % of the soil water capacity, have shown a significant increase of yield. Thus, in these treatments yield ranges from 17440.4 to 19630.0 kg/ha maize.

In treatments where irrigation was 40 % of the soil water capacity by adding nitrogen dose gave a small increase of yield as it happened in seed fresh weight and on the fresh weight of 1000 grains (Table 2).

According to Duncan multiple range test at 0.05 significance level, it seems that the treatments B2 and B3 were not significantly different, i.e. treatments at 40% of soil water capacity with either the addition of 160 kg N/ha (B2) or 240 kg N/ha (B3). All the other treatments were significantly different to each other. According to literature [7], the maximum leaf area index, number of grains per cob, grain yield and harvest index were achieved with the addition of 250 kg N/ha and eight irrigations while the highest biological yield was reported by the addition of 300 kg N/ha and eight irrigations.

Figure 2 presents the data regarding the yield percentage of maize fertilization and irrigation levels. The effect of nitrogen in maize is important to the total production. The addition of nitrogen, i.e. 160 and 240 kg/ha, with 70% of soil water capacity led to an increase in yield percentage around 65% and 86% respectively compared to soil with lack of nitrogen (N0-P10-K10).

Note that maize yield, as it seems to the previous measurements, affects the rate and soil moisture. This can be interpreted that fertilizers have a greater impact on crop yield, when they are combined with the appropriate level of irrigation. Similar results were also referred in literature [8-11].

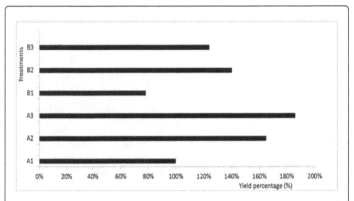

Figure 2: Effect of nitrogen levels and soil water capacity on maize yield percentage.

Conclusions

The nutritional conditions greatly influenced the nature and direction of the processes involved in developing plant. The addition of nutrients, regardless of the combination of each treatment and their dosage, positively affected plant growth and accumulation of fresh plant mass relative to soil with lack of nitrogen.

A high yield was observed with the addition of nitrogen in proportion equal to 160 and 240 kg/ha. Nitrogen was the main factor which increased maize yields maintaining the level of irrigation at 70 % of soil water capacity.

Comparing the high dose of nitrogen, i.e. 240 kg/ha and the different irrigation levels, it seems that the yield percentage increased (186%) with the increase of soil water capacity (70%). This can be interpreted that fertilizers have a greater impact on maize yield, when they are combined with the appropriate level of irrigation.

References

1. Romains HR (2001) Crop production in Tropical Africa. Directorate General for International Co-operation, Belgium.

2. Georgakopoulou A (2011) Influence of mycorrhiza presence on maize plants increase (*Zea mays* L.). Master's Thesis, Department of Biology, University of Patras, Greece.

3. Katsantonis N, Sfakianakis IN, Katziannias A, Katranis N, Georgiadis S, et al. (1988) Maize fertilization: II. Yield fruit, Nitrogen germination index and percentage of nitrogen fertilization, Agricultural Research.

4. Lubrication and Nutrition. Accessed 12/12-2014.

5. Molla A, Ioannou Z, Dimirkou A., Mollas S (2014) Reduction of nitrate nitrogen from alkaline soils cultivated with maize crop using zeolite-bentonite soil amendments. Int J Waste Resources 4: 155.

6.   Ogola JBO, Wheeler TR, Harris PM (2002) Effects of nitrogen and irrigation on water use of maize crops. Crops Research 78: 105-117.

7.   Hammad HM, Ahmad A, Azhar F, Khaliq T, Wajid A, et al. (2011) Optimizing water and nitrogen requirement in maize (Zea mays L.) under semi-arid conditions of Pakistan. Pak J Bot 43: 2919-2923.

8.   Pandey RK, Maranville JW, Admou A (2000) Deficit irrigation and nitrogen effects on maize in a Sahelian environment I. Grain yield and yield components. Agricultural Water Management 46: 1-13.

9.   Paolo ED, Rinaldi M (2008) Yield response of corn to irrigation and nitrogen fertilization in a Mediterranean environment. Field Crops Research 105: 202-210.

10.   Ramos TB, Gonçalves MC, Castanheira NL, Martins JC, Santos FL, et al. (2009) Effect of sodium and nitrogen on yield function of irrigated maize in southern Portugal. Agricultural Water Management 96: 585-594.

11.   Gehl RJ, Schmidt JP, Maddux LD, Gordon WB (2005) Corn yield response to nitrogen rate and timing in sandy irrigated soils. Agron J 97: 1230-1238.

14

Removal of Cationic Dye Methylene Blue from Aqueous Solution by Adsorption on Algerian Clay

Djelloul Bendaho*, Tabet Ainad Driss and djillali Bassou

Laboratory of organic chemical-physical and macromolecular Faculty of exact sciences, University DjilaliLiabès, Algeria

*Corresponding author: Djelloul Bendaho, Laboratory of organic chemical-physical and macromolecular Faculty of exact sciences, University DjilaliLiabès, FaubourgLarbi Ben m'hjdi P.O.Box-89, SidiBel-Abbès 22000, Algeria
E-mail: bendaho_djelloul@yahoo.fr

Abstract

The objective of this study was to demonstrate the potential of Tiout-Naama (TN) clay for removing a cationic Methylene blue (MB) dye from aqueous solutions which was used for the first time like an adsorbent. For this, the effect of several parameters such as contact time, adsorbent dose, pH and temperature have been reported. Nearly 30 min of contact time are found to be sufficient for the adsorption to reach equilibrium. The residual concentration of the dye is determined using UV/Vis Spectrophotometer at wavelength 664 nm. Langmuir and Freundlich isotherm models were used to describe adsorption data. The result revealed that the adsorptions of MB dye onto TN clay is the best-fit both Langmuir and Freundlich isotherms, further to understand the adsorption kinetics the adsorption data were analyzed by the second-order and the pseudo-second-order. The results show that the methylene blue adsorption follows pseudo-second–order kinetics.

Keywords: Dye; Adsorption; TN clay; Adsorption kinetics; Adsorption isotherms

Introduction

Water and soil pollution is a source of decay of the environment. Among these pollutants are the textile industry rejects which are heavily saturated by organic colorants. These colorants are usually used in excess to make the dye better, and consequently sewage is highly concentrated with colorants. In this objective, several methods are used to eliminate these colorants from industrial wastes Traditional process; such as chemical-physical treatments based on the addition of coagulant and flocculants (aluminium salts and polymers) [1] and biological process by the use of activation mud with a sufficient aeration [2], nevertheless these methods are not satisfying because of the weak colorant biodegradability [3]. The adsorption on active carbon is efficient but very expensive and has high operating costs due to the high price of the activated carbon and to the high water flow rate always involved [4]. In this way the majority of the processes are very selective according to the colorant categories to treat and some just move the pollution instead of removing it. It is necessary for the process to mineralize the colorant. Research is focussing on process using natural materials such as clay, agriculture waste residues [5] e.g wood, Waste. The different applications of the clay depend on their specific adsorption properties, the ion exchange and the surface nature. Due to these qualities, clay is used in different field, like in medical and pharmaceutical industries, organic molecule polymerization [6], and pollutant retention eg. Pesticides, organic colorants, heavy metals [7,8]. This paper presents the elimination of Methylene blue, an organic colorant used in textile industries, by adsorption on a natural material that is the clay of Tiout region in Algeria. This kind of clay fulfils all the environment protection conditions, and is very abundant and not expensive [9]. Kinetic and isotherm adsorption models are performed in order to better understand the organic colorant adsorption mechanisms [10].

Materials and Methods

Materials

The clay used in this work is montmorillonite type from Tiout region located south west Algeria . The clay undergoes a pretreatment by stirring it for two hours at room temperature. The obtained suspension is filtered, ground and heated untill obtaining a constant weight, then stored in a securely closed flask against the moisture.

In order to show the clay capacity of decoloration, we chose an organic colorant that it has methylene blue taken as a reference for pollutants because of its middle-size molecule and its several uses in tests, has a high solubility in water and it is one of the Thiazine colorants having the chimical formula $C_{16}H_{18}N_3ClS$ and molecular weight of 319.89 g/mole. The chemical structure is presented in Figure 1.

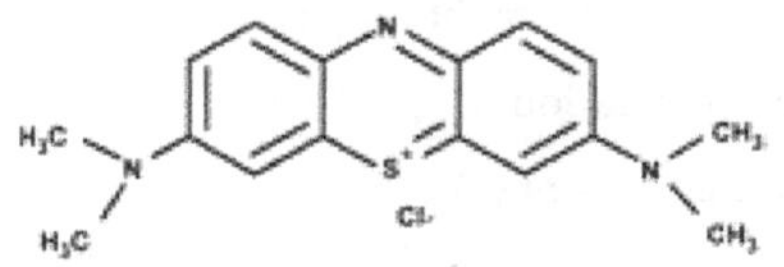

Figure 1: Chemical structure of methylene blue dye.

Chemical analysis showed the clay used is composed essentially of silica and alumina approximately 71.5% and of iron oxide 7.30% (Table 1), the presence of ions Na^+, Ca^+ and K^+ in the clay gives it a swelling type. The ratio $SiO_2/Al_2O_3=3.51$ reveals its montmorillonite characteristics.The mineralogical composition of the natural clay was determined by X-ray diffraction (XRD) using DRX.D8 ADVANCE BRUKER generator with copper anticathode ($\lambda CuK\alpha=1,5406$ Å).The Xray spectrum shows that the TN clay is a mixture of monmorillonite and impuretes of calcite and quartz (Figure 2).

Elements	SiO$_2$	Al$_2$O$_3$	Fe$_2$O$_3$	CaO	MgO	Na$_2$O	K$_2$O	SO$_3$	SO$_3$gyp	Cl	PF
% CCM	55.72	15.85	7.30	4.11	2.26	1.30	4.08	0.43	1.11	0.01	7.66

Table 1: Chemical composition of TN Clay.

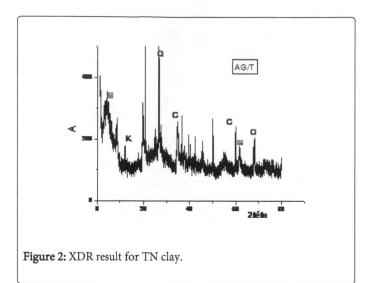

Figure 2: XDR result for TN clay.

Methods

In a series of 100 ml flasks, 80 mg of adsorbent (raw clay) is mixed with each 50 ml volume of Methylene blue aqueous solution having an initial concentration 25.59 mg/l. The adsorption tests are perfumed at natural pH, room temperature and magnetic stirring at 450 rpm at different time intervals. After equilibration, the aqueous phase was separated by centrifugation at 2500 rpm during 15 min. The concentration of the residual dye Cr was determined using SHIMADZU 1240, a spectrophotometer UV/Vis at 664nm of wavelength after 30 min. The equilibrium adsorption capacity q_e (mg/g) was calculated from the following equation:

$$q_t = (C_0 - C_r) . \frac{V}{m} \quad (1)$$

q_e: is the amount of dye adsorbed at equilibrium (mg/g)

C_0 and C_r are the initial and equilibrium concentrations of the dye, respectively, computed from the calibration curve (mg/l).

V: is the volume of the solution (l).

m: is the mass of the adsorbent (g).

Results and Discussion

Effect of contact time

The influence of time is achieved at natural pH of the solution for an initial concentration of 25.59 mg/1, with a mass of clay of 80 mg/1 and at room temperature. The amount of colorant adsorbed q_t at time t was determined by the following expression:

$$q_t = (C_0 - C_r) . \frac{V}{m} \quad (2)$$

Figure 3 shows the time course of adsorption equilibrium of Methylene Blue onto raw clay. The amount of dye adsorbed by adsorption on raw clay was found to be rapid at the initial period of contact time and then become slow and stagnant with increase in contact time; the mechanism of adsorbent removal can be described in migration of the dye molecule from the solution to the adsorbent particle and diffusion through the surface [11,12]. A 30 min contact was recommended for carrying out adsorption experiments.

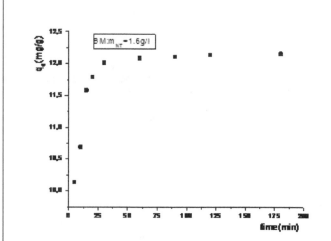

Figure 3: The effect of contact time on the adsorption of MB on TN clay.

Kinetics order

Adsorption kinetics is literary reviewed in different forms. In this paper kinetic laws of the second order and pseudo-second order are used. The second order kinetic model has the following formula.

$$\frac{dq}{dt} = k(q_e - q_t)^2 \quad (3)$$

Where is the rate constant of second-order adsorption has the following formula.

$$1/(q_e - q_t) = 1/q_e + k . t \quad (4)$$

For the pseudo-second-order model is given by the following equation.

$$\frac{dq_t}{dq_e} = k' . (q_e - q_t)^2 \quad (5)$$

Where is the rate constant of pseudo-second-order adsorption has the following formula.

$$\frac{t}{q_t} = 1 / (k' . q_e^2) + \frac{t}{q_e} \quad (6)$$

The kinetic data obtained using the second order and pseudo-second order is depicted in Figures 4 and 5.

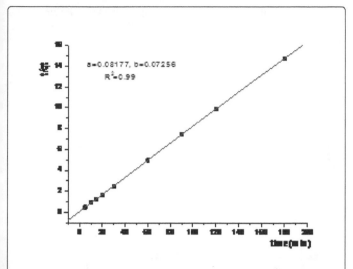

Figure 4: Plot of t/q_e versus t from the pseudo-second-order model.

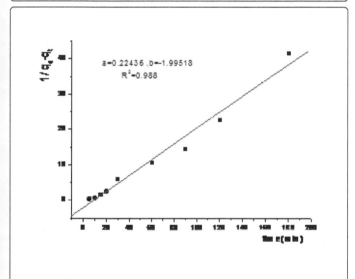

Figure 5: Plot $1/q_e-q_t$ versus t from the second-order model.

The q_e value calculated from the pseudo-second-order model is in accordance with the experimental q_e value. The correlation coefficient R^2 of the linear plot is very high. The result is shown in Table 2. The value of kinetic constant and qe indicates that the adsorption follow the pseudo-second order model. Several studies found that the kinetics of adsorption of dyes on clay supports obey to the pseudo-second-order [13-16].

Parameters	Pseudo-second-order
K' (g/mg min)	0.092
Q$_{ecal}$ (mg/g)	12,229
q$_{e\ épx}$ (mg./g)	12,183
R^2	0.99

Table 2: The pseudo- second -order parameters of MB adsorption into TN clay.

Effect of adsorbent mass

The effect of adsorbent mass on the adsorption capacity was studied using 50 ml of Methylene blue (25.591 mg/l) onto 1-4 g/l of adsorbent with shaking at room temperature for 30 min, the solid particles were removed and the remaining concentration of colorant in the filtrate was measured by using the UV/Vis spectrophotometer at 664 nm.

The effect of sorbent quantity on dye removal is illustrated in Figure 6. From the figure it can be seen that an increase mass of crude clay 1 g/l down to a value of 4 g/l causes a decreases in residual dye concentration. The increase in Methylene blue adsorption with the increase in adsorbent mass is attributed to increase in surface area of micro pores and the increase in availability of vacant adsorption sites. The same results were obtained by other authors [17,18].

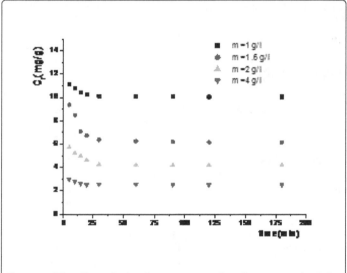

Figure 6: The effect of adsorbent mass on the adsorption of MB by TN clay.

Effect of pH

The influence of pH on dye removal was determined by performing the adsorption experiments at different initial pH of the solution (2-11) at room temperature. The pH of the solution was adjusted with HCl (0.1 N) or NaOH (0.1 N) solution by using a HANNA 210 pH-meter equipped with a combined pH electrode. The adsorption of Methylene blue onto clay is highly dependent on pH of the solution. The Figure 7 shows that the adsorption increased with increasing pH. The removal dye is low in the acid pH region because the hydrogen ions neutralize the negatively charged clay surface thereby decreasing the adsorption of the positively charged caution because of reduction in the force of attraction between adsorbate and adsorbent. The removal of dye is more at higher pH, because the surface of used clay is negatively charged. Therefore, the electrostatic attractive force between the colorant dye, which has a positive charge, and the adsorbent surface increases, and consequently, the rate of dye adsorption increases, the highest dye removal was detected in pH 9.

Similar results have been reported for the adsorption of Methylene blue on morocco clay [19,20].

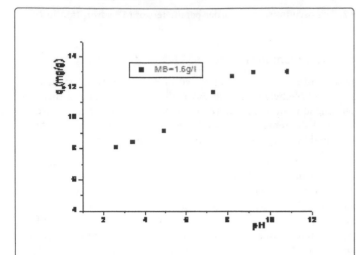

Figure 7: The effect of pH of the solution on the adsorption of MB on TN clay.

Effect of temperature

The amount of dye adsorbed on Tiout-Naama clay was determined at 20, 30, 40 and 60°C to investigate the effect of temperature. 80 mg of adsorbent was added to 50 ml dye solution with initial concentration of 25.59 mg/l. The contents in the flasks were agitated for 30 min.

The Figure 8 shows that as temperature increases from 20°C to 60°C, the adsorbed amount of dye at the same equilibrium concentration increased. Similar observations have been reported in the literature [21]. When temperature increased, the physical bonding between the organic compounds (including dyes) and the active sites of the adsorbent weakened. Besides, the solubility of Methylene bleu also increased with increase in temperature.

Thermodynamic parameters

The Gibbs energy is calculated from the given equation:

$$\Delta G = -RTLnKc \qquad (7)$$

Kc represented the ability of the retain the adsorbate and extent of movement of it within the solution. The value of K_c can be deduced from the following formula:

$$K_C = \frac{qe}{Ce} \quad (8)$$

Where:

q_e: is the amount of dye adsorbed at equilibrium (mg/g)

C_e: is the equilibrium concentrations of the dye in the solution.

The thermodynamic equation:

$$\Delta G = \Delta H - T\Delta S \qquad (9)$$

And the Vant'Hoff formula:

$$\Delta G = -RTLnKc \qquad (10)$$

Can be deduced the following formula:

$$lnKc = \frac{\Delta S}{R} - \frac{\Delta H}{R} \cdot \frac{1}{T} \qquad (11)$$

The values and can be obtained by plotting the versus 1/T (Figure 9).

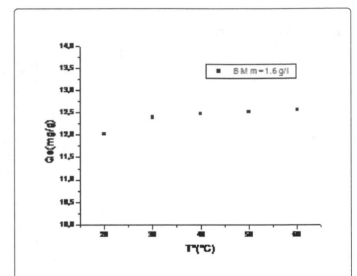

Figure 8: The effect of the temperature on the adsorption of MB on TN clay.

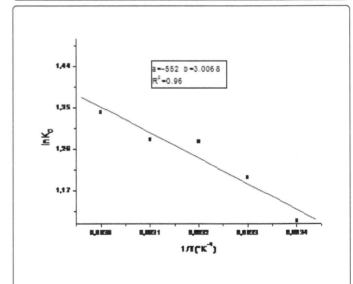

Figure 9: Plot of ln Kd versus 1/T for the estimation of thermodynamic parameters.

According to the thermodynamic parameters represented in Table 3, we realized that the ΔH enthalpy of the system is positive so the adsorption process on Tiout-Naana clay is endothermic, the low value of this energy (<40 Kj/mole) shows that it is a physical adsorption, the positive value of ΔS shows the attraction of the adsorption according to the dye, the negative value of ΔG indicates that the adsorption is done through a spontaneous and favourable process [22].

Adsorbant	Colorant	T('K)	ΔG KJ/mol	ΔH KJ/mol	ΔS mol.K	R²
TN clay	MB	20	-2.90	4.58	24.96	0.92
		30	-2.98	-	-	
		40	-3.23	-	-	
		50	-3.48	-	-	
		60	-3.72	-	-	

Table 3: Thermodynamic parameters of MB adsorption into TN clay.

Parameters	Langmuir	Freundlich
qmax (mg/g)	56.850	-
R2	0.99	0.99
b	2.94	-
1/n	-	0.759
LogKF	-	0.479

Table 4: Isotherm constants for adsorption of MB on TN clay.

Adsorption isotherm

Several laws have been proposed for the study of adsorption. They express the relation between the amount adsorbed and the concentration of aqueous solution at a specific temperature; the most commonly models used for adsorption are Langmuir and Freundlich isotherms those have been selected in this study.

The Langmuir isotherm is valid for monolayer adsorption onto a surface with a finite number of identical sites. The homogeneous Langmuir adsorption isotherm is represented by the following equation:

$$q_e = q_{max} \cdot \frac{bC_e}{(1 + bC_e)} \qquad (12)$$

Where q_e is the amount adsorbed at equilibrium (mg/g), Cethe equilibrium concentration (mg/l), b constant related to the adsorption energy (l/mg), and qmax the maximum adsorption capacity (mg/g).

The linear form of Langmuir equation may be written as

$$\frac{1}{q_e} = \frac{1}{q_{max}} + \frac{1}{b} \cdot \frac{1}{C_e} \qquad (13)$$

By plotting $(1/q_e)$ versus C_e, q_{max} and b can be determined if a straight line is obtained.

The Freundlich isotherm is an empirical equation assuming that the adsorption process takes place on heterogeneous surfaces, and adsorption capacity is related to the concentration of colorant at equilibrium. The heterogeneous Freundlich adsorption isotherm is represented by the following equation:

$$q_e = C_e^{1/n} \cdot K_F \qquad (14)$$

Where the KF is Freundlich constant related to the adsorption capacity (mg/g) and 1/n shows the adsorption intensity (l/mg).

The linear form of Freundlich equation may be written as

$$\log q_e = \frac{1}{n} \cdot \log C_e + \log K_F \qquad (15)$$

The values of and can be determined by plotting the versus if a straight line is obtained.

In the experiments of equilibrium adsorption isotherm, a fixed amount of 80 mg adsorbent is contacted with 50 ml of aqueous solutions Methylene bleu have different concentrations (10–250 mg/l). The adsorption was carried at room temperature while keeping all other parameters constant and the result is shown in Figures 10 and 11. Values for q_{max}, b, n and K_F, are summarised in Table 4. It can be seen, the result revealed that the adsorption of Methylene bleu dye onto raw clay was the best-fit both Langmuir and Freundlich isotherms. The n value is greater than 1 which indicates that the adsorption process is favourable [23].

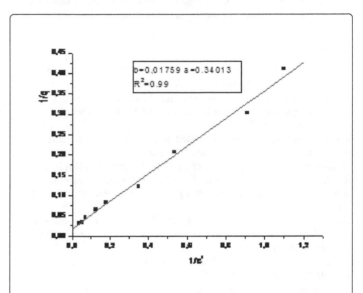

Figure 10: Langmuir plots for the adsorption of MB on TN clay.

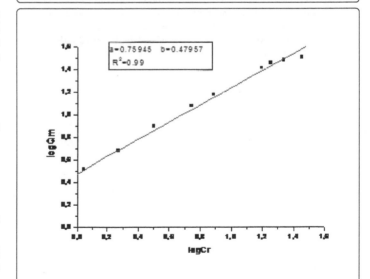

Figure 11: Freundlich plots for the adsorption of MB on TN clay.

Conclusions

The present study shows that the natural Algerian clay, abundant low-cost clay, can be used as sorbent for the removal of Methylene blue dye from aqueous solution.The value of kinetic constant and q_e indicates that the adsorption follow the pseudo-second order model.

Nearly 30 min of contact time are found to be sufficient for the adsorption to reach equilibrium. When the amount of crude clay increases from 1 g/l to a value of 4 g/l causes decreases in residual dye concentration. The results suggested that the adsorption capacity of organic compound colorant on raw clay adsorbent increased with increasing pH, rising the temperature induces a slight growth in the adsorption capacity, the negative value of ΔG indicates that the adsorption is done through a spontaneous and favourable process, the result revealed that the adsorption of Methylene bleu dye onto Tiout-Naama clay was the best-fit both Langmuir and Freundlich isotherms and the maximum adsorption capacity is in the order of 56.85 mg /g.

References

1. Annadurai G, Chellapandian M, Krishnan MRV () Adsorption of reactive dye on chitin. Environ Monit Asses 59: 111-119.

2. Alinsafi A, Khemis M, Pons MN, Leclerc JP, Yaacoubi A, et al. (2005) Electro-coagulation of reactive textile dyes and textile wastewater. Chem Eng Process 44: 464-470.

3. Khenifi A, Sekrane F, Kameche M, Deriche Z (2007) Adsorption study of an industrial dye by an organic clay. Adsorption 13: 149-158.

4. Chen S, Zhang J, Zhang C, Yue Q, Li Y, et al. (2010) Equilibrium and kinetic studies of methyl orange and methyl violet adsorption on activated carbon derived from Phragmites australis. Desalination 252: 149-156.

5. Robinson T, Chandran B, Nigam P (2002) Removal of dyes from an artificial textile dye effluent by two agricultural waste residues, corncob and barley husk. Environ Int 28: 29-33.

6. Ferrahi MI, Belbachir M (2005) Synthesis of cyclic polyesters of poly(oxybutyleneoxymaleoyl). J Polym Res 12: 167-171.

7. Zamzow MJ, Eichbaum BR, Sandgren KR, Shanks DE (1990) Removal of heavy metals and other cations from wastewater using zeolites. Sep Sci Technol 25: 1555-1569.

8. Bhattacharyya KG, Gupta SS (2006) Kaolinite, montmorillonite, and their modified derivatives as adsorbents for removal of Cu+2 from aqueous solution. Sep Purif Technol 50: 388-397.

9. Ho YS, Chiang CC (2001) Sorption studies of acid dye by mixed sorbents. Adsorption 7: 139-147.

10. Nagarethinam K, Mariappan MS (2001) Kinetics and mechanism of removal of methylene blue by adsorption on various carbons: a comparative study. Dyes Pigment 51: 25-40.

11. Gürses A, Doğar C, Yalçin M, Açikyildiz M, Bayrak R, et al. (2006) The adsorption kinetics of the cationic dye, methylene blue, onto clay.J Hazard Mater 131: 217-228.

12. Wibulswas R (2004) Bach and fixed bed sorption of methylene blue on precursor and QUACs modified montmorillonit. Sep Purif Technol 39: 3-12.

13. Almeida CA, Debacher NA, Downs AJ, Cottet L, Mello CA (2009) Removal of methylene blue from colored effluents by adsorption on montmorilloniteclay. J Colloid Interface Sci 332: 46-53.

14. DoÄŸan M, Alkan M, TÃ¼rkyilmaz A, Ozdemir Y (2004) Kinetics and mechanism of removal of methylene blue by adsorption onto perlite.J Hazard Mater 109: 141-148.

15. Ho YS, Chiang CC, Hsu YC (2001) Sorption kinetics for dye removal from aqueous solution using activated clay. Separation Science and Technology 36: 2473-2488.

16. Ferrero F (2010) Adsorption of Methylene Blue on magnesium silicate: kinetics, equilibria and comparison with other adsorbents.J Environ Sci (China) 22: 467-473.

17. Tahir SS, Rauf N (2006) Removal of a cationic dye from aqueous solutions by adsorption onto bentoniteclay.Chemosphere 63: 1842-1848.

18. Mohan D, Singh KP, Singh G, Kumar K (2002) Removal of dyes from wastewater using fly ash, a low-cost adsorbent. Ind Eng Chem Res 41: 3688-3695.

19. Fil BA, Yilmaz MT, Bayar S, Elkoca MT (2014) Investigation of adsorption of the dyestuff astrazon red violet 3rn (basic violet 16) on montmorillonite clay. Braz J Chem Eng 33: 171-182.

20. Tsai WT, Hsu HC, Su TY, Lin KY, Lin CM, et al. (2007) The adsorption of cationic dye from aqueous solution onto acid-activated andesite.J Hazard Mater 147: 1056-1062.

21. Richards S, Bouazza A (2007) Phenol adsorption in organo-modified basaltic clay and bentonite. Applied Clay Science 37: 133-142.

22. El Ouardi M, Alahiane S, Qourzal S, Abaamrane A, Assabbane A, et al. (2013) Removal of Carbaryl Pesticide from Aqueous Solution by Adsorption on Local Clay in Agadir. American Journal of Analytical Chemistry 4: 72-79

23. Tsai WT, Chang YM, Lai CW, Lo CC (2005) Adsorption of Basic Dyes in Aqueous Solution by Clay Adsorbent from Regenerated Bleaching Earth. Applied Clay Science 29: 149-154.

15

White Shrimp (*Litopenaeus vannamei*) Culture using Heterotrophic Aquaculture System on Nursery Phase

Supono[1]*, Johannes Hutabarat[2], Slamet Budi Prayitno[2] and YS Darmanto[2]

[1]*Department of Aquaculture, Faculty of Agriculture, Lampung University, Indonesia*

[2]*Department of Fisheries, Faculty of Fisheries and Marine Science, Diponegoro University, Indonesia*

*Corresponding author: Supono, Department of Aquaculture, Faculty of Agriculture, Lampung University, Indonesia
E-mail: supono_unila@yahoo.com

Abstract

Heterotrophic aquaculture system is an environmental friendly shrimp culture that has a huge potency to improve yields of *Litopenaeus vannamei*. Biofloc grown in a heterotrophic aquaculture system that can be used as an alternative feed for shrimp due to its high nutrition. Biofloc contains bacterial protein and polyhydroxybutyrate that are able to enhance growth. Biofloc also contains bacteria that have peptidoglycan and lipopolysaccharide on their cell walls. The aim of the research was to study the effect of heterotrophic aquaculture system on culturing of *Litopenaeus vannamei* during nursery phase. The experiment was arranged in split plot design in three replicateses. The treatments consisted of two factors namely various densities and different aquaculture systems. The aquaculture systems were autotrophic and heterotrophic aquaculture system, while densities were 1,000, 1,500, and 2,000 PLm^{-3}. The result showed that there was no significant interaction between densities and aquaculture system toward the growth rate, protein efficiency ratio and yield of *Litopenaeus vannamei*. The heterotrophic aquaculture system was able to increase the yield of *Litopenaeus vannamei* on nursery phase. However heterotrophic aquaculture system did not significantly affect growth rate and protein efficiency ratio of *Litopenaeus vannamei*. While, the density significantly affected survival rate and yield of *Litopenaeus vannamei*.

Keywords: Environmental friendly; *Litopenaeus vannamei*; Heterotrophic; Biofloc

Introduction

The intensive development of the shrimp culture has been accompanied by an enhancement in environmental impact. In the autotrophic aquaculture system, waste of shrimp culture can be a serious problem both in culture pond and in the environment. The effluent as uneaten feed, faces, and excretions is mostly inorganic nitrogen (mobile nitrogen) in form of ammonia and nitrite due to high protein content in feed (30-40%). Nitrogen in feed is only 25% that is recovered in the shrimp on harvest while about 75% is released into the pond ecosystem, mostly as TAN [1]. Ammonia and nitrite in culture pond is toxic to shrimp. Inorganic nitrogen built up in ponds, is controlled by algae and nitrification. Inorganic nitrogen is converted to organic nitrogen to build algae cell. This process is limited by the rate carbon assimilation by algae. Nitrification is a slow process and need a few weeks to complete it [2].

In order to minimalize impact of the shrimp culture effluent, it is impotant to develope shrimp farming with zero water exchange. The effluent produced by shrimp should be recycled in the pond before releasing to the surrounding environment. The method that can be applied to overcome the problem is heterotrophic aquaculture system (biofloc system). A carbon source namely sugar, molasses, and starch is added into culture pond increasing ratio of C:N to immobilize inorganic nitrogen and to stimulate the growth of heterotrophic bacteria to form biofloc.

The shrimp cultivated in heterorophic system has higher price than that in autotrophic system due to an environmental friendly product.

Heterorophic system in aqculture has huge potential to increase growth and survival rate of shrimp. Heterorophic system is capable to decrease cost production from feed because biofloc can be an alternative feed for shrimp that has high nutrition content [2].

Biofloc dominated by bacteria is high protein content and able to produce polyhydroxybutyrate as reserve energy and carbon [3]. Polyhydroxybutyrate is capable to accelerate the animal growth and to inhibit pathogenic vibrio in intestine tract. The objective of this experiment was to study the effect of heterotrophic system on the performance of on nursery *Litopenaeus vannamei* phase.

Materials and Methods

The experiment was arranged in split plot design with two factors in three replicates. The treatments consisted of aquaculture system (as sub plot) and density (as main plot). The aquaculture systems were heterotrophic system (HS) and autotrophic system (AS) while stocking densities of *Litopenaeus vannamei* were 10, 15, and 20 PL in 10 litres container equivalent to 1,000, 1,500, and 2,000 PLm^{-3}.

Eighteen plastic containers were filled with sterile saline water. Water salinity was adjusted to osmolarity of *Litopenaeus vannamei* haemolymph at intermolt phase. Osmolarity of *Litopenaeus vannamei* haemolymph was measured with method conducted by Anggoro and Muryati [4]. Nine containers were used to culture *Litopenaeus vannamei* in heterotrophic system. Shrimp feed and glucose was added to get C:N ratio of 21 [2]. Bacterium of Bacillus cereus (10^6 CFU/ml) was inoculated into media to stimulate heterotrophic system. Each container was equipped by aeration and was installed on the container bottom to maintain dissolved oxygen min. at 4 mg/l and water movement. After 10 days of growing biofloc, *Litopenaeus vannamei* postlarvae (PL 17) with an average body weight of 11 ± 1.0 mg and average length of 1.36 ± 0.15 cm was stocked into culture media.

Shrimp culture was conducted in 30 days. Shrimp were fed by formulated feed of 38% protein on feeding rate of 5%. Zero water exchange was applied in heterotrophic system. The growth and survival rate of *Litopenaeus vannamei* were calculated with methods conducted by Far et al. [5]. Protein efficiency ratio was analyzed with a method described by Hoffman and Falvo [6]. All data were further analyzed statistically using Two-Way-Anova after testing of normality, homogeneity, and additivity using SPSS statistical software. Statistical significance of differences required that p<0.05.

Results and Discussion

Haemolymph osmolarity of *Litopenaeus vannamei*

Haemolymph osmolarity of PL 11- *Litopenaeus vannamei* on premolt phase was 933,89 mOsm/l H₂O, equivalent to 32‰, meanwhile haemolymph osmolarity of *Litopenaeus vannamei* on intermolt phase was 861,00 mOsm/l H₂O, equivalent to 29.5‰ (Table 1).

Phase	Osmolarity (m Osm/l H2O)	Salinity (‰)
Premolt	1. 933.89	32
	2. 933.90	32
	3. 933.87	32
	Average	32
Intermolt	1. 861.02	29.5
	2. 860.98	29.5
	2. 860.98	29.5
	Average	29.5

Table 1: Osmolarity haemolymph of *Litopenaeus vannamei.*

According to the data, applied salinity of media was 30‰. Osmorality of haemolymph on premolt phase and intermolt phase is the best range for growth (4).

Water quality

Dissolved oxygen, total ammonia nitrogen (TAN), and pH of culture media of heterotrophic system and autotrophic system are performed at Figures 1, 2 and 3, respectively. The range of dissolved oxygen (DO) in culture media was 5.0–8.0 mg/l, TAN was 0.01-0.05 mg/l, and pH was 6.8-7.2. The values were in range for growing *Litopenaeus vannamei* [5].

Shrimp Performance

In general, performance of *Litopenaeus vannamei* in heterotrophic aquaculture system was better than that in autotrophic one. Specific Growth Rates (SGR), survival Rate (SR), protein efficiency ratio (PER), and yield of *Litopenaeus vannamei* at harvest are summarized in Table 2.

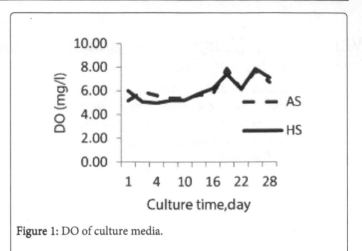

Figure 1: DO of culture media.

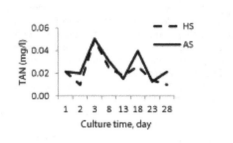

Figure 2: TAN of culture media.

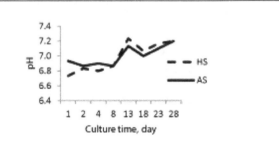

Figure 3: pH of culture media.

Density (PL/m³)	System	SGR (%)	SR (%)	PER	Yield (g/m³)
1000	AS	12.75 ± 2.08	60.0 ± 10.0	0.83 ± 0.37	256.6 ± 107.8
	HS	13.07 ± 1.27	73.3 ± 5.8	1.08 ± 0.35	29.9 ± 101.93
1500	AS	12.85 ± 2.08	60.0 ± 6.7	0.82 ± 0.37	378.1 ± 85.7
	HS	14.04 ± 0.80	66.7 ± 6.7	1.25 ± 0.26	571.4 ± 114.5
2000	AS	12.90 ± 2.08	41.7 ± 10.4	0.55 ± 0.10	344.3 ± 59.2

| | HS | 13.15 ± 2.08 | 50.0 ± 10.0 | 0.73 ± 0.19 | 450.8 ± 113.6 |

Table 2: Performance of *Litopenaeus vannamei*.

According to statistical analytic (ANOVA), there was no significant interaction between densities and aquaculture system to performance of *Litopenaeus vannamei*. The use of heterotrophic system did not enhance the growth rate and protein efficiency ratio (Figures 4 and 5). However survival rate and yield of *Litopenaeus vannamei* significantly increased in heterotrophic system (Figures 6 and 7). Stocking densities significantly affected the survival rate and yield of *Litopenaeus vannamei*.

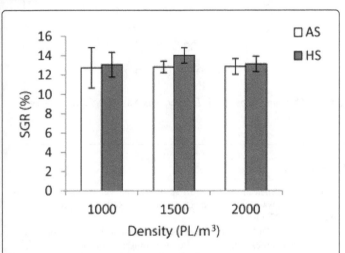

Figure 4: Specific growth rates (SGR) of *Litopenaeus vannamei*.

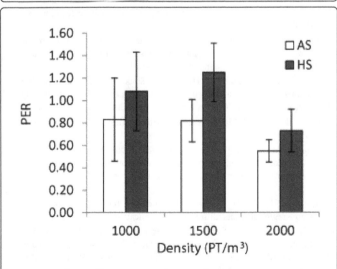

Figure 5: Protein efficiency ratio (PER) of *Litopenaeus vannamei*.

Discussion

Ammonia in pond is produced as a major end product of the metabolism due to high content protein of feed and is excreted as ammonia across the gill of shrimp [7]. No water exchange applied in heterotrophic system was able to control TAN in aquaculture system.

By adding organic carbon source to the water, it forces the bacteria to immobilize any inorganic nitrogen present in the pond. Inorganic nitrogen was recycled in the culture pond resulting in microbial protein biomass needed for cell growth and multiplication. At high ratio of C:N, heterotrophic bacteria will assimilate ammonium nitrogen directly from water metabolized to cell biomass. The addition of carbon source is most effective method in decreasing inorganic nitrogen mostly TAN [7] and is often more stable and reliable than algal uptake or nitrification [1].

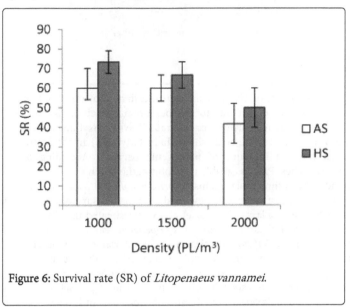

Figure 6: Survival rate (SR) of *Litopenaeus vannamei*.

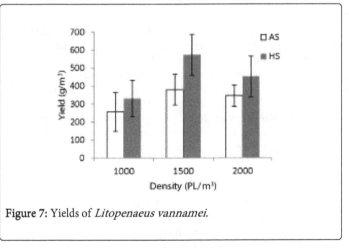

Figure 7: Yields of *Litopenaeus vannamei*.

Specific growth rates and protein efficiency ratio of *Litopenaeus vannamei* in heterotrophic aquaculture system tended to be better than those in autotrophic system. Biofloc, formed in heterotrophic system, can be benefited by shrimp as an alternative feed (Figure 8). Biofloc contains bacterial protein [8] and polyhydroxybutyrate [3] produced by bacteria. Biofloc was consumeable due to the appropriate size for shrimp. Polyhydroxybutyrate is the most dominant polymer and is useful in aquaculture. The advantages of PHB are an energy reserve for fish, digestible in intestine, increasing unsaturated fatty acid, and increasing growth of fish [1]. Biofloc also contains methionine, lysine, vitamins and minerals, especially phosphorus

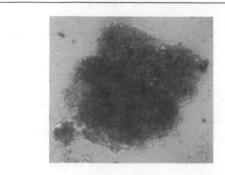

Figure 8: Biofloc.

Survival rate of *Litopenaeus vannamei* in heterotrophic system experienced enhancement due to biofloc contains bacteria. Bacteria are capable to produce polyhydroxybutyrate. Polyhydroxybutyrate will release 3-hydroxy butyric acid (short chain fatty acid) in the gastro intestinal tract as inhibitor of pathogenic bacteria. According to several researches, PHB is capable to inhibit pathogen in the intestinal tract and to be antimicrobial against *Vibrio, E. coli,* and *Salmonella,* to control pathogen of *Vibrio harveyi,* and to enhance survival rate of *Artemia franciscana* larvae [9]. Far et al. [5] investigated that *Bacillus* is able to increase survival rate of *Litopenaeus vannamei* and to decrease luminous *Vibrio* densities in the pond water. Bacteria also contain peptydoglycan and lipopolysaccharide on their cell wall. Peptydoglycan and lipopolysaccharide are immunostimulant being capable to inccrease nonspecific immunity of shrimp. The substances influence prophenoloxidase activity and phagocytosis of hyaline cells [10].

The yields of *Litopenaeus vannamei* in heterotrophic system enhanced significantly compared to autotrophic one. The enhancement was affected by growth (Figure 4) and survival rate (Figure 6). According to polynomial orthogonal analysis, optimal density of *Litopenaeus vannamei* in autotrophic system on nursery phase was 1612 PL/m^{-3} yielded 466 g/m^{-3}, while in heterotrophic system, optimal density was 1638 yielded 639 g/m^{-3} (Figure 9). Heterotrophic system was capable to increase yield of *Litopenaeus vannamei* of 37% compared to autotrophic system.

Conclusions

Heterotrophic aquaculture system can be an alternative method to culture *Litopenaeus vannamei* in pond. The system is environmental-friendly one overcoming the problem in shrimp culture related to deterorietion of environment quality. Pond water quality was controlled resulting in good growth, high survival rate, and increased yield of *Litopenaeus vannamei* on nursery phase. Heterotrophic system aquaculture was capable to increase the yield of *Litopenaeus vannamei* on nursery phase. However heterotrophic system did not significantly affect the growth rate and protein efficiency ratio of *Litopenaeus vannamei*. The density affected the survival rate and yield of *Litopenaeus vannamei*.

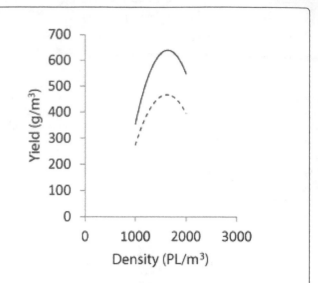

Figure 9: Optimal density of *Litopenaeus vannamei* on nursery phase.

References

1. Crab R, Avnimelech Y, Defoirdt T, Bossier P, Verstraete W (2007) Nitrogen removal techniques in aquaculture for a sustainable production. Aquaculture 270: 1–14.

2. Avnimelech Y (2009) Biofloc Technology–A Practical Guide Book. The World Aquaculture Society Baton Rouge Louisiana United State.

3. De Schryver P, Sinha AK, Kunwar PS, Baruah K, Verstraete W, et al. (2010) Poly-beta-hydroxybutyrate (PHB) increases growth performance and intestinal bacterial range-weighted richness in juvenile European sea bass, Dicentrarchus labrax. Applied Microbiology and Biotechnology 86: 1535-1541.

4. Anggoro S, Muryati (2006) Osmotic respons of tiger shrimp (Penaeus monodon Fab.) juvenile and adult at various level of molting stages and salinity. Buletin Penelitian dan Pengembangan Industri 1: 59-63.

5. Far HZ, Saad CRB, Daud HM, Harmin SA, Shakibazadeh S (2009) Effect of Bacillus subtilis on the growth and survival rate of shrimp (Litopenaeus vannamei. African Journal of Biotechnology 8: 3369-3376.

6. Hoffman JR, Falvo MJ (2004) Protein - Which is Best? J Sports Sci Med 3: 118-130.

7. Ebeling JM, Michael B. Timmons JJ, Bisogni (2006) Engineering analysis of the stoichiometry of photoautotrophic, autotrophic, and heterotrophic removal of ammonia–nitrogen in aquaculture systems. Aquaculture 257: 346–358.

8. Hargreaves JA (2013) Biofloc Production System for Aquaculture. Southern Rregional Aquaculture Center Publication No: 4503.

9. Crab R, Lambert A, Defoirdt T, Bossier P, Verstraete W (2010) The application of bioflocs technology to protect brine shrimp (Artemia franciscana) from pathogenic Vibrio harveyi. J Appl Microbiol 109: 1643-1649.

10. Yeh ST, Li CC, Tsui WC, Lin YC, Chen JC (2010) The protective immunity oh white shrimp Litopenaeus vannamei that had been immersed in the hot water extract of Gracilaria tenuistipitata and subjected to combined stresses of Vibrio alginolyticus injection and temperature change. Fish & Shellfish Immunology 29: 271-278.

16

Rapid Waste Composition Studies for the Assessment of Solid Waste Management Systems in Developing Countries

Max J Krause and Timothy G Townsend*

Environmental Engineering Sciences, University of Florida, Gainesville, FL 32611, USA

***Corresponding author:** Timothy G Townsend, Environmental Engineering Sciences, University of Florida, Gainesville, FL 32611, USA
E-mail: ttown@ufl.edu

Abstract

A methodology for the rapid assessment of waste composition was assessed by examining municipal solid waste from five rural communities throughout Central America and the Caribbean. Target waste components were minimized and a sieve-shaker table was employed to maximize the quantity of waste that could be sorted in an efficient and timely manner. Food waste (along with other fine materials) was the largest component by weight, but plastics represented a major fraction. To illustrate potential utility of composition study results, the data were used to estimate the methane generation potential, L_0, of each municipality's waste stream. While the approach does not provide the statistical rigor of more standardized waste composition methodologies, the technique does provide a tool for rapid assessment of local waste characteristics.

Keywords: Waste composition; Food waste; Landfill; Central America; Haiti; Methane potential

Introduction

Adequate solid waste management systems (SWMS) have lagged in developing countries behind other infrastructure needs. Rural municipalities, in particular, do not have the resources to construct and maintain the infrastructure that supports the use of conventional waste collection vehicles. While rural municipalities are often legally responsible for municipal solid waste (MSW) collection for all residents, the burden often falls to informal businesses. Without formal contracts or established liabilities, collectors often dispose of the waste wherever they can [1,2]. In many cases, an unmanaged dump site is established over several years or decades. While this creates a single, isolated source of pollution, it also concentrates the liquid and gaseous emissions. Furthermore, as technology advances and social habits change, waste composition changes. Whereas most of the waste dumped twenty years ago in rural areas may have been predominantly organic and biodegradable, the proliferation of plastics, electronics, and other environmentally persistent materials within current waste streams warrant modern SWMS.

To address the problems posed from sub-standard waste disposal, rural communities are often urged to implement more sustainable SWMS, both by national regulations and by outside parties. Challenges to implementing more sustainable practices include limited financial resources and the need to match an appropriate technology to the region's specific waste characteristics. Accurate waste composition data can be a crucial tool for selecting an appropriate waste management approach; the relative fractions of food waste, plastics, metals, and other components dictate the viability of technologies such as composting, anaerobic digestion, thermal energy recovery, and recycling. Such data are often lacking in developing countries [3–6], and even rarer in rural areas [7,8]. The cost and nature of the studies can be prohibitive to small and local governments, as they are typically performed over a period of multiple days or weeks, and sampling events are repeated throughout the year [9–11]. The largest direct cost of a composition study is attributed to labor, which typically requires 4 to 12 operators as well as a supervisor [9,12].

The work presented in this paper was motivated by challenges the authors faced as part of several efforts providing assistance to rural communities on waste management issues. Some estimate of the composition of the local waste stream was desired, but time and resources were not available for a complete waste composition study. Over several trips, a protocol was developed that allowed a quick evaluation of a municipality's waste stream in a single sampling event. A description and assessment of this procedure are presented within to provide benefit to others interested in similar activities. Results of composition studies from Costa Rica, Guatemala, Haiti, and Honduras are described, and the utility of the results are illustrated by estimating the MSW methane potential, L_0 (m^3 CH_4/Mg MSW) of the different communities and through evaluating the occurrence of plastics in each waste stream. Other challenges with respect to waste collection and management in the subject communities are also discussed to provide a broader context of issues to the reader.

Materials and Methods

Site Descriptions

Rapid waste composition studies were performed on six waste streams in five rural towns across Central America from 2010 to 2013 as part of an on-going effort by the authors to assess current waste management practices in developing countries and assist local communities. As illustrated in Figure 1, the locations included Nosara, Costa Rica; Tactic and San Juan Chamelco, Guatemala; Cabaret, Haiti; and Villa de San Francisco, Honduras. The waste streams examined in the study were market or residential waste as indicated in Table 1. The local markets were the primary sources of commerce within the communities. As such, large amounts of waste were generated weekly at these central locations.

Town	Country	Waste Stream	Site ID
Cabaret	Haiti	Market waste	CAB
San Juan Chamelco	Guatemala	Market waste	SJM
San Juan Chamelco	Guatemala	Residential	SJR
Tactic	Guatemala	Residential	TAC
Nosara	Costa Rica	Residential	NOS
Villa de San Francisco	Honduras	Residential	VSF

Table 1: Towns and waste streams evaluated in this research.

In Cabaret, Haiti, much of the market waste was piled together and burned on site. San Juan Chamelco had bins located throughout the central market area for public waste disposal. San Juan Chamelco and Tactic used several waste haulers (private and municipal) to collect residential waste once per week. Nosara had a single, private hauling service that collected waste from all areas of the town. The town was highly segregated between tourist hotels, restaurants, and local residences. Villa de San Francisco had a single waste hauler (municipal) that collected waste weekly.

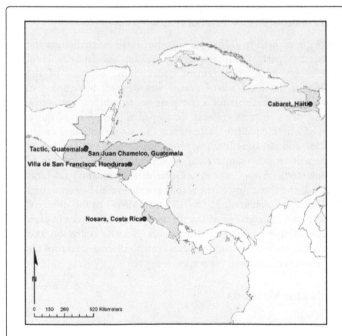

Figure 1: Geographic locations where rapid waste composition studies were performed.

For the composition studies, waste was collected from the hauler's truck directly or arrangements were made to retain a truckload from the previous day at the landfill site. Rapid waste composition studies were performed at the landfills in Cabaret, Nosara, and Villa de San Francisco and off-site in San Juan Chamelco and Tactic. In San Juan Chamelco and Tactic, the municipality provided facilities that protected sorters from the sun and allowed haulers to drop off waste throughout the day.

The objectives of the waste sorts for each location were similar. The amount of organic matter subject to biological decomposition was of interest; this material could contribute to methane gas production when placed in a landfill and would be amenable to treatment via composting or anaerobic digestion. A second interest was an assessment of potential recovered resources (recyclable materials) present, as well as possible materials of concern (hazardous wastes); some materials offer real economic value, while other materials may have potential for pollution [13].

Rapid waste composition methodology

A standard method for waste composition analysis currently exists and additional guidelines have been established and published [9–12]. Generally, a representative sample of waste (90 – 140 kg) is taken from a collection truck and manually sorted into several categories (e.g., plastics) and subcategories (e.g., polyethylene terephthalate, PET). A great number of different waste categories are often sorted. The components are weighed and the relative fractions of waste are determined from the combined weight of the total waste sample. Samples from a large number of waste collection vehicles are sorted, with the number of samples based on statistical calculations that consider the components being categorized and the desired confidence level [9].

In the current study, a methodology was needed to provide a general sense of the waste stream components in a relatively short period of time. Goals included processing a relatively large amount of waste into a few basic categories. To assist in this objective, a sieve-shaker table was constructed at each location with dimensional lumber and a sliding, wire mesh top for categorizing waste components. The tables were approximately 2 m by 1 m by 1 m (length by width by height) with the surface of the table consisting of steel mesh layer (square grid sizes of approximately 6.5 cm^2 (1 in^2); see Figure 2). The mesh table was designed to slide back and forth to agitate the waste sample, maximizing the amount of material that passed through the sieve; fine materials less than size of the mesh fell through the table surface onto a plastic tarp. The objective of the screening operation was to quickly remove small pieces that were difficult and time-consuming to categorize; these materials often did not represent components of particular value for resource recovery. The fine fraction was visually assessed to estimate an organic content; it was not sorted and was weighed in bulk. This reduced the amount of time required per waste load and allowed for greater amounts of waste to be sorted throughout the day.

Loads were sorted into six to nine categories as shown in Table 2 and weighed to the nearest 0.5 kg. Biodegradable and inert categories were chosen based on the need to have distinct, uniquely identifiable categories while also being inclusive of common and uncommon materials. "Plastics" incorporated resin codes 1 to 7 for both rigid and film plastics. "Food waste" consisted of both food and soiled paper that would not be suitable for recycling. If an item was not immediately identifiable, it was placed in the "Other" category. The amount of time for sampling did not allow for critical sub-categorization, however, site-specific categories were created if one particular waste component was found in significant quantity.

Figure 2: The sieve-shaker table designed and used for rapid waste composition studies.

Biodegradable Fraction	Inert Fraction
Paper Products	Plastics
Food waste	Metals
Rubber, Leather, and Textiles	Glass
Yard waste	Other
Wood	

Table 2: Waste categories of a rapid waste composition study.

Garbage bags were selected at random from the truck or landfill stockpile. Bags were opened onto a receiving tarp and waste was spread out to first look for hazardous components. Once the waste was deemed free of hazards by the supervisor, waste sorters would add waste to the sieve-shaker table. After shaking, waste was sorted from the table and placed into new garbage bags. Once a bag was filled with a single component it was closed and weighed. Categorized samples were weighed in bags with hanging scales (range 0 to 40 kg). The weight was recorded in the respective waste category and a new bag was designated for that category. Researchers were able to sort 250 to 780 kg of waste in a single day, the maximum allotted time for waste characterization.

Maintaining well-trained labor is always a concern for waste composition studies [9]. Waste sorters must be aware of the target categories to avoid contamination and erroneous results. Safety is another major concern. In this study, all sorters wore rubber and textile gloves over waterproof nitrile gloves. Long pants and long-sleeve shirts were recommended as wells as hats to protect from the sun. Sorters also wore safety glasses and, in some cases, dust masks.

Methane Potential of MSW

Waste composition data are useful for the development of accurate greenhouse gas (GHG) inventories and life-cycle assessments of SWMS [14,15]. Landfills are a major contributor to GHG emissions and dumping is the most common method of waste disposal in Central America [16]. Unmanaged dumps emit much of the generated methane into the atmosphere, unlike sanitary landfills that incorporate regular cover soil and gas collection and control systems for treatment or energy recovery. To illustrate the utility of the waste composition study results, the L_0 of each community's waste stream was estimated. The US EPA landfill gas production model requires the assumption of an ultimate methane potential, L_0 (m^3 CH_4/Mg-MSW), of the waste [17]. Past research has reported the methane potential of individual waste components (measured using laboratory experiments) [18–20] and others have coupled these data with composition study results to estimate a L_0 value for a specific waste stream [21,22]. In the present study, methane potential data from Owens and Chynoweth [19] and Jeon et al. [20] were used with the composition results to estimate L_0 for each community's waste stream.

Results and Discussion

Rapid waste composition studies observations

Because municipal participation in waste collection is not mandatory, it was thought that not all sources of waste generation would be captured by the rapid composition studies. Field observations were made during waste collection and disposal at all locations to determine if waste streams other than MSW were entering the landfill. Dump site assessments were conducted in each of the towns. In most cases, the dumping areas had developed over decades, with the site being selected based on proximity to the municipality and availability rather than engineering criteria.

Because much of the garbage sorted in Haiti was previously burned before disposal at the dump site, the fines collection tarp required weighing several times. When the fines collection tarp was being weighed, all sorting was required to pause until the tarp could be emptied and replaced in order to ensure all fines were collected from the table. Thus, in cases where fine material is abundant, the sorting process may be slowed because of frequent re-weighing of the collection tarp. However, the minimization of categories helps to offset the time attributed to identifying waste components and any such breaks in sorting activity.

Worker safety was paramount during waste composition studies. Sorters were outfitted with safety equipment to prevent ingestion, inhalation, or dermal contact of waste materials. In Central America particularly, the heat and sun are of concern as well. Because the studies rely on obtaining fresh MSW samples, at times the collection trucks would not arrive on site until mid-morning. Thus, the majority of the sorting took place through midday and the afternoon. Worker fatigue posed a challenge to productivity and multiple breaks were provided throughout the day; water was made available to sorters at all times.

The ability to complete the study in a single day significantly lowered the cost of the research. For these composition studies, many of the sorters were student volunteers. If additional laborers were required they were compensated with approximately $20 USD/day, higher than the average daily wage in any of these countries. Table 3 itemizes the costs of a rapid waste composition study, including materials and labor. This is significantly less than typical composition studies [9].

Item		USD ($)		
		Cost	Quantity	Total
Table	lumber	$80	1	$80
	2 m x 1 m mesh	$20	1	$20
	6 cm nails (pack)	$5	1	$5
	3 m x 2 m tarp	$20	2	$40
Labor		$20	8	$160
Materials	Work gloves	$12	4	$48
	Nitrile gloves	$18	4	$72
	Safety glasses	$6	10	$60
	Dust masks	$36	1	$36
	Garbage bags (pack)	$25	2	$50
	Hanging scales (0 – 40 kg)	$12	2	$24
Total				$595

Table 3: Materials and labor cost estimate of a rapid waste composition study.

All of the collection vehicles observed in this study were flatbed trucks with wooden railings. Some of the trucks had hydraulic lifts to tilt the bed but none had any compaction mechanism. This allowed the researchers to select samples from all areas of the truck. The absence of compaction also decreased potential contamination of plastics and paper products by food wastes and liquids which can skew composition data [12].

Rapid waste composition studies data

Waste composition data for the six waste streams (as well as the US presented for comparison) are provided in Figure 3. Food scraps dominated most of the waste streams, with the highest amounts found in San Juan Chamelco (70% residential, 62% marketplace) and Tactic (69%), Guatemala. Nosara, Costa Rica had the lowest food waste fraction (22%), likely because of the higher tourist population which produced more packaging waste (paper and plastics) than the other towns. Paper products and cardboard constituted a large portion of the waste, but without available recycling markets or size-reduction equipment for composting, these materials were destined for the landfill.

In Villa de San Francisco, Honduras, a large amount of textiles were found during the composition study and observed at the dump site. This was not typical of other locations; a specific reason for this occurred was not determined. Yard waste and wood (from construction and demolition debris) were not widely observed or recorded in any of the waste streams of this study. These materials are more likely to be used at the home for cooking or simply discarded adjacent to the source. Yard wastes and wood are more resistant to anaerobic decay and thus may not greatly contribute to potential landfill methane emissions [19,23].

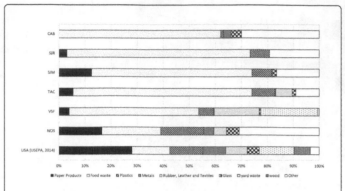

Figure 3: Waste composition data from 5 rural towns (6 waste streams) and the United States for comparison.

Small, light-weight items (less than 6.5 cm^2 (1 in^2)) would not typically be targeted as a recoverable commodity at a waste recovery operation, other than potentially being included as part of the biologically treated waste fraction (i.e., that destined for composting or anaerobic digestion). The sieve-shaker table allowed for maximum sorting efficiency of bulky or heavy material while allowing residual pieces to fall through to a collection tarp below. The fines were grouped into a single category that was visually assessed, weighed and initially categorized separately. Generally, the fine fraction was found to be highly organic (food scraps) or soil-like in nature (dust, street sweepings) and were subsequently added to the "Food waste" category. Small non-biodegradable items, such as bottle caps, were determined to be negligible to the total weight of the "Fines" category.

As expected, many types of plastics (resin codes 1 to 7) were found within the waste streams. Plastics contributed 17% of the weight to the waste stream of Nosara, but only 7%, 8%, 9%, 6%, and 0.4% to San Juan Chamelco residential, San Juan Chamelco market, Tactic, Villa de San Francisco, and Cabaret, respectively. The amount of plastic measured by weight in Nosara, Costa Rica during this study was higher than the estimated plastic fraction of the US waste stream [24]. Again, this can be attributed to the tourists that frequent the area, consuming and discarding single-use plastic bottles and bags. Because plastics are much less dense than other items such as cardboard, or food waste, the 17% by weight constitutes a considerable volume fraction of the waste stream.

Large amounts of PET drink containers were observed in all waste streams. PET is a valuable recyclable commodity and was weighed separately during the composition studies. In Guatemala, a significant fraction of polystyrene was found during the study. It was categorized and weighed separately and subsequently included in the "Plastics" category. Plastic film was common in most of the waste streams and many of the large garbage bags contained smaller bags of waste. This was later found to be a method of preventing vectors from being attracted to waste that was stored in a home or building that lacked air conditioning.

The largest product category of US MSW is "Containers and Packaging," which includes paper and plastic [24]. It was hypothesized that the percentage of plastics in a waste stream would relate to a country's economic status. The percentage of plastics found within the waste stream was compared to the gross domestic product purchase power parity (GDP PPP) per capita of the respective countries [25]. A trend was found that shows a direct relationship between GDP per

TAC	Guatemala	5	69	6	0	101
NOS	Costa Rica	17	22	5	0	72
VSF	Honduras	4	50	17	22	104

capita and the fraction of plastic within the waste stream (See Figure 4). Two additional studies examining household waste composition in rural Mexico [7] and Serbia [5] were included to better assess this trend. The trend line does not include the US datapoint, which is the US EPA's national average, not specifically rural generation [24].

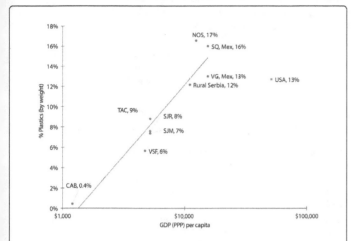

Figure 4: Percentage of plastics found within a rural waste stream compared to the country's GDP per capita.

Table 4: The methane potential, L_0, of 6 Central American waste streams based on waste composition data.

Conclusions

Rapid waste composition studies were performed on waste from five rural communities in Costa Rica, Guatemala, Haiti, and Honduras. Time and resource constraints did not permit an extensive waste composition study as might be typical in developed countries, but some general characterization of the waste was needed. The protocol described herein provides a means for interested parties to collect waste composition data in a single sampling event. By characterizing only a limited number of categories (six to nine) and using a shaker-table to quickly screen out difficult to identify components, the methodology allows for rapid data collection while also providing valuable information regarding relative amounts of biodegradable waste, recyclable materials, and components destined for landfill. Observations of collection and disposal sites indicated that the waste collected for the studies was representative of the overall waste stream.

Food waste (plus fines) accounted for 50% or greater of the waste streams in San Juan Chamelco, Tactic, Cabaret, and Villa de San Francisco. Paper and plastics were significant fractions but varied by location. Nosara was found to have notably greater percentage of plastics (17%) – this was attributed to single-use bags and bottles. To illustrate the utility of the waste composition data, the results were coupled with component-specific methane yield data to estimate site-specific L_0 values; the results fell within the default ranges for each country in the US EPA's landfill emission model (Haiti is not included in the model). A direct relationship was found between the percentage of plastics within the waste stream and a country's per capita GDP. This may be of value to other municipalities that lack waste composition data, but are interested in enacting a recycling program for plastics. Some plastics have a high value in the recycling markets, but many of the films and other packaging materials lack viable recycling markets and will be landfilled.

Some valid concerns of the proposed methodology remain. A single sampling event cannot replace a full, long-term waste composition study. This study does not consider sample sizes and only considers the total amount of waste that can be sorted in a single day. Seasonal variations cannot be accounted for without additional data. The sampling event could be supplemented by information from the municipality or waste hauler to extrapolate waste generation rates. While uncertainty exists within any waste composition study, these data serve to inform the municipalities, the public, and researchers in geographic areas where such data do not exist, and allow more informed decision-making regarding solid waste management in developing countries.

Calculation of methane emissions from waste composition data

The L_0 of each waste stream was calculated using component-specific data and are presented in Table 4. Calculated methane potentials of the waste streams ranged from 72 to 104 m^3 CH_4/Mg, within the range of default values given in the US EPA's Central America Landfill Gas Model [17] and similar to the US default value (100 m^3 CH_4/Mg MSW) [26]. Machado et al. [21] calculated L_0 for fresh waste and excavated landfill samples from Brazil, determining L0 of that site to be 67 m^3 CH_4/Mg MSW. L_0 is dependent of the biodegradable fraction of MSW, but it is also dependent of moisture content because it is a volume per "as-received" mass coefficient. Food waste can be largely made of water, which contributes to the weight of the waste, but does not contribute to methane potential. Paper products have high volatile solids and low moisture contents, which could increase the methane potential of that waste stream, as shown in Table 4. Haiti, having the second highest food waste fraction, had the second lowest methane potential, 73 m^3 CH_4/Mg MSW. This was largely attributed to the difficulties in obtaining fresh MSW samples before they were burned. Partially burned samples were taken multiple times throughout the day, but in all samples, some portion of the waste had already been combusted. Better coordination with the local waste haulers would be needed to better represent unburned MSW.

Acknowledgements

This work was made possible by the students of the Solid and Hazardous Waste Management Capstone courses (2010–2013) as well as donations from private parties, including HDR, Jacobs Engineering,

| Site ID | Country | Percent by weight of waste stream (%) | | | | L_0 |
		Paper Products	Food Waste	Rubber, Leather, and Textiles	Yard Waste	(m^3 $CH_4/$ Mg MSW)
CAB	Haiti	0	62	0	0	73
SJR	Guatemala	3	70	0	0	90
SJM	Guatemala	12	62	0	0	102

Golder Associates, Jones Edmunds and Associates, CH$_2$MHill, and Koogler & Associates. The authors would like to recognize Antonio Yaquian-Luna for his contribution to this work. The authors would also like to acknowledge the continued support of the University of Florida Engineering School of Sustainable Infrastructure and Environment.

References

1. Guerrero LA, Maas G, Hogland W (2013) Solid waste management challenges for cities in developing countries. Waste Manag 33: 220-232.

2. Yeomans JC (2008) Development of an Integrated Waste Management Plan for Ordinary Solid Waste in Rural Communities in Latin America. Tierra Tropical 4: 1–25.

3. UN-Habitat (2011) Collection of Municipal Solid Waste Key issues for Decision-makers in Developing Countries. UN-HABITAT Nairobi 38.

4. Badgie D, Samah M, Manaf L (2012) Assessment of Municipal Solid Waste Composition in Malaysia: Management Practice and Challenges. Polish Journal of Environmental Studies 21: 539–547.

5. Vujic G, Jovicic N, Redzic N, Jovicic G, Batinic B. et al. (2010) A fast method for the analysis of municipal solid waste in developing countries - case study of Serbia.Environmental Engineering and Management Journal 9: 1021–1029.

6. Henry RK, Yongsheng Z, Jun D (2006) Municipal solid waste management challenges in developing countries--Kenyan case study. Waste Manag 26: 92-100.

7. Taboada-González P, Armijo-de-Vega C, Aguilar-Virgen Q, Ojeda-Benítez S (2010) Household solid waste characteristics and management in rural communities. The Open Waste Management Journal 3: 167–173.

8. Laines Canepa JR, Zequeira Larios C, Valadez Treviño MEM, Garduza Sánchez DI (2012) Basic diagnosis of solid waste generated at Agua Blanca State Park to propose waste management strategies. Waste management and Research, 30: 302–310.

9. McCauley-Bell P, Reinhart DR, Sfeir H, Ryan BO (1997) Municipal Solid Waste Composition Studies. Practice Periodical of Hazardous Toxic and Radioactive Waste Management 1: 158–163.

10. Dahlén L, Lagerkvist A (2008) Methods for household waste composition studies. Waste Manag 28: 1100-1112.

11. ASTM International (2008) ASTM D5231 - 92(2008) Standard Test Method for Determination of the Composition of Unprocessed Municipal Solid Waste. ASTM International Conshohocken PA.

12. Sfeir H, Reinhart DR, McCauley-Bell PR (1999) An Evaluation of Municipal Solid Waste Composition Bias Sources. Journal of the Air & Waste Management Association 49: 1096–1102.

13. Ojeda-Benítez S, Aguilar-Virgen Q, Taboada-González P, Cruz-Sotelo SE (2013) Household hazardous wastes as a potential source of pollution: a generation study. Waste Manag Res 31: 1279-1284.

14. Finnveden G, Johansson J, Lind P, Moberg A (2005) Life cycle assessment of energy from solid waste part 1: general methodology and results. Journal of Cleaner Production 13: 213–229.

15. US EPA (2006) Solid Waste Management and Greenhouse Gases. A Life-Cycle Assessment of Emissions and Sinks (3 edtn).

16. World Bank (2012) What a Waste: A Global Review of Solid Waste Management.

17. US EPA (2007) User's Manual Central America Landfill Gas Model. Washington DC.

18. Eleazer WE, Odle WS, Wang YS, Barlaz MA (1997) Biodegradability of Municipal Solid Waste Components in Laboratory-Scale Landfills. Environmental Science and Technology 31: 911–917.

19. Owens J, Chynoweth DP (1993) Biochemical methane potential of municipal solid waste (MSW) components. Water Science and Technology 27: 1–14.

20. Jeon EJ, Bae SJ, Lee DH, Seo DC, Chun SK, et al. (2007) Methane generation potential and biodegradability of MSW components. Sardinia 2007 Eleventh International Waste Management and Landfill Symposium Cagliari Italy.

21. Machado SL, Carvalho MF, Gourc JP, Vilar OM, do Nascimento JC (2009) Methane generation in tropical landfills: simplified methods and field results. Waste Manag 29: 153-161.

22. Cho HS, Moon HS, Kim JY (2012) Effect of quantity and composition of waste on the prediction of annual methane potential from landfills. Bioresour Technol 109: 86-92.

23. Kim SK, Lee T (2009) Degradation of lignocellulosic materials under sulfidogenic and methanogenic conditions. Waste Manag 29: 224-227.

24. US EPA (2014) Municipal Solid Waste Generation, Recycling and Disposal in the United States: Facts and Figures for 2012. 2012 MSW Characterization Reports.

25. CIA (2013) Country comparison GDP (purchasing power parity). The World Factbook.

26. US EPA (1998) Municipal Solid Waste Landfills. Compilation of Air Pollutant Emission Factors. Stationary Point and Area Sources (AP-42). US EPA, Research Triangle Park NC.

17

Environmental Health Monitoring: A Pragmatic Approach

Savariar Vincent

Loyola Institute of Frontier Energy, Chennai, India

*Corresponding author:Dr. Vincent, Dean of Research, Loyola College & Director, Loyola Institute of Frontier Energy, Chennai - 600 034, Tamil Nadu, India
E-mail: svincentloyola@gmail.com

Abstract

Eco benign environment, wherein human race survives and evolves, comprises the interaction of biotic and abiotic systems. Maintaining such environment in view of drastically changing human practices due to increased industrialization, urbanization and modernization is an urgent necessity for the global human community. The responsibility to protect the environment as well as to make it suitable for sustaining human life with well-being propels research-based implementation of best industrial and environmental practices. The nature of strategies and research designed for arriving eco-benign solutions to the challenges of environmental protection widely varies with different fields of industry. It is important to note that the intervention of environmental friendly strategies for making hazard and disease free environment is inevitable in all spheres of life, and the research on green solutions envisaged for management of environmental resources is the need of hour. Hence, in this review, we highlight three socially important environmental research areas such as impact of toxicity of metal pollutants in aquatic biological systems with special reference to fish as a bioindicator, mosquito control by bioinsecticides, and management of disease and disaster management by Geographical Information Systems (GIS) that are of our areas of contribution. Though these areas are viewed as different domains, they share an unique feature among themselves i.e. the crisis in water management affecting environment and health. The review also attempts to discuss about the possible biointervention solutions to address the challenges in water management in terms of pollution.

Keywords: Metal toxicity; Aquatic health; Biomonitoring; Metallothionein; Bioinsecticides; GIS; Drainage

Changing Quality of Aquatic Life

Quality of life in aquatic system is determined by the characteristics of biotic resources and their interactions that are complex and dynamic. Naturally, there is a self-balance between the biotic resources and their interactions both in quantitative and qualitative aspects and the balance is affected when new elements are introduced to the system. Especially, the elements introduced into the system have toxic properties; the aquatic system is under threat of quality degradation that seriously affects global biodiversity. Aquatic system today is polluted by industrial effluents comprising organic and synthetic chemical substances, heavy metals, dyes, oil etc; house hold wastes such as detergents and drainage; e-wastes, pesticides from agricultural lands, spillage of oil from ships etc. These pollutants of both biodegradable and non-biodegradable change the water qualities such as colour, odour, surface tension, thermal properties, conductivity, density, pH etc [1]. The acidic and alkali pollutants destroy most invertebrates and microorganisms [2]. Due to increased anthropogenic activities, aquatic organisms are continuously exposed to the elevated levels of metal concentration that the levels not previously encountered. The metal contaminants in aquatic systems usually remain either insoluble or suspension form and finally tend to settle down at the bottom are taken up by the organisms. The progressive and irreversible accumulation of these metals in various organs of marine creatures leads to metal related physiological impairing in the long run because of their toxicity and persistence, thereby endangering the aquatic biota and other organisms [3]. Metals and pesticides, in particular, have an inclination to accumulate, undergo food chain magnification and exhibit chronic toxicity [4]. It is also important to monitor the bioaccumulation of these metals and pesticides in a living system (fish) and assess the possible impact on human health due to consumption [5]. There is also a need to intervene with options such as reduction in the use of heavy metals, replacement of toxic metals by nonmetallic substances in process and product, metal recovery and recycling and reducing the metal release in process as well as from product through leaching to minimize the damage by metal toxicity caused to the aqua-biotic system. Development and use of novel bioinsecticides will not only eliminate the ill effects of biomagnification but also helps to sustain the environment for bioproductivity.

Metal Toxicity in Aqua-Biotic Systems

Aquatic organisms readily absorb heavy metals through respiratory surfaces, mucus covering gills, where it gradually diffuse into to the binding sites, adsorption through body surface and cell wall where it diffuse through cell membrane and accumulate in cells during ion exchange. The metal which has a relatively high density and toxic at low quantity is referred as 'heavy metal', e.g., arsenic (As), lead (Pb), mercury (Hg), cadmium (Cd), chromium (Cr), thallium (Tl), etc. Some 'trace elements' are also known as heavy metals, e.g., copper (Cu), selenium (Se) and zinc (Zn). They are essential to maintain the body metabolism, but they are toxic at higher concentrations. Bioaccumulation of metals reflects the amount ingested by the organism, the way in which the metals are distributed amongst the different tissues and the extent to which the metal is retained in each tissue. Also, metal accumulation depends upon other factors such as species, age, exposure time, temperature, salinity, metal and physiological factors [6,7]. Heavy metal accumulation dominantly occurs in metabolically active tissues such as liver and kidney [8]. There are many studies on fish response to metal contamination and fish is generally recognized as one of the most sensitive indicator for environmental livability studies for assessing any change in aquatic system due to pollutants [9,10]. In fish tissues, metals form complexes and may then be readily excreted. In some cases, however, the metals may alter the biochemical composition of tissues, rendering it unfit for human consumption.

XenoticPotential of Zinc And Nickel on Aquatic Life

Zinc and nickel are nutritionally essential trace elements for all organisms, and hence either deficiency or toxicity symptoms arise respectively when too little or too much accumulation occur. Effluents of textile, paint and tanning industries were reported to contain 120 to 160 mg/L of zinc and nickel metals [11]. Zinc is an integral part of certain enzymes, e.g., carbonic anhydrase, carboxypeptidase, and several hydrogenases. It was reported that zinc concentrations exceeding 330 mg/L may be lethal to fish [12]. The study also reported that the general trend of zinc accumulation in various tissues of Silver carp fish *Hypophthalmichthys molitrix* was found to be gills (460 mg/L) < liver (513 mg/L) < intestine (562 mg/L) < kidney (828 mg/L) after exposed to sublethal concentration of 6 mg/L for 30 days. It was found that zinc was most toxic in soft water of pH 4-6 and 8-9 and that excess zinc was precipitated in mucus and epithelial layer of gills that resulted in mortality due to damage of the gill epithelium [13]. Histopathological effects in gills, liver, intestine and kidney tissues of Silver carp after zinc sulphate exposure revealed that all the above tissues were affected due to zinc accumulation. Pathological features such as hypertrophy, necrosis, vacuolization, disintegration of cells, pyknotic nuclei and haemolysis were observed in all tissues with slight variations in features [14].

In a study, it was reported that the exposure of Silver carp to Zn and Ni at the sublethal concentrations of 6.8 mg/L and 5.7 mg/L respectively resulted in the synthesis of stress proteins in the gill, liver and kidney tissues with the reduction in the synthesis of other proteins compared with that of protein profile of unexposed fish controls [15,16]. Other heavy metals such as cadmium also was known to evoke the expression of stress proteins in fish and marine invertebrates [17]. Ni is introduced into the hydrosphere by removal from the atmosphere, by surface run-off by discharge of industrial, municipal and mining waste, and also following natural erosion of soils and rocks [18]. Ni has not been considered a broadscale global contaminant; however, ecological changes, such as a decrease in the number and diversity of species, have been observed near Ni emitting sources [19]. Fresh water levels of Ni are typically about 1-10 µg/L in unimpact areas [20] and Ni concentrations in highly contaminated fresh waters may reach as high as several hundred to 1000 µg/L [21].

The kidney plays a principal role in the accumulation, detoxification, and excretion of Ni and it is considered to be a target organ for Ni toxicity [22]. The assessment of genotoxic potential of Ni and Zn after exposing Silver carp at the above stated sublethal concentrations revealed chromosomal aberrations such as polyploidy, dicentric chromosome, acentric fragment, variation in the length of chromosomal arms, chromatid break, chromosomal clumping and centromeric fusion [23]. The frequency of these aberrations were higher in fish exposed to Nickel Chloride when compared Zinc Sulphate. Toxic metals exhibiting genotoxic effects were known to form reactive oxygen species as well as electrophilic free radical metabolites that interact with DNA to cause disruptive changes [24]. Ni could bind to DNA and proteins in cells in vitro and to chromatin *in vivo*. Such binding to macromolecules could be correlated to the

ability of Ni compounds to interfere with DNA synthesis and to induce slight increases in chromosomal alterations, as well as its mutagenic action [25,26]. It was reported that Ni-induced abnormal DNA repair is a mechanism for carcinogenesis [27].

Mode of Chromium Toxicity and its Bioremediation Mechanism

Chromium compounds have widespread uses in steel production, wood preservation, leather tanning, metal corrosion inhibition, paints and pigments, metal plating and other industrial applications. Trivalent and Hexavalent forms are the dominant oxidation states of chromium that exist in the environment. Cr (VI), a carcinogen, is highly toxic to all forms of life. Cr (III), an essential micronutrient for many higher organisms, is relatively insoluble in water and 100 times less toxic than Cr (VI) [28]. In our early study, we investigated the effect of Cr on acid (ACP) and alkaline phosphatase (ALP) activities from various tissues such as gill, intestine, liver and kidney after exposing Indian major carp fish *Catla catla* to the sublethal concentration in the range of 20 to 32 mg/L [29]. It was found that though both ACP and ALP in these tissues were inhibited to the extent of 40 to 70%, intestine was the most affected tissue and higher inhibition was observed for ALP in these tissues than ACP. Besides, it was observed that decline in the function of phosphatases due to Cr inhibition resulted in lysosome injury reflecting cellular stress. ALP is involved in transphophorylation reactions and its inhibition might be related to the ability of toxicant to alter cell configuration by binding to the membrane system.

In another study, we reported that exposure of fish (*Catla catla*) to sublethal concentration of Cr resulted in reduction in overall metabolism due to depletion of food utilization parameters such as consumption, assimilation, gross and net production, biomass and metabolic rates [30]. Elevated levels of egestion were observed in the study that reflected poor energy assimilation.

We have also reported that untreated tannery effluents having high Cr concentration when loaded to the aquatic system, they affected gross (GPP) and net primary productivity (NPP) of major aquatic macrophytes such as *Hydrilla verticillata* and *Ceretophyllum demersum* due to creation of metal influenced imbalance in the photosynthetic machinery [31]. An increase in the respiratory rate of macrophytes in the study suggested increased metabolism and extra energy demand under effluent stress condition. Primary production in aquatic system is the most important biological phenomenon on which an array of life depends for energy either directly or indirectly.

Studies were also conducted on macrophytes such as *Caldesia paranassipolia* for assessing its potential of the affinity to Cr and its use on Cr bioremediation as the aquatic plant has the potential to survive under various ecological conditions and has the ability to uptake heavy metals [32]. The findings of the study suggested that the absorption of Cr was about 75% when the plant was exposed to water having Cr concentration of 10000 ppm for 15 days. The morphology of the plant after exposed to Cr in the stated condition did not reveal much change except slight leaf withering and weakening of stem.

In an another Cr bioremediation study, the potential of microbes such as alkalophilic *Bacillus* sp. was assessed for their ability to reduce Cr (VI) since detoxification under alkaline conditions derive importance due to alkaline nature of many industrial effluents [33]. The chromium resistant bacteria isolated from tannery effluent contaminated soil was found to grow and reduce Cr (VI) to Cr (III) up to 100% at an alkaline pH 9.0 as evidenced by XPS and FT-IR spectra. The chromate reductase assay confirmed that constitutive membrane bound enzymes mediated the reduction, and the rate of reduction was reduced when the pH is reduced to below 7.0 due to secretion of acidic metabolites of growth. Several Cr (VI) detoxification studies mediated by bacteria were reported at neutral/near-neutral pH; but, very few studies were reported under alkaline condition [34,35]. Cr (VI) was found to affect colony morphology and sporulation ability in *Bacillus sp.* which may also affect Cr bioremediation process. Confocal Laser Scanning Microscope (CLSM) studies revealed changes in the production of exopolymeric substances of bacteria exposed to Cr and these substances are essential for absorption of Cr [28].

Arsenic and Cadmium Toxicity – A Threat to Aquatic System

The arsenic (As) is commonly present in air, water, soil and all living tissues. It is at the 20th abundant element in the earth's crust, 14th in seawater and 12th in human body. It is reported as a carcinogen, and causes foetal death and malformations in many mammal species. Contamination of aquatic environment by arsenicals and their impact on the aquatic organisms has now emerged as a serious environmental problem [36]. In aquatic systems, inorganic arsenic can occur in both As^{3+} and As^{5+} oxidation states. The level of As^{3+} to As^{5+} depends on the influences of pH, metal sulfide and sulfide ion concentrations, iron concentration, temperature, salinity, and the distribution of the biota in an aquatic ecosystem.

Generally, the inorganic As compounds are more toxic and carcinogenic than organic compounds, and trivalent As (arsenites, As $^{+3}$) are more toxic than pentavalent As (arsenates, As^{+5}). Most of the As compounds are used in manufacture of agricultural products such as insecticides, herbicides, fungicides, algaecides, wood preservatives, and growth stimulants for plants and animals. The atmospheric emissions from smelters, coal-fired power plants and arsenical herbicide sprays; water contaminated by mine tailings, smelter wastes and natural mineralization; and diet, especially from consumption of marine biota, all cause As toxicity [37].

In freshwater fish, As can be present in two different oxidation states, as arseno-sugars and arseno-lipids, these two differ in their toxicity which is supposed to be responsible for the pathophysiology of arsenic [38]. Arsenobetaine and arsenocholine are non-toxic organic forms of arsenic present in fish. The majority of total arsenic in fish tissue is present as arsenobetaine [39]. In a study on the spotted snakehead fish *Channa punctatus*, it was observed that when the high concentration (2 mM) of sodium arsenite (NaAsO) affected these fishes, they died within 3 h. The chromosomal DNA of liver cells were fragmented which indicated that NaAsO might have caused death of those cells through apoptosis [40]. It was reported that As severely affected liver of tilapia (*Oreochromis mossambicus*) while treating at sublethal concentration of 28 ppm. The liver was the most affected organ and the histopathological studies of liver tissue showed focal lymphocytic and macrophage infiltration, congestion, vacuolization and shrinkage of hepatocytes, necrosis and nuclear hypertrophy [41]. In a study, we evaluated the toxic effects of arsenic trioxide (Ar_2O_3) to freshwater fish *C. carpio* at different median lethal concentration levels and assessed its influences on the expression of metallothionein (MT) activity in liver and kidney [42]. It was found that the acute exposure resulted in instantaneous death of fish because of As-induced increases in mucus production, causing suffocation, or direct detrimental effects on gill epithelium besides increasing MT protein expression. Oxidative stress-induced apoptosis is a possible mechanism of arsenic toxicity in a zebra fish *Danioreri oliver* cell line [43].

Cadmium (Cd) is a transition heavy metal widely used in manufacture of semiconductors, integrated circuits, electroplating, and photoelectric appliances. The metal is released into the environment during manufacturing processes such as etching, wet polishing, and cleaning operations, which may produce much potentially hazardous waste. Cd was found to exhibit lethal effect at 0.05 mg/L concentration on fresh water swamp shrimp *Macrobrachium nipponense* [44]. It was reported that Cd increased the expression of oxidative stress biomarkers such as catalase, Glutathione-S-peroxidase, glutathione reductase and lipid peroxidase in response to increase in reactive oxygen species (ROS) in liver and gills of fish *Gambusia holbrooki* [45]. Blood being a sensitive indicator of stress, and white cells being involved in the immune defence of the organism, the state of health of fish *Catla catla* was found to be stressed under the toxic influence of Cd [46]. It was found that there was depletion in monocytes and lymphocytes but an elevation in neutrophils and thrombocytes in blood suggesting a reduction in the activation of immune defense mechanisms.

Metallothioneins– The Scavenger and Savior Proteins

Toxic metals are ubiquitous in our environment, and heavy metals such as cadmium, lead, and mercury have no essential biochemical roles, but exert diverse, severe toxicities in multiple organ systems as they bind in tissues, create oxidative stress, affect endocrine function, block aquaporins, and interfere with functions of essential cations such as magnesium and zinc. Discharge of heavy metals to water streams potentially affects different stages of the aquatic food chain as aquatic flora and fauna tends to accumulate metals. The metal accumulation does not necessarily indicates deleterious effects since organisms have possibilities to protect themselves from metal toxicity by increased excretion, differential allocation among organs and by binding the metals intracellularly [47]. Stress indices or stress biomarkers are used to evaluate the effects of heavy metal pollutants on marine organisms (bioindicators). One such important stress biomarker for assessing heavy metal toxicity is Metallothioneins (MTs), which are low molecular weight (6 to 10 kDa), cysteine-rich (20-30%), metal binding proteins whose synthesis represents a specific response of the organisms to pollution by heavy metals such as Cu, Zn, Cd and Hg, both under laboratory and field conditions [48]. They also play an important role in detoxification processes implicating toxic heavy metals. Glutathione is another potent chelator involved in cellular response, transport, and excretion of metal cations and is a biomarker for toxic metal overload.

Fish species that differ in their feeding strategies and/or detoxification capacities might accumulate metals to a different extent. Bottom dwelling species such as the gudgeon (*Gobio gobio*) will be exposed to water, sediment and food whereas pelagic species are mainly exposed via water and food. As a consequence, given the different way of exposure, both metal accumulation and MT-induction might differ among different species, which might be responsible for differences in sensitivity to metals [49]. It was reported that MT's role in detoxification was organ specific in common carp *Cyprinus carpio* and he reported that gill, kidney and liver tissues could accumulate Cd and induce MTs differently [50]. In *H. mylodon*, liver was more sensitive to metal ions than kidney and gill, and the increase of hepatic MTmRNA was dose-dependent. Induction of hepatic MTs by Cd was time-dependent but transient [51]. The above findings were also supported by one of our study that assessing the toxicity of Ar2O3 on fresh water fish *C. caprio*, we found that As elevated the expression of MTs in different tissues though liver tissues showed high MT content [42]. In another study, we reported expression of MT in response to time- and dose-dependent accumulation of Cd in various tissues of fresh water cat fish of *Clarias gariepinus* when the fish was treated with $CdCl_2$ in the concentration range of 5 to 20 ppm [52].

MTs as Biomarker for Monitoring the Metal Toxicity in Aquatic System

Owing to its highly inducible expression during exposures to various heavy metals, MTs have been paid much attention as a potential biomarker to monitor the heavy metal pollution of aquatic ecosystem, a major receptor of pollutants especially with relatively high amount of heavy metals. The isolation and characterization of MTs from bioindicator aquatic organisms should have implications for understanding the biological response to pollutants and for environmental biomonitoring. To date the function of MTs has been focused on its role in metal transport, mineral nutrition, metal detoxification and detoxification of other chemicals (xenobiotics). Other possible functions of MT such as its role in embryonic development, cell differentiation and preventing carcinogenesis via inhibiting metal induced ROS generation require identification and characterization of MTs from other metal binding proteins, and a few reports are available on the studies. Hence, we developed a software program "G primer" to compute primers for MT genes [53]. Given a nucleotide sequence, the program identifies the functionally active regions of MT for gene amplification. Primer selection in Gprimer is specifically built for MT genes without intronic region occurrence. The program has built-in database of functionally active sites, where MT genes are classified based on their respective phylum. To characterize MTs, we developed a computer program "ID3 algorithm" wherein isolated and purified MT proteins were subjected to conditions such as (1) protein with low molecular weight/high molecular weight, (2) proteins with metal content/without metal content, (3) aromatic amino acids/without aromatic amino acids and (4) sulphur content/without sulphur content. The derived conditions at every step were trained in an expert system (ID3 algorithm) [54]. The conditions were then formulated into an IF–THEN–ELSE algorithm and translated into VISUAL BASIC language statements. The developed software solution proposes to categorize MT proteins without aromatic amino acids and high metal content and the solution can be expanded to other types of proteins with specific known characteristics.

BiotoxicEffects of Heavy Metals

The biotoxic effects of heavy metals are observed in the body when they are consumed above the bio-recommended limits. Although individual metals exhibit specific signs of their toxicity, the following have been reported as general signs associated with cadmium, lead, arsenic, mercury, zinc, copper and aluminium poisoning: gastrointestinal (GI) disorders, diarrhoea, stomatitis, tremor, allergic dermatitis, bronchitis, hemoglobinuria causing a rust–red colour to stool, ataxia, paralysis, vomiting and convulsion, depression, and pneumonia when volatile vapours and fumes are inhaled [55]. The nature of effects could be toxic (acute, chronic or sub-chronic), neurotoxic, carcinogenic, mutagenic or teratogenic, and these effects vary among organisms as well as with the oxidation state of metals. The tolerance limits of some metals are shown in Table 1. Even when

the organisms are exposed to heavy metal concentration that is much lower than the permissible limits in their living environments, the metals are accumulated in their body due to continued exposure and xenobiotic food chain biomagnifications [56]. Therefore, a common characteristic of toxic metals is the chronic nature of their toxicity. Hence, elucidating the mechanistic basis of heavy metal interactions in physiological system is essential for assessing health risk.

Heavy metal	Max. conc in air (mg/m3)	Max. conc in soil (mg/Kg or ppm)	Max. conc in drinking water (mg/L)	Max. conc in aquatic water supporting life (mg/L or ppm)
Cd	0.1-0.2	85	0.005	0.008
Pb	NA	420	0.01	0.0058
Zn	NA	7500	5	0.0766
Hg	NA	<1	0.002	0.05
Ca	5	Tolerable	50	Tolerable > 50
Ag	0.01	NA	0	0.1
Cr	0.005	NA	0.1	0.02
As	NA	NA	0.01	NA

Table 1: Maximum contamination levels for heavy metal concentration in air, soil and water according to United States of Environmental Protection Agency (USEPA) and Occupational Safety and Health Administration's permissible exposure limits (OSHA PEL).

Air, Water and Soil Pollution Study in Industrial Units Using Environmental Flow Diagram

Industry is a major consumer of natural resources and a major contributor to the overall pollution load including the heavy metals and organics. More than 60% of annual greenhouse gas emissions are related to the industrial activities in the world (transportation fuels and distribution 25.3%, power stations 21.3% and industrial process 16.8%) [57]. According to the OECD (Organization for Economic Cooperation and Development) estimates, industries account for about one-third of global energy consumption, and >10% of the total fresh water withdrawal thus creating a negative environmental impact in terms of energy crisis and lack of green processes. The degradation of surface and groundwater quality is due to the release of untreated or partially treated industrial effluents or urban wastes into a water body. Ecological flow diagram shows the release of heavy metals into the biosphere comprising air, soil and aquatic environmental systems due to natural and anthropogenic processes, their accumulation in flora and fauna, and the flow cycle of heavy metals (Figure 1).

Studies such as computational and mathematical models as well as designs on environmental management such as Bayesian and DAVID influence diagrams, Hasse diagram (ProRank), and multivariate statistical techniques were reported on the pollution load in air, soil and water and environmental management are available separately but they lack the comprehensive approach in addressing the net pollution effect on the entire biosphere [58]. Besides, these studies were focusing more on the forecasting and management using conventional methods rather than addressing real-time challenges and input-output auditing of pollution load.

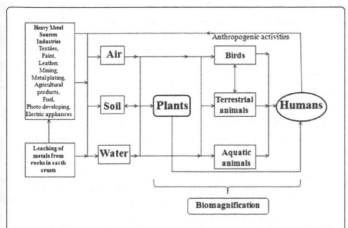

Figure 1: Ecological flow diagram of heavy metals impacting the environment and living systems.

It is important to have information on the quality and quantity of pollution elements generated from the pollution sources to reduce as well as to eliminate the risk of pollution to the living systems. Environmental flow diagrams (EFD) are useful for determining the sources of a range of pollutants, flow of pollutants from sources to acceptor environments, and analyzing the impact of energy optimization solutions to reduce pollutants in process and in field. The diagrams are made based on the energy reference system (RES) and the process flow diagram (PFD) for each industrial firm or unit. RES represented all energy levels such as extraction, collection, primary and final processing, separation, conversion, storage, transmission, distribution, loading, and end-users of energy carriers. An user friendly EFD software to analyze the pollution flow and impact from industries was developed comprising the features of Health, Safety and Environment (HSE) aspects of operational units for providing detailed knowledge of pollution level in industrial areas and the nearby environment [57,58].

Heavy metal pollution in water using multivariate statistical techniques was assessed in industrial areas in India [59]. This study used multivariate statistical techniques for evaluation and interpretation of the data with a view to get better information about the water quality and designs some remedial techniques to prevent the pollution caused by hazardous toxic elements in future. Soil magnetometry was used for mapping particulate pollution loads in urban forests and it was reported that very low soil pH favored the release of heavy metals and other toxic elements into the soil environment, and through the soil, directly into the forest ground flora and underground water system [60]. However, these studies were unidirectional and several factors of pollution aspects were not taken into consideration. The emerging new technology platforms that use computational EFD softwares will help to deal better with environmental challenges in civil and energy engineering projects than the conventional cumbersome methods [61]. EFD softwares were useful in examining pollution indicators such as greenhouse gases and air pollutants in atmosphere and BOD, COD, heavy metals, organics, and total hardness in water and soil environments. Methods of air pollution estimation in EFD include sampling of emission sources, emission factors available in international resources, engineering calculations, and process simulation. In addition, mode of evaluating water and wastewater qualities in EFD include sampling of industrial wastewater in operational area, comparison with national and international standards, and detecting pollutants that are above environmental standards.

Development of Strategic Approach for Vector Control

Mosquitoes are known vectors of several disease-causing pathogens, which affect millions of people worldwide causing huge morbidity and mortality. *Aedes aegypti* is known to transmit viral infections such as dengue, yellow fever and chikungunya; *Aedes albopictus* is known to carry both dengue and chikungunya viruses and can co-transmit the pathogens. Malarial parasites are carried by *Anopheles stephensi*, and filarial disease by *Culex quinquefasciatus*. To prevent or to reduce the burden of mosquito-borne diseases and improve public health, it is necessary to control them. Vector control, which includes both anti-larval and anti-adult measures, constitutes an important aspect of any mosquito control programs. In recent years, however, mosquito control programs have been suffering from failures because of the ever-increasing insecticide resistance of these vectors [62]. Besides, the recalcitrant chemical insecticides that are currently used in several tropical countries impact severely the both terrestrial and aquatic life systems due to biomagnification and food chain entry.

Biological control of mosquitoes using plant based bioinsecticides has become one of the most important alternatives to prevent development of these vectors as well as reduce the trend of insecticide resistance among them. In addition to the efficiency, they are relatively safe, degradable and readily scalable compared to the problems associated with synthetic insecticides. Several groups of phytochemicals such as alkaloids, steroids, terpenoids, essential oils and phenolics from different plants have been reported for their mosquitocidal activities and several of these are toxic secondary metabolites. When mosquitoes feed on the metabolites, they develop non-specific effects on a wide range of molecular targets. These targets range from proteins (enzymes, receptors, signaling molecules, ion-channels and structural proteins), nucleic acids, biomembranes, and other cellular components [63]. This in turn, affects mosquito physiology in several ways such as abnormality in the nervous system e.g., inhibition of acetylecholinestrase (by essential oils), GABA-gated chloride channel (by thymol), hormonal balance disruption, mitotic poisoning (by azadirachtin), disruption of the molecular events of morphogenesis and alteration in the behaviour and memory of cholinergic system (by essential oil) .

Though many studies on plant extracts against mosquito larvae have been conducted around the world, we present our efforts on the development of a range of ecofriendly bioinsecticides in this review [64-66, 68-72]. The methanolic extract of leaves from *Acalypha alnifolia* exhibited larvicidal activity against three important mosquitoes such as malarial vector, *Anopheles stephensi*, dengue vector, *Aedes aegypti* and Bancroftian filariasis vector, *Culex quinquefasciatus* [64]. The larval mortality was observed after 24 h exposure. The early fourth instar larvae of *A. stephensi* had LC_{50} and LC_{90} values in the range of 125 to 200 ppm and 390 to 460 ppm, respectively. The *A. aegypti* had LC_{50} and LC_{90} values in the range of 150 to 200 ppm and 380 to 480 ppm, respectively. The *C. quinquefasciatus* had LC_{50} and LC_{90} values in the range of 140 to 200 ppm and 390 to 460 ppm. This was the first report on the larvicidal activities of South Indian plant extracts against three species of mosquito vectors. The whole plant extracts of *Leucas aspera* and *Bacillus sphaericus* was found to be effective against the first to fourth instar larva and pupa of the malarial vector, *Anopheles stephensi* [65]. *A. stephensi* had LC_{50} in the range of 9 to 11% and 12% for larvae (for

4 instars) and pupae, respectively while treating with ethanolic extract of *L. aspera*; the vector had LC_{50} in the range of 0.051% to 0.062% and 0.073% for larvae and pupae, respectively while treating with *B. spaericus*. The use of *B. sphaericus* as a potential biolarvicides is limited due to the development of resistance by the target mosquito species. However, there are no reports on resistance against plant based bioinsecticides.

The larvicidal and pupicidal effects of leaf extract of *Carica papaya* were tested against chikungunya and dengue vector, *Aedes aegypti;* the IC_{50} values for larval instars (I to IV) ranges between 50 to 82 ppm, and for pupae, it was 440 ppm [66]. When the extract was combined with bacterial insecticide, spinosad, the treatment increased the pupicidal efficacy i.e., LC_{50}: 107 ppm. The larvicidal and pupicidal activities were attributed to carpain and papain phytochemicals in the extract. The extracts of *Carica papaya* leaves not only derive importance in anti-vector properties but also antiviral activities especially against arboviruses such as dengue [67]. The combination of ethanol extract from *Acalypha alnifolia* leaves and fungal insecticide, *Metarizhium anisopliae* was evaluated for larvicidal and pupicidal properties against the malaria fever mosquito, *Anopheles stephensi* and the formulation was found to be more effective in exhibiting anti-mosquito activities than the individual active components [68,69]. Another study explored the effects of *Jatropha curcas* leaf extract and *Bacillus thuringiensis israelensis* on larvicidal activity against the lymphatic filarial vector, *Culex quinquefasciatus* [70]. Combination of plant and microbial insecticides not only shows efficiency in vector control at different stages but also reduces the occurrence of insecticidal resistance.

The effects of solvent extracts of *Jatropha curcas, Hyptissu aveolens, Abutilon indicum*, and *Leucas aspera* were tested against third instar larvae of filarial vector, *Culex quinquefasciatus*, and the larval density was reduced in the range of 60 to 99% when the sewage water was treated with the extracts [71]. The extracts containing the phytocomponents such isovitexin, vitexin, α-amyrin, β-sitosterol, stigmasterol, and campesterol could be responsible for the larvicidal properties of the plants. *Sphaeranthus indicus, Cleistanthus collinus* and *Murraya koenigii* leaf extracts were tested against the third instar larvae of filarial vector, *Culex quinquefasciatus* and the extracts were found to exhibit mortality at 250 to 1000 ppm [72]. The efficacy of several essential oils exhibiting larvicidal and knock down effects against three important mosquito vectors was studied, and it was found that calamus oil, cinnamon oil, citro- nella oil, clove oil, eucalyptus oil, lemon oil, mentha oil and orange oil exhibited 100% larvicidal activity at 1000 ppm and 100% knockdown effect at 10% concentration [73]. The anti-mosquitocidal efficiency of the oils were further improved to 4 folds by preparing a formulation of combining these eight oils in appropriate proportions along with natural Camphor [74,75]. To summarize, an insecticide does not need to cause high mortality on target organisms in order to be acceptable but should be eco-friendly in nature. Synergistic approaches such as application of mosquito predators with botanical blends and microbial pesticides will provide a better effect in reducing the vector population and the magnitude of epidemiology.

Surveillance for Environmental Management Systems

Vector borne viral diseases such as dengue, chikungunya, Japanese encephalitis (JE) and parasitic diseases such as malaria and filariasis cause considerable disease burden in human population due to their high virulence and pathogenic properties. In tropical countries including India, the chikungunya affected areas overlap with Dengue endemic areas and provide opportunities for mosquitoes to become infected with both viruses [76]. In areas, where both viruses co-circulate, they can be transmitted together causing co-infections with varied atypical clinical manifestations with frequent outbreak recurrence [77,78]. It is a major challenge to the health system in adopting or implementing vector and viral control activities via addressing the problems of water management. Hence, control of these infections through effective water management is gaining not only greater clinical significance but also avoiding the formation of vector breeding sites.

Presently, the problems in viral and vector control activities are not effectively addressed due to reasons such as absence of curative therapy and effective safe mosquitocides, rapid urbanization with development of industrial areas, poor water management including storm water drain, inadequate sustainable environmental and waste management programmes, demographic behavior, presence of non-degradable tyres and plastic containers, garbage and increasing public agglomerations result in providing an ideal environment of breeding sites for mosquitoes. The situation demands a constant and effective viral-vector surveillance system and a vigil on water management strategies in urban cities. There is often a correlation between the potential vector breeding sites and endemicity of diseases. Yet, there is no rational way to approach the problem in view of virological and entomological monitoring with respect to mapping, which is envisaged to evolve new directions in both vector and viral control activities. Geographical Information System (GIS) has become a major tool for public health professionals, epidemiologists and environmental scientists to track the status and distribution of health indicators. GIS maps are used to identify the distribution of diseases and its variations over space and time. GIS in the past have been used primarily for the production of digital amps for resource management and management of land information systems. However, in recent times, it has been used in determining health indicators through geosurveillance of diseases and their patterns, city planning and arriving strategies for resource and waste management to prevent epidemics [79].

In vector borne disease management, creation of cluster mapping using GIS can indicate prevalence, incidence, identification of risk factors, prediction of outbreaks, potential niches of viral and vector activities, control measures implemented and the extent of efficiency of implementation etc [80]. Besides, GIS enables to conduct (i) periodic zone-wise entomological survey for the identification of mosquito larval breeding sites and sources, (ii) vector population density during different seasons in urban areas, (iii) isolation of breeding sites of mosquitoes in the zones near human habitats using mosquito indices, (iv) analyze the spread of different dengue serotypes and chikungunya infections in various zones of urban areas and correlate with vector distribution, and (v) topographic and climatic disease data stratification on the basis of location, intensity, type and time of reporting of symptoms to help the health administrators to assess the morbidity levels and disease pattern in any part of the country [81].With this technology, health departments and policy makers can view hot spots of vector and viral activities to initiate disease control and prevention activities. We have reported the use of GIS for forecasting of chikungunya epidemics via correlating the rainfall with the disease outbreak in thirty districts of Tamil Nadu and forecast for chikungunya outbreak [82]. Smoothing methods were adopted to filter the variability in the dataset; we found that the districts with the potential of having low, moderate and high outbreak intensity were mapped and correlated with rainfall.

In another study, we identified high prevalent zones of water and vector borne diseases in Chennai Metropolitan, which is the capital city of Tamil Nadu and the outbreak intensity of such diseases was correlated with topographical and climatological factors such as latitude, altitude, zonal areas, temperature, humidity, rainfall, wind velocity and direction, weather and season etc. It was observed that challenges in management of storm water drain could be the reason behind such outbreaks and we proposed that GIS could effectively address the challenges of storm water drain management in urban cities based on the Chennai model [79].

Integrated Information System for Disease and Disaster Management

The main role of disease surveillance is to predict, observe, and minimize the impact caused by sudden outbreak, epidemic, endemic and pandemic situations, disasters, as well as knowledge empowerment on factors contributing to such circumstances. Also, strengthening of Health Surveillance System in urban cities has become essential in the context of growing population, climatic, environmental and ecological interactions, rapid industrialization and urbanization and changing socio-economic profiles [83]. At present, there are some basic systems available to manage disease surveillance, but there remains a huge gap in availability of an Integrated and alert system at the field level. Capturing outbreak data from the field on a timely basis is a highly critical requirement for effective surveillance. More importantly, the systems available presently are not providing appropriate data on a timely basis to the decision makers or health authorities. This has caused lacunae in the efforts of Public Health Administration departments while managing sudden outbreaks of highly infectious vector borne diseases [84,85]. Effective surveillance system is necessary for management of any Public health hazards.

Due to the lack of Integrated Information System available in the public space for recording, analyzing and reporting on emerging diseases and disasters, we developed IDS (Integrated Decision Support) Online system that could meet the need for effectively managing the emerging infectious disease by providing a detailed demographic profile on the incidence of infectious diseases [86]. It is a web based GIS enabled Information system that collects, manages and analyzes the clinical data pertaining to infectious diseases and geographical factors, and provides time to time information to Public Health Authorities. This initiative would bridge the gap between the actual scenarios at the field level and the decision making process at the management level. Also, the data so available would be useful for healthcare professionals and researchers for further research and effective management of future outbreaks. Besides, the system could act as an Effective Reporting Mechanism, which will aid in disaster management.

Drain Water Management and Health Concerns

Loading of contaminants to surface waters, groundwater, sediments, and drinking water occurs via two primary routes: (1) point-source pollution and (2) non-point-source pollution [87]. Point-source pollution originates from discrete sources whose inputs into aquatic systems. Examples of point-source pollution include industrial effluents, municipal sewage treatment plants and combined sewage-storm-water overflows, resource extraction (mining), and land disposal sites. But, non-point-source pollution originates from poorly defined sources such as agricultural runoff (pesticides, pathogens, and fertilizers), storm-water and urban runoff, and atmospheric deposition (wet and dry deposition of persistent organic pollutants such as polychlorinated biphenyls (PCBs), mercury and heavy metals) and this type of pollution occurs over broad geographical scales. Globally, contamination of water by chemicals and pathogens poses the most significant health threat to humans, and there have been countless numbers of disease outbreaks and poisonings throughout history due to exposure to untreated water or lack of efficient drain water management system. Proper surface and subsurface drainage to remove excess water in a safe and timely manner plays an important role in controlling water related diseases. Careful control and appropriate reuse of drainage water can help protect the environment and optimize the use of water resources. Establishing an effective drain water management system requires understanding of the primary sources of toxic contaminants in surface waters and groundwater, the pathways through which they move in aquatic environments, factors that affect their concentration and structure along the many transport flow paths, and the relative risks that these contaminants pose to human and environmental health. Understanding the sources, fate, and concentrations of chemicals/pollutants in water, in conjunction with assessment of effects, not only forms the basis of risk characterization, but also provides critical information required to render decisions regarding regulatory initiatives, remediation, monitoring, and management .

The drainage system is an essential part of living in a city or urban area, as it reduces flood damage by carrying water away. When the excess accumulation of water occurs due to natural or industrial phenomena, some water naturally seeps into the ground. The rest makes its way through drainage systems, into rivers, canals and streams and eventually into the bays, or directly to the bays through storm water beach outlets. In areas with houses, commercial and industrial units and roads, there is a need to create alternative ways for this water to drain away. Large amounts of water can build up quickly during heavy rain and storms, and without adequate networked drainage system, this flows towards low-lying land, causing flooding, damage and pollution associated safety risks.

Downstream beneficial uses of any surface water body to which drainage water is added must be protected. For example, the discharge of saline drainage water into a river or lake is not advisable when that surface water body is being used for domestic or agricultural water supplies. However, it is acceptable to discharge drainage water into a large freshwater body but it is necessary to determine the assimilative capacity of the receiving water and identify the constituents in the drainage water to determine the 'safe level' or discharge requirements for the drainage water. The discharge requirements should specify the maximum allowable concentration of each element of pollution concern and the volume of drainage water discharge that is acceptable. There may be a significant or little difference between the quality of the drainage water and that of the receiving water. The dilution capacity of the receiving water varies from place to place and from time to time depending on numerous local conditions and the upstream uses of the receiving water. The discharge of drainage water of a higher quality than the receiving water is generally acceptable.

Improper use as well as inadequate maintenance of drainage structures (road drainage ditches, septic tanks, drainage canals in irrigation schemes, and also drainage water treatment and disposal facilities) are often associated with environmental health problems.

The health concerns associated with drainage water management can be grouped in three groups: (i) vector-borne diseases; (ii) faecal/orally transmitted diseases; and (iii) chronic health issues related to exposure to agrochemical residues such as pesticides, insecticides etc [88]. Besides, there is often a lack of adequate domestic water supplies and sanitation facilities. Hence, drainage canals or drainage water treatment and disposal facilities are often used for washing, drinking and untreated effluents or other wastes by marginalized sections that facilitates disease transmission. Silting, uncontrolled aquatic weed growth, slow water flow or stagnant pools associated with the resulting wetlands offer ideal breeding conditions for mosquitoes, pathogenic microorganisms such as enteric bacteria, parasites, enteric viruses, parasites, helminthes etc. The issues related to drainage systems and its associated health issues create more challenges to the health, hygiene and sanity measures of developing and under developing countries than developed nations.

Drainage, Water Logging and Salinity

Drainage may be practiced in conjunction with irrigation for protecting environment from pollutants and obtaining maximum land as well as agriculture productivity. There are several concerns about the sustainability of drainage and irrigation projects besides the water quality issues related to the disposal of drainage water. There are also problems with land degradation due to irrigation induced salinity and water logging. There are instances where saline or high nutrient drainage water caused damage to aquatic ecosystems. Several countries are facing water logging and salinity problems that are intensified by a range of factors including the use of wastewaters for irrigation, unsuitable cropping patterns, heavy rains and floods, lack of inadequate drainage structures, unsustainable management decisions, irrigation systems without paying attention to their adverse impacts on soil and quality of water resources etc [89]. Drainage continues to be a vital and necessary component of agricultural production systems since excess water/water with high salinity in the crop root zone soil is injurious to plant growth. Crop yields are drastically reduced on poorly drained soils or prolonged water logging as plants eventually die due to lack of oxygen in the root zone [90]. In order to enhance the net benefits of drainage systems, more attention needs to be given to the water quality impacts of drainage water disposal. An efficient drainage system removes salts added to the soil brought in by irrigation water. Various problems such as the increase in soil salinity, water logging, and water table rise into the crop root zone are encountered when an efficient drainage system is not provided in an irrigated area [91].

Drainage Systems Useful for Managing Soil Environmental Crisis

Various drainage systems have been used to drain the excess water both on the surface and between the soil horizons in order to keep the water table below a certain level and they comprise surface and subsurface drainage infrastructures [92].

Surface Drainage Systems

Drainage of stationary water on soil surface that can increase water table in the soil profile and harm plant root zone is accomplished by providing surface drainage (surface runoff) system. Surface drainage is often achieved by land forming and smoothing to remove isolated depressions, or by constructing parallel ditches. Ditches and furrow bottoms are gently graded and discharge into main drains at the field boundary. Although the ditches or furrows are intended primarily to convey excess surface runoff, there is some seepage through the soil to the ditches, depending on the water table position. This could be regarded as a form of shallow subsurface drainage. Surface drainage is especially important in humid regions on flat lands with limited hydraulic gradients to nearby rivers or other disposal points.

Subsurface Drainage Systems

In contrast to surface drainage system that removes excess water accumulated in the cropped area, the excess water and harmful salt solutions in the soil profile can be removed using subsurface drainage system [91]. These underground systems comprised of drain tile or tubing designed to lower the water table by subsurface flow. The downstream ends of the laterals are normally connected to a collector drain. Subsurface drainage is more effective in salt cleaning in soil profile than surface drainage [93]. Subsurface drainage provides several intangible benefits such as improvement in soil health due to the increased aeration of the soil, increased responses to fertilizer use, reduced mineral imbalances in the soil, reduction in salinity etc. In the areas where subsurface drainage system is not constructed, no efficient surface drainage system can be obtained in the soil profile even if surface drainage system is constructed [94].

Horizontal Drainage Systems

These systems are used in irrigated arid and semi-arid regions for rescuing saline and waterlogged lands to maintain long-term salt and water balances in the crop root zone. Salinity and water logging of lands are due to buildup of the water table, deep percolation of normal excess water and canal seepage. Buried horizontal pipe drains are installed deeply in arid regions when compared to humid regions for controlling salinity. During irrigation, excess quantity of water against the plant evapotranspiration is applied and this additional quantity of water applied is known as the leaching fraction. Naturally occurring as well as applied salts are then leached from the root zone by this water, and removed from the field via the pipe drains. Deeper drain installation ensures that salts do not rise too rapidly to the soil surface due to capillary action. The amount of irrigation water to be removed is generally less in arid than in humid regions [90].

Vertical Drainage Systems

These systems use tube-wells to control water logging and salinity and the system is widely adopted by South East Asian countries such as India, Pakistan, Sri Lanka etc. The primary goals of tube-wells are almost similar to that of horizontal drains in terms of extracting groundwater for irrigation. As a result of pumping, the water table is lowered, and salinization due to capillarity is minimized. This situation is ideal where the groundwater is not very brackish or saline, and is therefore suitable for irrigation. In areas where the groundwater is highly saline, the pumped water may be too saline for irrigation, unless mixed with fresher or less saline water. Where the groundwater is too saline for crop production, it must be disposed of.

The design of horizontal and vertical drainage systems is based on the layout, depth and spacing of the drains and is generally carried out using subsurface drainage parameters such as the depth of drain, water table and soil, hydraulic conductivity of the soil and drain discharge [95]. Though several studies on the designs of horizontal and vertical drainage systems were described, these studies were performed separately on these two systems without comparing the efficacy of them using anisotropic soils. However, comparative study on these two systems for their suitability in anisotropic soils using EnDrainWin and WellDrain softwares for drain spacing and well spacing, respectively was reported [95,96]. The study results showed that horizontal drainage systems were better than vertical drainage systems in terms of higher spacings between drains thus reducing number of drainage and cost. However, vertical drainage systems due to the lower changes in well spacing in different anisotropic soils were suitable for conditions that soil hydraulic conductivity was likely to change. In another study, the En Drain software was used to analyze the drainage parameters on the changes of drain discharge in subsurface drainage systems as scrutiny of inflow to the drains is essential for improving soil environmental conditions [97].

Often, agricultural land drainage consists of a combination of both surface and subsurface systems. At the field scale, subsurface drain pipes and field ditches normally exit to an open main or collector drain. At the regional level, the latter then empties into a river or canals. In some instances, depending on the character of the hydrological basin, main drains may dispose of drainage water to an evaporation pond, to a wetland, or to a saline agriculture/agriculture-forestry system. In practice, it is often difficult to differentiate between surface and subsurface drainage because the outflow in drainage ditches or canals is usually a combination of both surface and subsurface flow. The relative proportion of surface and subsurface flow in the total drainage volume depends on many factors. These include rainfall intensity, land surface roughness and slope, vegetation, soil permeability, and ditch or drain tubing spacing and depth. Developing countries, in the last half century, experience rapid strides in the changing practices such as conversion of agricultural lands to construct house and industrial units that impact the efficiency of functioning of drainage, water resource management and agriculture systems. Sustainability in public health and agricultural productivity demands macroeconomic policies on the use of land and resources due to the limited availability of water resources and its vulnerability to exploitation and contamination. Actual agricultural yield as percentage of potential yield was higher for North America, Western and Central Europe, South America and North Africa (50-60%) than Central America, Eastern Europe and Sub-Saharan Africa (≤30%), and the wide variation in the yields across continents and nations suggest the need for agricultural water management through policies and incentives [98]. Studies that examined land use policies and agricultural water management in different regions of the world including the Africa in the past half century suggested the need for constructing irrigation and drainage facilities as well as implementation of effective land use policies for improving water resources and crop productivity [98,99]. These studies were conducted to analyse data on several indices such as anthropological, agricultural, economical and irrigation and drainage indices obtained from Food and Agricultural Organization (FAO) databases and World Bank Group. Interestingly, these studies revealed drastic variations in the production of crops, provided list of strengths and weakness and emphasized the avoidance of trial and error policies to sustain agricultural productivity and water resources.

GIS Application in Drainage Master Planning and Water Resource Management

The scope of sustainable management of water resources in drainage systems and agriculture concerns the responsibility of policy makers, technologists, users and public to ensure that water resources are allocated efficiently and equitably and used to achieve socially, environmentally and economically beneficial outcomes [100]. In changing anthropogenic domain, there exists a need to update water conservation systems and improve agricultural practices with less chemo-insecticides and pesticides based on water resource information [101]. Urban drainage master planning, in recent years, depends heavily on spatial-based data to establish the characteristics and performance of both major and minor drainage systems. This data includes the topological and geophysical characteristics of the storm sewer and open channel drainage network, the runoff catchments, and overland flow paths and potential hazard areas associated with flooding during extreme runoff events. In the lack of adequate data and information on the drainage system for an area or a region, remotely sensed satellite image based GIS tools have become of great role in analysis. GIS has become an indispensable, multi-purpose tool for infrastructure master planning that enables sophisticated desktop analyses, efficient simulation model pre- and post-processing, and effective data management and communication, all through the use of intrinsic functionality and readily available on-line utilities. Its applications, in urban planning and water resource management, covers storm and sanitary drainage systems, water distribution and resources such as surface as well as ground water, morphometry of streams, watershed planning, regional groundwater protection, a wide range of agricultural systems, rain harvesting systems, climatic conditions across regions, recycled wastewater, and desalinated water. Besides, features on the thematic presentation of real time data facilitate implementation of appropriate schemes for construction of drainage structures and agricultural water management aspects including irrigation water management in rain-fed agriculture; management of floods, droughts, and drainage; and conservation of ecosystems and associated cultural and recreational values. GIS tools for water resource management function on the following platforms using separate softwares: (a) Data management - includes the tasks of reviewing the geospatial images obtained by remote sensing, editing of catchment and infrastructure data, data on hydraulic features and hydraulic model simulation; (b) analysis - using custom applications developed for water balance analysis and core GIS functions such as geoprocessing, surface modeling and hydrologic analysis of major drainage systems for spatial analysis; and (c) presentation - presenting study findings using thematic mapping and three dimensional models.

The use of GIS was reported to explore the temporal and spatial variations in river water quality and to estimate the influence of watershed land use, topography and socio-economic factors on river water quality and various pollutants [102]. In another study, GIS was used to determine the levels of pollution by organic pollutants, salts, metals and microbial indicators in fresh and saline/brackish environments. In this study, surface quality maps for dissolved oxygen, nitrogen and phosphate contents, metals, and total coliforms were developed to highlight hot-spot areas of pollution [103]. It was reported that geospatial data along with GIS was useful in analyzing the topological elements of drainage systems and in inducing their geomorphologic and hydrologic characteristics [104].

Water resource management is crucial for sustaining the environment as well as the survival of life. It is necessary to find faster and more effective methods to identify and manage sources of pollution and minimize the levels of surface and ground water contamination. In some cases, the contamination has already occurred and the primary focus is clean up. GIS contributes solutions to the above problems and further proves useful in adopting water resource policy for promoting a more efficient and equitable allocation of natural and community resources.

Conclusion

Aquatic ecosystem is in threat due to anthropogenic activities that change the quality of water to a great extent. The researches that we had undertaken in this perspective were presented in this review and the findings revealed the factors and mechanism in causing damage to such a precious ecosystem. The review also gives some clue and scope to the research community for finding new enviro-friendly sustainable solutions for the complex problems associated with water contaminants. This review also encourages researchers working in environmental biology to explore novel eco-management strategies to preserve water ecosystem as water is the elixir of life.

Acknowledgements

I thank the funding agencies of Govt of India such as Department of Science and Technology (DST), Ministry of Environment and Forests (MoEn), University Grants Commission (UGC), Department of Information Technology (DIT) for their support to undertake the need based research. I am also grateful to the Management of Loyola College for their encouragement. Our special thanks to Dr. S. Sivasubramanian, Scientist for his great help in preparation of the review.

References

1. Dash MC (1994) Fundamentals of Ecology. Tata McGraw-Hill, New Delhi, India.

2. Sharma BK, Kauai H (1994) Environmental Chemistry. Goel Pub House, Meerut, India.

3. Zou E (1997) Effects of sublethal exposure to zinc chloride on the reproduction of the water flea, Moinairrasa (Cladocera). Bull Environ ContamToxicol 58: 437-441.

4. James R, Sampath K, Selvamani P (1998) Effect of EDTA on reduction of copper toxicity in Oreochromismossambicus (Peters). Bull Environ ContamToxicol 60: 487-493.

5. Kotze P, Du Preez HH, Van Vuren JHJ (1999) Bioaccumulation of copper and zinc in Oreochromismossambicus and Clariasgariepinus from the Olifants River, Mpumalanga, South Africa. Water SA 25: 99-110.

6. Abreu SN, Pereira E, Vale C, Duarte AC (2000) Accumulation of mercury sea bass from a contaminated lagoon (Ria de Aveiro, Portugal). Mar Pollut Bull 40: 293.169.

7. Govind P, Madhuri S (2014) Heavy metals causing toxicity in animals and fishes. Res J Animal, Veterinary and Fishery Sci 2: 17-23.

8. Dallinger R, Kutzky H (1985) The importance of contaminated food for the uptake of heavy metals by rainbow trout (Salmogairdnen) a field study. Oecologia 67: 82-89.

9. Arockiadoss T, Vincent S, Xavier FP, Nagaraja KS, Selvanayagam M (1998) pH-based conductivity studies on fish in a contaminated environment. Bull Environ ContamToxicol 61: 645-649.

10. Gohil MN, Mankodi PC (2013) Diversity of fish fauna from downstream zone of river Mahisagar, gujarat state, India. Res J Animal, Veterinary and Fisheries Sci 1: 14-15.

11. Ambrose T, Vincent S, Cyril Arun Kumar L (1994). Susceptibility of the fresh water fish Gambusiaaffinis (Bird and Girad), Sarotherodonmossambicus (Peters) and Cirrhinusmrigala (Ham.) to zinc toxicity. Indian J Environ Toxicol 4: 29-31.

12. Athikewavan S, Vincent S, Velmurugan B (2006) Accumulation of zinc in the different tissues of the Silver carp, Hypophthalmichthysmolitrix. Poll Res 25(1): 47-49.

13. Everall NC, Mcfarlane AA, Sedgwick RW (1989) The effects of water hardness upon the uptake, accumulation and excretion of zinc in the brown trout, Salmotrutta L. J Fish Biol 35: 888-892.

14. Athikesavan S, Vincent S, Velmurugan B (2006) Histopathological effects of zinc sulphate on gill, liver, intestine and kidney tissues of Hypophthalmichthysmolitrix. Indian J Environ Toxicol 16: 27-31.

15. Athikesavan S, Vincent S, Velmurugan B, Janardhanan S (2006) Impact of nickel and zinc on protein profile of the Silver carp, Hypophthalmichthysmolitrix. Asian J Microbiol Biotech EnvSci 8: 147-149.

16. Athikesavan S, Vincent S, Velmurugan B (2006) Investigation of acute toxicity of zinc sulphate in Silver carp (Hypophthalmichthysmolitrix). Aquacult 7: 331-336.

17. Veldhuizen-Tsoerkan MB, Holwerda DA, Van der Mast CA, Zandee DI (1990) Effects of cadmium exposure and heat shock on protein synthesis in gill tissue of sea mussel, Mytilusedulis. Comp BiochemPhysiol 96: 419-426.

18. Athikesavan S, Vincent S, Ambrose T, Velmurugan B (2006) Nickel induced histopathological changes in the different tissues of freshwater fish, Hypophthalmichthysmolitrix (Valenciennes). J Environ Biol 27: 391-395.

19. Nickel. In Environmental Health Criteria (1991) IPCS (International Programme on Chemical Safety).

20. Ambient water quality criteria for nickel (1980) USEPA (US Environmental Protection Agency). EPA Report 440/5-80-060. p206.

21. Eisler R (1998) Nickel hazards to fish, wildlife, and invertebrates: a synoptic review. US Geological Survey, Biological Science Report: 1998-0001, U.S. Fish and Wildlife Service, USA. p76.

22. Ptashynski MD, Klaverkamp JF (2002) Accumulation and distribution of dietary Ni in Lake Whitefish (Coregonusclupeaformis). Aquatic Toxicol 58: 249-264.

23. Athikesavan S, Vincent S, Velmurugan B (2005) Genotoxic effect of nickel chloride and zinc sulphate on fish Hypophthalmichthysmolitrix. J Indian Fish Assoc 32: 111-117.

24. Alarifi S, Ali D, Alakhtani S, Al Suhaibani ES, Al-Qahtani AA (2014) Reactive oxygen species-mediated DNA damage and apoptosis in human skin epidermal cells after exposure to nickel nanoparticles. Biol Trace Elem Res 157: 84-93.

25. Morita H, Umeda M, Ogawa HI (1991) Mutagenicity of various chemicals including nickel and cobalt compounds in cultured mouse FM3A cells. Mutat Res 261: 131-137.

26. Sarkar B (1995) Metal replacement in DNA-binding zinc finger proteins and its relevance to mutagenicity and carcinogenicity through free radical generation. Nutrition 11: 646-649.

27. Hartmann M, Hartwig A (1998) Disturbance of DNA damage recognition after UV-irradiation by nickel(II) and cadmium(II) in mammalian cells. Carcinogenesis 19: 617-621.

28. Mary MS, Gopal J, Tata BVR, Rao TS, Vincent S (2008) A confocal microscopic study on colony morphology and sporulation of Bacillus sp. World J MicrobiolBiotechnol 24: 2435-2442.

29. Vincent S, Cyril Arun Kumar L, Mani T, Ambrose T, Selvanayagam M (1995) Heavy metal chromium induced alterations on phosphatases activity in the Indian major carp, Catla catla (HAM). Arch Hydrobiol 135: 283-287.

30. Vincent S, Cyril Arun Kumar L, Ambrose T (1996) Impact of heavy metal chromium on bioenergetics of the Indian major carp, Catla catla (HAM). Poll Res 15: 273-275.

31. Ambrose T, Vincent S, Cyril Arunkumar L (1994) Impact of tannery effluent on primary production in the aquatic macrophytesHydrillaverticillata and Ceretophyllumdemersum. Geobios 21: 89-92.

32. Vincent S, Cruz MM, Thomas AL (2001) Bioremediation of chromium by the aquatic macrophyteCaldesiaparanassipolia (L) Parl. Poll Res 20: 75-77.

33. Mary Mangaiyarkarasi MS, Vincent S, Janarthanan S, SubbaRao T, Tata BV (2011) Bioreduction of Cr(VI) by alkaliphilic Bacillus subtilis and interaction of the membrane groups. Saudi J BiolSci 18: 157-167.

34. Ye Q, Roh Y, Carroll SL, Blair B, Zhou J, et al. (2004) Alkaline anaerobic respiration: isolation and characterization of a novel alkaliphilic and metal-reducing bacterium. Appl Environ Microbiol 70: 5595-5602.

35. Stewart DI, Burke IT, Mortimer RJG (2007) Stimulation of microbially mediated chromate reduction in alkaline soil water systems. Geomicrobiol J 4: 655-669.

36. Allen T, Rana SV (2004) Effect of arsenic (AsIII) on glutathione-dependent enzymes in liver and kidney of the freshwater fish Channapunctatus. Biol Trace Elem Res 100: 39-48.

37. Govind P, Madhuri S, Shrivastav AB (2014) Fish Cancer by Environmental Pollutants, 1st edn. Narendra Publishing House, Delhi, India.

38. Bears H, Richards JG, Schulte PM (2006) Arsenic exposure alters hepatic arsenic species composition and stress-mediated gene expression in the common killifish (Fundulusheteroclitus). AquatToxicol 77: 257-266.

39. Ciardullo S, Aureli F, Raggi A, Cubadda F (2010) Arsenic speciation in freshwater fish: focus on extraction and mass balance. Talanta 81: 213-221.

40. Das S, Unni B, Bhattacharjee M, Wann SB, GangadharRao P (2012) Toxicological effects of arsenic exposure in a fresh water teleost fish, Channapunctatus. African J Biotechnol 11: 4447-4454.

41. Ahmed MK, Habibullah-Al-Mamun M, Parvin E, Akter MS, Khan MS (2013) Arsenic induced toxicity and histopathological changes in gill and liver tissue of freshwater fish, tilapia (Oreochromismossambicus). ExpToxicolPathol 65: 903-909.

42. Kovendan K, Vincent S, Janardhanan S, Saravanan M (2013) Expression of metallothionein in liver and kidney of freshwater fish Cyprinuscarpio var. communis (Linn) exposed to arsenic trioxide. Amer J SciInd Res 4: 1-10.

43. Seok SH, Baek MW, Lee HY, Kim DJ, Na YR, et al. (2007) Arsenite-induced apoptosis is prevented by antioxidants in zebrafish liver cell line. Toxicol In Vitro 21: 870-877.

44. Yang JL (2014) Comparative acute toxicity of gallium(III), antimony(III), indium(III), cadmium(II), and copper(II) on freshwater swamp shrimp (Macrobrachiumnipponense). Biol Res 47: 13.

45. Nunes B, Caldeira C, Pereira JL, Gonçalves F, Correia AT (2014) Perturbations in ROS-related processes of the fish Gambusiaholbrooki after acute and chronic exposures to the metals copper and cadmium. Environ SciPollut Res Int .

46. Vincent S, Ambrose T, Selvanayagam M (1994) Influence of the heavy metals cadmium and chromium on leukocytes of fresh water fish, Catla catla (Ham.) Indian J Environ Toxicol 4: 45-47.

47. Van Campenhout K, Infante HE, Hoff PT, Moens L, Goemans G (2010) Cytosolic distribution of Cd, Cu and Zn, and metallothionein levels in relation to physiological changes in gibel carp (Carassiusauratusgibelio) from metal-impacted habitats. Ecotox Environ Safe 73: 296-305.

48. Mieiro CL, Bervoets L, Joosen S, Blust R, Duarte AC, et al. (2011) Metallothioneins failed to reflect mercury external levels of exposure and bioaccumulation in marine fish--considerations on tissue and species specific responses. Chemosphere 85: 114-121.

49. Bervoets L, Knapen D, De Jonge M, Van Campenhout K, Blust R (2013) Differential hepatic metal and metallothionein levels in three Feral fish species along a metal pollution gradient. PLoS One 8: e60805.

50. De Smet H, De Wachter B, Lobinski R, Blust R (2001) Dynamics of (Cd,Zn)-metallothioneins in gills, liver and kidney of common carp Cyprinuscarpio during cadmium exposure. AquatToxicol 52: 269-281.

51. Wang WC, Mao H, Ma DD, Yang WX (2014) Characteristics, functions, and applications of metallothionin in aquatic vertebrates. Marine Pollution 1: 34.

52. Rose S, Vincent S, Meena B, Suresh A, Mani R (2014) Metallothionein induction in fresh water catfish Clariasgariepinus on exposure to cadmium. International J Pharmacy Pharm Sci 6: 377-383.

53. Thomas PJ, Anand T, Suresh P, Janardhanan S, Vincent S (2006) Designing specific oligonucleotide primers for metallothionein genes. Indian J Biotechnol 5: 120-122.

54. Praveen B, Vincent S, Murty US, Krishna AR, Jamil K (2005) A rapid identification system for metallothionein proteins using expert system. Bioinformation 1: 14-15.

55. Duruibe JO, Ogwuegbu MOC, Egwurugwu JN (2007) Heavy metal pollution and human biotoxic effects. Int J Physical Sci, 2(5), 112-118.

56. Tchounwou PB, Yedjou CG, Patlolla AK, Sutton DJ (2012) Heavy metal toxicity and the environment. EXS 101: 133-164.

57. Valipour M, Mousavi SM, Valipour R, Rezaei E (2013) A new approach for environmental crises and its solutions by computer modeling. The 1st International Conference on Environmental Crises and its Solutions, Kish Island, Iran.

58. Valipour M, Mousavi SM, Valipour R, Rezaei E (2012) Air, water, and soil pollution study in industrial units using environmental flow diagram. J. Basic. Appl. Sci. Res 2: 12365-12372.

59. Krishna AK, Satyanarayanan M, Govil PK (2009) Assessment of heavy metal pollution in water using multivariate statistical techniques in an industrial area: A case study from Patancheru, Medak District, Andhra Pradesh, India. J Hazard Mater 167: 366-373.

60. Magiera T, Strzyszcz Z, Rachwal M (2007) Mapping particulate pollution loads using soil magnetometry in urban forests in the Upper Silesia Industrial Region, Poland. Forest Ecology and Management 248: 36-42.

61. Valipour M, Mousavi SM, Valipour R, Rezaei E (2013) Deal with environmental challenges in civil and energy engineering projects using a new technology. J Civil Environ Eng 3: 127.

62. Thomsen EK, Strode C1, Hemmings K1, Hughes AJ1, Chanda E2, et al. (2014) Underpinning sustainable vector control through informed insecticide resistance management. PLoS One 9: e99822.

63. Ghosh A, Chowdhury N, Chandra G (2012) Plant extracts as potential mosquito larvicides. Indian J Med Res 135: 581-598.

64. Kovendan K, Murugan K, Vincent S (2012) Evaluation of larvicidal activity of Acalyphaalnifolia Klein ex Wild. (Euphorbiaceae) leaf extract against the malarial vector, Anopheles stephensi, dengue vector, Aedesaegypti and Bancroftianfilariasis vector, Culexquinquefasciatus (Diptera: Culicidae). Parasitol Res 110: 571-581.

65. Kovendan K, Murugan K, Vincent S, Barnard DR (2012) Studies on larvicidal and pupicidal activity of Leucasaspera Willd. (Lamiaceae) and bacterial insecticide, Bacillus sphaericus, against malarial vector, Anopheles stephensi Liston. (Diptera: Culicidae). Parasitol Res 110: 195-203.

66. Kovendan K, Murugan K, Naresh Kumar A, Vincent S, Hwang JS (2012) Bioefficacy of larvicdial and pupicidal properties of Carica papaya (Caricaceae) leaf extract and bacterial insecticide, spinosad, against chikungunya vector, Aedesaegypti (Diptera: Culicidae). Parasitol Res 110: 669-678.

67. Sharma N, Mishra D (2014) Papaya leaves in dengue fever: is there scientific evidence? Indian Pediatr 51: 324-325.

68. Murugan K, Kovendan K, Vincent S, Barnard DR (2012) Biolarvicidal and pupicidal activity of Acalyphaalnifolia Klein ex Willd. (Family: Euphorbiaceae) leaf extract and Microbial insecticide, Metarhiziumanisopliae (Metsch.) against malaria fever mosquito, Anopheles stephensi Liston. (Diptera: Culicidae). Parasitol Res 110: 2263-2270.

69. Kovendan K, Murugan K, Vincent S, Barnard DR (2012) Efficacy of larvicidal and pupicidal properties of Acalyphaalnifolia Klein ex Willd. (Euphorbiaceae) leaf extract and Metarhiziumanisopliae (Metsch.) against Culexquinquefasciatus Say. (Diptera: Culicidae). J Biopest 5: 170-176.

70. Kovendan K, Murugan K, Vincent S, Kamalakannan S (2011) Larvicidal efficacy of Jatrophacurcas and bacterial insecticide, Bacillus thuringiensis, against lymphatic filarial vector, Culexquinquefasciatus Say (Diptera: Culicidae). Parasitol Res 109: 1251-1257.

71. Kovendan K, Murugan K, Panneerselvam C, Mahesh Kumar P, Amerasan D, et al. (2012) Laboratory and field evaluation of medicinal

plant extracts against filarial vector, Culexquinquefasciatus Say (Diptera: Culicidae). Parasitol Res 110: 2105-2115.

72. Kovendan K, Arivoli S, Maheshwaran R, Baskar K, Vincent S (2012) Larvicidal efficacy of Sphaeranthusindicus, Cleistanthuscollinus and Murrayakoenigii leaf extracts against filarial vector, Culexquinquefasciatus Say (Diptera: Culicidae). Parasitol Res 111: 1025-1035.

73. Manimaran A, Cruz M, Muthu C, Vincent S, Ignacimuthu S (2012) Larvicidal and knockdown effects of some essential oils against Culexquinquefasciatus Say, Aedesaegypti (L.) and Anopheles stephensi (Liston). Advances in Bioscience and Biotechnology 3:855-862.

74. Manimaran A, Cruz MMJ, Muthu C, Vincent S, Ignacimuthu S (2013) Repellent activity of plant essential oils formulation against three diseases causing mosquito vectors. J AgricTechnol 9: 845-854.

75. Manimaran A, Cruz M, Muthu C, Vincent S, Ignacimuthu S (2013) Larvicidal and growth inhibitory activities of different plant volatile oils formulation against Anopheles stephensi (Liston), Culexquinquefasciatus Say and Aedesaegypti (L.). International Journal of Phytotherapy Research 3: 38-48.

76. Chahar HS, Bharaj P, Dar L, Guleria R, Kabra SK, et al. (2009) Co-infections with chikungunya virus and dengue virus in Delhi, India. Emerg Infect Dis 15: 1077-1080.

77. Vazeille M, Mousson L, Martin E, Failloux AB (2010) Orally co-Infected Aedesalbopictus from La Reunion Island, Indian Ocean, can deliver both dengue and chikungunya infectious viral particles in their saliva. PLoSNegl Trop Dis 4: e706.

78. Taraphdar D, Sarkar A, Mukhopadhyay BB, Chatterjee S (2012) A comparative study of clinical features between monotypic and dual infection cases with Chikungunya virus and dengue virus in West Bengal, India. Am J Trop Med Hyg 86: 720-723.

79. Sanjeevi Prasad S (2007) Public health issues related to storm water drain in Chennai (Zone 2-3) using GIS as a tool for management. Ph.D thesis. University of Madras, India.

80. Palaniyandi M (2012) The role of remote sensing and GIS for spatial prediction of vector-borne diseases transmission: a systematic review. J Vector Borne Dis 49: 197-204.

81. Bouzid M, Colón-González FJ, Lung T, Lake IR, Hunter PR (2014) Climate change and the emergence of vector-borne diseases in Europe: case study of dengue fever. BMC Public Health 14: 781.

82. Chakkaravarthy VM, Vincent S, Ambrose T (2011) Novel approach of Geographic Information systems on recent outbreaks of chikungunya in Tamilnadu, India. J Environ SciTechnol 4: 387-394.

83. Thakur JS (2006) Integrated Disease Surveillance - A Key Step to improve public health in India. Indian J Community Medicine 31: 215.

84. Zeng D, Chen H, Lynch C, Eidson M, Gotham I (2004) Infectious disease informatics and outbreak detection. Medical Informatics 359-395.

85. Kant L (2008) Combating emerging infectious diseases in India: orchestrating a symphony. J Biosci 33: 425-427.

86. Antony Raj G, Vincent S, Muthumariappan M (2011) Development of an integrated decision support system (IDS Online) for an effective disease surveillance and disaster management. Int J Pharma and Biosciences 2: B452-458.

87. Ritter L, Solomon K, Sibley P, Hall K, Keen P, et al. (2002) Sources, pathways, and relative risks of contaminants in surface water and groundwater: a perspective prepared for the Walkerton inquiry. J Toxicol Environ Health A 65: 1-142.

88. Fritsch MS (1997) Management of agricultural drainage water quality: Water Reports 13. Food and Agriculture Organization of the United Nations. Madramootoo CA, Johnston WR, Willardson LS (Eds.). Natural Resources and Environment, Italy.

89. Valipour M (2014) Drainage, waterlogging, and salinity. Archives of Agronomy and Soil Science 60: 1625-1640.

90. Madramootoo CA (1997) Management of agricultural drainage water quality: Water Reports 13. Food and Agriculture Organization of the United Nations. Madramootoo CA, Johnston WR, Willardson LS (Eds.).

91. Fayrap A, Koc C (2012) Comparison of drainage water quality and soil salinity in irrigated areas with surface and subsurface drainage systems. Agricultural Research 1: 280-284.

92. Fayrap A, Tonkaz T, Kiziloglu FM (2010) Spatial distribution patterns of ground water levels and salinity in Igdir Plain. In: International soil sciences congress on management of natural resources to sustain soil health and qualityâ€™ OndokuzMayis University, Samsun, Turkey. p286.

93. Bahceci I, Nacar AS (2008) Subsurface drainage and salt leaching in irrigated land in south east Turkey. Irrig Drainage 57: 1-11.

94. Dunn SM, Mackay R (1996) Modelling hydrological impacts of open ditch drainage. J Hydrol 179: 37-66.

95. Valipour M (2012) A Comparison between Horizontal and Vertical Drainage Systems (Include Pipe Drainage, Open Ditch Drainage, and Pumped Wells) in Anisotropic Soils. IOSR Journal of Mechanical and Civil Engineering 4: 7-12.

96. Valipour M (2013) Comparison of Different Drainage Systems Usable for Solution of Environmental Crises in Soil. In: The 1st International Conference on Environmental Crises and its Solutions, Kish Island, Iran.

97. Valipour M (2013) Scrutiny of Inflow to the Drains Applicable for Improvement of Soil Environmental Conditions. In: The 1st International Conference on Environmental Crises and its Solutions, Kish Island, Iran.

98. Valipour M, Ahmadi MZ, Raeini-Sarjaz M, Sefidkouhi MA, Shahnazari A, et al. (2014) Agricultural water management in the world during past half century. Arch Agron Soil Sci 1-12.

99. Valipour M (2014) Land use policy and agricultural water management of the previous half of century in Africa. Appl Water Sci 1-29.

100. Sustainable management of water resources in agriculture, OECD 2010. DOI 10.1787/9789264083578-en.

101. Valipour M (2013) Need to Update of Irrigation and Water Resources Information According to the Progresses of Agricultural Knowledge. Agrotechnol S10: e001.

102. Chen J, Lu J (2014) Effects of land use, topography and socio-economic factors on river water quality in a mountainous watershed with intensive agricultural production in East china. PLoS One 9: e102714.

103. Wilbers GJ, Becker M, Nga LT, Sebesvari Z, Renaud FG (2014) Spatial and temporal variability of surface water pollution in the Mekong Delta, Vietnam. Sci Total Environ 485-486: 653-65.

104. Saud MA (2012) Use of remote sensing and GIS to analyze drainage system in flood occurrence, Jeddah - Western Saudi coast, drainage systems, Prof. Muhammad SalikJavaid (Ed.).

18

Occurrence and Significance of Secondary Iron-rich Products in Landfilled MSWI Bottom Ash

Saffarzadeh A[1]* and Takayuki Shimaoka[1]

[1]*Department of Urban & Environmental Engineering, Kyushu University, Fukuoka, 819-0395, Japan*

***Corresponding author:** Saffarzadeh A, Associate Prof. (Ph.D, D.Eng) Department of Urban & Environmental Engineering, Kyushu University, Fukuoka, 819-0395, Japan
E-mail: amir@doc.kyushu-u.ac.jp

Abstract

Incineration is one of the most effective techniques for the treatment of both municipal and hazardous wastes. Via this technique, the majority of toxic substances are expected to be stabilized in the durable matrix of the end-of-process bottom ash products. These products consist of a variety of glassy/crystalline components including primary Fe-rich phases that may undergo alterations when exposed to natural environment. In the present research, the impact of natural weathering on the behavior of primary Fe-rich phases, their alteration, and the formation of the relevant secondary products in the weathered bottom ash samples of a (mono) landfill site was systematically investigated. Samples of various ages (1-20 yrs) were collected from four locations of the landfill in 2009. Optical microscopy, SEM-EDX, XRD and XRF examinations were applied in order to document the footprints of weathering processes. Using these techniques, we understood that several secondary (newly-formed) products (amorphous or crystalline) have been developed, including goethite (α-FeOOH), lepidocrocite (γ-FeOOH), hematite (Fe_2O_3), magnetite (Fe_3O_4), iron oxide (FeO), and Fe-rich Ca-Si and Ca-Al-Si gel phases. They occurred under variable environmental conditions as the weathering products of the primary iron-rich phases. The strong affinity of these secondary phases with heavy metals of environmental significance such as Zn, Cu, Pb, and Ni was also identified. This suggests that the development of secondary Fe-rich products can partially contribute to the reduction of heavy metals release to the surrounding environments. However such phenomena may have inhibitory effect on the utilization of bottom ash as recycled aggregates.

Keywords: Environment; Heavy metals; Landfill; MSW incineration bottom ash; Natural Weathering; Secondary Fe-rich products

Introduction

Incineration technology has been adopted as an effective strategy for the treatment of municipal and hazardous wastes in many communities at different scales with highest share among the developed countries. This treatment technique results in the generation of various residues including bottom ash and fly ash as the major solid outputs totally ranging from 4-10% by volume and 15-20% by weight of the original quantity of waste [1]. They predominantly consist of glassy phase and significant concentration of hazardous components comparing with the source materials. Among MSW incineration products, bottom ash is the most significant by-product that accounts for 85 to 95% of all the residues produced in the course of combustion [2].

Municipal solid waste incineration (MSWI) bottom ash is considered as biologically and chemically reactive residues whose long-term behavior and evolution have been comprehensively studied by several workers over the past two decades [3-6]; therefore its efficient management always remains one of the most controversial environmental topics. Among different treatment techniques, natural ageing and weathering has been considered as one of the most cost-effective methods of treatment for the chemical stabilization of MSWI bottom ash. Many researchers have been investigating the natural or accelerated aging of bottom ash, a process which could also be applied to fly ash disposal [7-10].

In the incineration bottom ash a variety of primary iron-rich phases are present that behave differently when influenced by natural weathering processes. These phases may originate from the waste stream (e.g. metal iron and Fe-rich minerals) or may be formed in the course of incineration (e.g. magnetite spinels and metal inclusions). In either case, such phases become unstable and convert to secondary (or tertiary) products when exposed to natural weathering. The iron-related reactions and products may trigger problems by generating unnecessary heat, gases, and leachate or cause staining when bottom ash is used as cement aggregate or when placed in the landfill.

These phenomena initiate immediately upon bottom ash characterization and persist for unlimited duration. To ad-dress such problems, a thorough understanding from the properties of the primary and secondary Fe-rich phases in the ash products is a requirement. The main goal of the current research was to study both fresh and weathered bottom ash with emphasis on the characterization of secondary (neo-formed) Fe-rich products in the weathered samples of a MSWI (mono) landfill site in the north east of the US. The impact of natural weathering on the behavior of Fe-rich phases, their alteration, and the formation of the relevant secondary products was systematically investigated. In the meantime, the strong affinity of these secondary phases with heavy metals of environmental significance such as Zn, Cu, Pb, and Ni was also identified. This paper presents one major task of a large multipurpose project by specifically focusing on the characterization of secondary Fe-rich species in the landfilled incineration ash residues.

Iron in the Incineration Residues- An Overview

Iron is one of the big 8 elements in the Earth's crust, being the fourth most abundant element (after oxygen, silicon and aluminum) at about 5% by weight [11]. Thanks to the combination of low cost and high strength, iron is the most used of all the metals. Iron is a very good conductor of both heat and electricity. It is also strong, ductile and malleable. Therefore, it is extensively mined, used and recycled throughout the world across an infinite variety of domains, from home to industry.

Metal iron may originate from various sources including household and municipal waste, curbside trash, industrial solid waste, and iron production waste. As a result of extensive utilization of iron in daily life, it is expected that a considerable fraction of this element enters the waste course and eventually incinerators as non-separable end-of-life material. Non-metallic iron also exists in a variety of inorganic components such as fine glass, ceramic, stone and mineral particles inherited from the waste stream that partially supply iron to the incineration residues. Hence, it is considerably enriched in the MSWI bottom ash ranging from 5-15% [2] that is comparable with its crustal abundance as mentioned earlier. Metal iron and iron-bearing components of the waste convert to primary phases during the incineration and quenching processes that can alter to the secondary Fe-hydrate products (chiefly amorphous) and finely crystalline Fe-rich phases as a result of ageing or natural weathering phenomena. Formation of the secondary phases may continue or be inhibited upon ash disposal or recycling.

Materials and Methods

A field survey was conducted in September 2009 in order to collect naturally weathered MSWI bottom ash of different ages from different depths of Franklin (mono) landfill site in the state of New Hampshire, United States. Sampling was conducted within three consecutive days by excavating the deposited ash at four distinct locations A (1 yr), B (10 yrs), C (13–14 yrs), and D (20 yrs) (Figure 1). At each location, samples were collected from six levels- at depths 0, 0.5, 1, 2, 3, and 4 m for locations A, B, C and 0, 0.5, 1, 1.5, 2, and 3 m for location D; all hereinafter referred to as points. Therefore, totally 24 sets of samples were collected. Several parameters including temperature, Eh and pH were measured on site at each point. In order to identify the characteristics of the primary Fe- rich minerals, we collected fresh (intact) bottom ash from three incineration plants in the US and Japan as well.

The collected ash samples from all sources were mixed manually, dried at ambient temperature and made into respective sub-samples using coning and quartering technique [12]. Sub-samples were sieved using a 2 mm screen in the laboratory to separate out the coarser grain intact particles from the fine-grained fraction. Several intact bottom ash particles were carefully picked out of the two size fractions for standard petrographic polished thin section preparation.

The thin sections were finely polished by diamond paste of 1 μm for both polarized light and electron microscopy. They were observed through different optical modes (plane polarized light-PPL, cross polarized light-XPL, and reflected light-RL). Scanning electron microscopy coupled with quantitative energy-dispersive X-ray microanalysis (SEM-EDX) was performed for detailed, high-resolution grain-specific mineralogical and chemical data of various samples. With the emerging high-performance software technologies SEM-EDX has become a standard method for the characterization and semi-quantitative analysis in various disciplines including material, mineral, environmental and biological sciences [13-15].

Backscattered electron (BSE) and characteristic X-ray images were taken by a scanning electron microscope, and the qualitative and qu0061ntitative spot analyses of the samples were conducted under accelerating voltages of 15-25 kV, the working distance (WD) of 17 mm, a beam current of 3 to 5 nA, and a probe size of 5 in a high vacuum atmosphere with spectral acquisition time of 100-300 s at desired magnifications.

Selected samples were pulverized for bulk analyses. Semi-quantitative bulk chemical analyses of the powdered samples were completed by using a Rigaku RIX3100 X-Ray fluorescence spectrometer (XRF) [16]. Powder X-Ray diffraction (XRD) analysis was also practiced in a Rigaku Multiflex diffractometer using CuKα radiation at 30kV voltage and 40 mA current in order to identify the existing crystalline phases (both primary and secondary) [17].

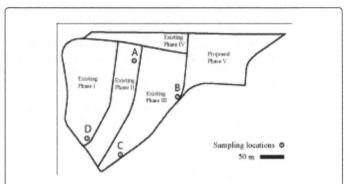

Figure 1: A schematic of the Franklin mono-landfill site representing the sampling locations (A, B, C and D).

Results and Discussion

Primary iron-rich constituents of MSWI bottom ash

MSWI bottom ash may contain a considerable amount of metal-rich phases that are essentially made up of iron, aluminum and copper. These elements (Fe and Al in particular) are concentrated in a variety of phases in the final ash products. Based on our microscopic and micro analytical experiments, it is possible to divide the metal-rich constituents of bottom ash into two groups: 1) metallic fragments (up to 3 vol%) originated from the waste source, and 2) primary metal-rich minerals (up to 7 vol%) that have been formed during the incineration process [18]. The first group is the remnant of larger metallic scraps typically with irregular shape ranging from several microns to a fraction of millimeter in size set in the bottom ash particle. Although they are essentially made up of Fe, Al and Cu alloys, they may enclose minor amounts of other metals and non-metals (Pb, Sb, Ni, Si, S, P etc). Such waste-derived metals assume to have been partially melted and contributed to the formation of the glassy matrix and primary metal-rich phase (second group) of the ash particles.

The members of the second group can be subdivided into metallic minerals (particularly magnetite spinels) and metal-rich inclusions. The magnetite spinel family is the most abundant primary iron oxide minerals within the incineration melt glass products. They present well-developed crystalline habits as well as irregular dendritic shape that are indicative of melt super-cooling and rapid quenching (Figure

2a). These minerals have been shaped up as tightly-packed clusters of euhedral to subhedral microcrystals in the silicate glass matrix of the ash. The metal-rich inclusions as the other member of the second group are commonly found as discrete phases that are indicative of complete immiscibility with their host silicate glass (Figure 2b).

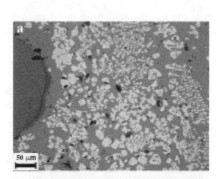
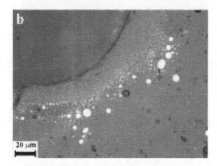

Figure 2: a) Microcrystals of magnetite spinel as a primary iron-rich phase (light gray) set within the glassy matrix (dark gray) of the incineration ash- RL; b) The spherical metal-rich inclusions (bright) as a primary iron-rich phase embedded in the glassy matrix (dark gray) of the incineration ash- RL.

Their sizes vary from submicron to several microns and are sphere-shaped. Metallic iron (up to 90%) and copper (up to 10%) are the major components of the metal-rich inclusions; however some other elements such as P, Si, Pb, Zn, Sb, Sn, Ni, and S might also be present at variable but lower concentrations.

Alteration of iron particles and interconversion of iron-rich phases are the most dominant weathering phenomena particularly in the older ash deposits of the landfill leading to the formation of secondary oxide and hydroxide iron species. Metallic minerals (spinels) and metal-rich inclusions display somewhat identical behavior as documented through the present study.

Occurrence of Secondary Iron-rich Products

Oxidation of metallic iron and iron-rich minerals at the presence of water and dissolved oxygen extensively occurs in nature or through industrial activities at broad Eh-pH ranges. The weathering processes convert the primary iron-rich phase to hydrate/oxide products through hydration/oxidation at different rates. As mentioned earlier, a variety of primary iron-bearing substances are present in the incineration bottom ash that may behave differently under the prevailing ash alkaline environment. Following our comprehensive examinations, the primary iron phases have been altered to a combination of secondary products including goethite partly with

lepidocrocite (as hydrate phases), magnetite and hematite (as oxide phases), and several intermediate phases mixed with silicate and carbonate species.

Phases such as goethite and lepidocrocite are normally formed under oxidizing conditions as weathering products of iron-bearing minerals such as siderite, magnetite, pyrite, etc [19] through inorganic-chemical or organic precipitation from various kinds of Fe-bearing solutions. The Fe-hydrates are the most dominant phases present in various zones of the landfill. Figure 3 shows the bulk iron concentration in the sampling locations (A through D); each column corresponds to the average iron concentration of all six sampling points in each location. Figure 4 presents the stability fields of several Fe-rich compounds as a function of redox potential (Eh) against pH. The Eh and pH data obtained from every sampling point in the field was plotted in the diagram. The entire population is distributed in a linear arrangement and presents a broad range of pH from weakly to strongly alkaline and a narrow range of positive redox potential (oxic). Such environment theoretically favors the stability of hydrate iron compounds.

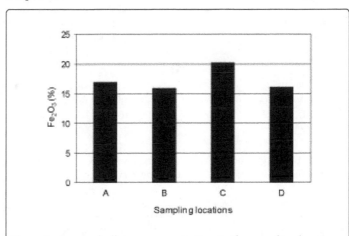

Figure 3: Average bulk iron concentration in the sampling locations (A to D).

An obvious trend in the transformation of primary Fe-bearing substances into secondary hydrate products and their distribution could not be delineated in the landfill, though evidences are indicative of higher concentrations of newly-formed Fe-rich species in the older ash deposits. Extensive formation of hydrate Fe-rich minerals, particularly goethite, can be corresponded to the weak to strong alkaline conditions (7.5- 11.5) and relatively oxidizing environment (average Eh of 0.3 V in locations A-C) (Figure 4). Under such conditions, most metallic iron and Fe-rich minerals are expected to become destabilized leading to the incipient formation of thermodynamically more stable hydrate Fe-rich species (mostly amorphous).

The oldest ash deposits (location D) is the richest zone with respect to the accumulation of Fe-rich hydrate phases. Although the average Eh in this location ($\sim$ 0.2 V) is slightly below the other locations, the higher average pH ($\sim$ 11) at all sampling depths of location D might have accelerated the transformation reactions. Schwertmann and Murad [20] suggest that the formation of goethite is strongly pH-dependant and goethite is the only phase to form at pH 12. Highly alkaline environment (and oxic condition) also favors the conversion of primary magnetite (Fe_3O_4) to goethite phase as confirmed by He and Traina [21]. In addition, the role of burial time as another important factor should not be overlooked in the completion of such processes in location D. The iron hydroxide precipitates present a broad variety of morphologies such as massive, crusty, fibrous, fanshaped (Fig. 5a), twisted (Fig. 5b) and convoluted habits. Such associations are usually amorphous or poorly crystalline. In some cases, the present crystals are aligned as parallel filaments or criss-crossed lamellar intergrowths.

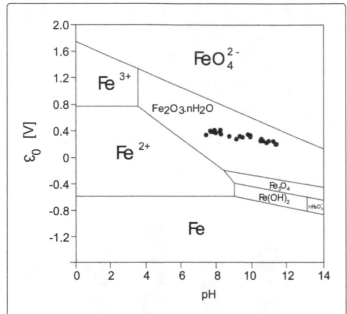

Figure 4: Eh-pH diagram of Fe-rich components at 25°C and 1 bar. Plots of the field-collected data display a linear arrangement in the diagram.

Although the Fe-hydrates were the most dominant phases in our landfill system, secondary magnetite (and partially hematite) are also present but at lesser quantities in several samples particularly from older locations C and D. Magnetite is a common mineral formed at elevated temperatures and pressures in igneous and metamorphic rocks; almost no magnetite is known to be formed inorganically in the biosphere [22]. It is therefore assumed that they have been formed in the landfill through the (biologicallycontrolled) dehydration/reduction of the secondary Fe-hydrate products. They have grown as individual euhedral to subhedral microcrystals (box-shaped or polygonal face) or aggregates of numerous crystals presenting linear or spherical orientations (Fig. 5c). Such morphologies display similarities with biologically synthesized magnetite from other works [23-24].

There were also evidences from the conversion of massive goethite to euhedral microcrystalline magnetite in a number of samples (Fig. 5d). This may indicate a local shift to more reducing condition by the solutions that locally transport ferrous iron in the landfill system at the scale of microenvironment that can be interpreted by the reaction no. 1. However there might be other alternatives for the production of Fe-rich species such as goethite, lepidocrocite, ferrihydrite, hematite and even magnetite. These species can all be produced from ferrous iron solutions by the oxidation of iron followed by hydrolysis via the basic reaction no.2 [25]. This reaction can be tuned in a manner to be consistent with the formation of various iron species.

$$2FeOOH + Fe^{2+} \rightarrow Fe_3O_4 + 2H^+ \quad (1)$$

$$2Fe^{2+} + 2H_2O \rightarrow 2Fe_3+ + H_2 + 2OH^- \quad (2)$$

It should be noticed, however, that not all the primary Fe-rich phases have been affected by the dominating environmental condition in the ash deposits, since a significant amount of them have wholly or partly remained intact. This might have been resulted from variations in the chemistry of both host solid and percolating solutions at microscopic scale that has locally influenced the exchange reactions. The melt glass phase also protects a considerable amount of metal inclusions from weathering, although the end stage of the glass durability is not yet known. In addition, as observed in several occurrences, the newly-formed phases might have produced stable oxide-hydroxide films over the unstable phases as a barrier to the transport of reacting species. Cornell and Schwertmann [25] provide reasons for the decelerated corrosion rate of metallic iron based on recent opinion under similar condition. The extremely heterogeneous components of bottom ash particles and the irregular distribution of microfracturing networks are amongst other major reasons that can partly justify the disproportionate alteration of primary Fe-rich substances.

Entrapment of Heavy Metals

In order to evaluate the influence of the secondary Fe-rich phases on the entrapment (scavenging) of environmentally hazardous metals, over 35 points from those phases in the polished thin sections of various samples were analyzed by EDX of which 24 sets of data with higher accuracy were selected. The data are indicative of a wide range of compositional variations in the secondary Fe-rich species due to the presence of cations such as, Si, Ca, Al and other elements in the system; thereby a standard chemical formula cannot be reproduced for them. In most cases, however, based on oxygen concentration and the total amount of major cations (Σ Fe, Si, Ca, Al) - aided by microscopic observation and XRD results- it was possible to propose several non-stoichiometric compositions close to goethite (lepidocrocite), hematite, ferrous iron oxide, magnetite and inter-mediate versions for the existing species. Goethite is the best example of an isomorphously substituted iron oxide with maximum observed caption substitution [25].

From the current research, it was distinguished that the newly-formed Fe-rich phases play substantial role in the entrapment of most available heavy metals (in particular Cu, Ni, Zn and Pb). This might have come into effect through a series of mechanisms including co-precipitation, complexation, adsorption and ion exchange with the solutions. Cu shows the highest affinity with the secondary Fe-rich phases- it was detected at almost 70% of the measurements (17 points) at moderately considerable quantities (0.31-5.14 wt %). The amount of Pb enriched in the secondary Fe-rich phases is particularly high (0.82-9.91wt %); although it was not frequently detected in such phases through our measurements. Correspondingly, the amount of Ni concentrated in these phases was about 0.13-0.9 wt%, and that of Zn varied between 0.48-1.15 wt% at various analytical points.

Based on the microanalytical data, it is possible to suggest the adsorption selectivity of heavy metals in the following order: Cu>Ni >Zn>Pb onto the Fe-rich oxyhydroxides.

The problem, however, is that we do not know what portion of the total heavy metals inventory in the MSWI ash is bound to active sorption sites in equilibrium with the landfill leachate.

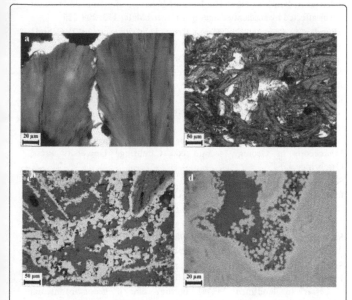

Figure 5: Photomicrographs of secondary Fe-rich products in the landfilled MSWI bottom ash formed during the weathering processes; semitranslucent orange-reddish fan-shaped (a) and fibrous (b) goethite from location D and C of the landfill, respectively- transmitted light (PPL), euhedral to subhedral microcrystals of magnetite- RL (c), and conversion of goethite to euhedral magnetite crystals (center) -RL (d). Both images "c" and "d" are from the location C of the landfill

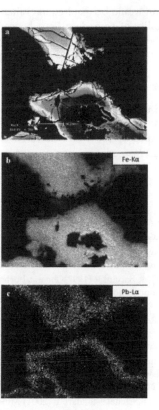

Figure 6: BSE image of a secondary iron hydrate aggregate (goethite) (a), and characteristic X-ray images of Fe, and Pb (b and c) from location B of the landfill. The areas with higher heavy metal concentration appear brighter in the image "c".

The SEM-EDX imagery technique was very useful for identifying the distribution pattern and elemental mapping of heavy metals in the secondary phases. Fig. 6a displays the backscattered electron (BSE) image of a secondary Fe-hydrate product (goethite) from the location B of the landfill. Figures 6b and 6c are the characteristic X-ray images of the zones that are rich in Fe (densely concentrated) and Pb, respectively. The heavy metals possess stronger effect in backscattering the electrons and therefore the dense population of brighter spots represents higher concentration of heavy metals particularly in the margins (Figure 6c).

Conclusions

We designed a systematic approach in order to evaluate the inorganic chemistry and mineralogical properties of MSWI bottom ash products when exposed to natural weathering in a full-scale MSWI mono (landfill) site. These products consist of a variety of glassy/crystalline components including primary Fe-rich phases that may undergo alteration when exposed to natural environment. The current research focused on the importance and role of mineralogy as a practical tool in the identification of newly-formed (secondary) Fe-rich products upon landfilling.

Samples of various ages (0–20 yrs) were collected from four locations (A to D) of the landfill site. The novelty of the current research is that it could confirm the formation and transformation of secondary Fe-rich phases and their role in the entrapment of heavy metals using direct microscopic observation and micro beam measurements. In the meantime, bulk analytical techniques including XRD and XRF were employed as useful complementary methods.

The techniques presented here are directly applicable to a broad range of combustion residues from other sources. We summarize our achievements into the following categories:

- Depending on the prevailing physicochemical conditions in an incineration ash landfill, alteration reactions may initiate and persist for days or for years.
- The alkaline and oxic conditions in the landfill trigger the formation of various Fe-rich oxyhydroxides.
- Several secondary products (amorphous or crystalline) were identified, including goethite (α-FeOOH), lepidocrocite (γ-FeOOH), hematite (Fe_2O_3), magnetite (Fe_3O_4), iron oxide (FeO), and Fe-bearing Ca-Si and Ca-Al-Si gel phases that have been developed under variable environmental conditions as the weathering products of primary iron-rich phases.
- The formation rate of the secondary Fe-rich constituents is considerably time-dependent. They are more dominant in the older ash deposits particularly in the locations C and D of the landfill.
- Such altered phases present different stability fields and different behaviors with respect to heavy metal entrapment capacity.
- A strong correlation between the secondary Fe-rich products and heavy metals of environmental significance such as Zn, Cu, Pb, and Ni was also identified. This suggests that the development of such products partially contribute to the reduction of heavy metals release to the landfill leachate.

- Further investigation would be required to evaluate the formation condition and stability of secondary Fe-rich products, and their impacts on both the long-term stabilization of heavy metals and the utilization potential of bottom ash as recycled aggregates.

Acknowledgement

Financial support to this project was provided through a Grant-in-Aid for Japan-USA joint research project (approved by Gakushinkyo-2-9 on April 1. 2010) and Research Grant-in-Aid for Sustainable Society and Waste Management funded by the Ministry of Environment (K22078).

References

1. RenoSam, Rambøll (2006)The most efficient waste management system in Europe. Waste-to-energy in Denmark, Technical report.

2. Chandler AJ, Eighmy TT, Hartlén J, Hjelmar O et al. (1997) Municipal solid waste incinerator residues, Amsterdam: Elsevier Science B.V.

3. Belevi H, Stämpfli DM, Baccini P (1992)Chemical behavior of municipal solid waste incinerator bottom ash in monofills, Waste Manage.Res.10: 153-167.

4. Zevenbergen C, Reeuwijk LP, Bradley JP, Comans RNJ et al. (1998)Weathering of MSWI bottom ash with emphasis on the glassy constituents. J. Geochem. Explor 62: 293-298.

5. Chimenos JM, Fernández AI, Miralles L, Segarra M, Espiell F (2003) Short-term natural weathering of MSWI bottom ash as a function of particle size. Waste Manag 23: 887-895.

6. Piantone P, Bodénan F, Chatelet-Snidaro L (2004) Mineralogical study of secondary mineral phases from weathered MSWI bottom ash: implications for the modeling and trapping of heavy metals. Appl. Geochem 19:1891-1904.

7. Meima JA, Comans RNJ, (1997) Geochemical modeling of weathering reactions in municipal solid waste incinerator bottom ash. Environ. Sci. Technol 31:1269-1276.

8. Meima JA, Comans RNJ (1999) The leaching of trace elements from municipal solid waste incinerator bottom ash at different stages of weathering. Appl. Geochem 14: 159-171.

9. Marchese F, Genon G (2009) Full scale tests of short-term municipal solid waste incineration bottom ash weathering before landfill disposal. Am. J. Environ. Sci 5: 569-576.

10. Polettini A, Pomi R (2004) The leaching behavior of incinerator bottom ash as affected by accelerated ageing. J Hazard Mater 113: 209-215.

11. Abundance of Elements in Earth's Crust, HyperPhysics (2012) Georgia State University.

12. IUPAC (1990) Analytical Chemistry Division, Commission on Analytical Nomenclature. "Nomenclature for sampling in analytical chemistry. Pure Appl. Chem. 62: 193–1208.

13. Kang E, Park I, Lee YJ, Lee M (2012) Characterization of atmospheric particles in Seoul, Korea using SEM-EDX. J Nanosci Nanotechnol 12: 6016-6021.

14. Reed SJB (2005) Electron Microprobe Analysis and Scanning Electron Microscopy in Geology, second ed., Cambridge University Press, Cambridge, UK.

15. Schatten H (2013) Scanning Electron Microscopy for the Life Sciences, first ed., Cambridge University Press, Cambridge, UK.

16. Beckhoff B, Kanngießer B, Langhoff N, Wedell R, Wolff H (2006) Handbook of Practical X-Ray Fluorescence Analysis, Springer.

17. Suryanarayana C, Norton MG (1998) X-ray Diffraction: A Practical Approach. Microsc Microanal 4: 513-515.

18. Saffarzadeh A, Shimaoka T, Wei Y, Gardner KH, Musselman CN (2011) Impacts of natural weathering on the transformation/neoformation processes in landfilled MSWI bottom ash: a geoenvironmental perspective. Waste Manag 31: 2440-2454.

19. Deer WA, Howie RA, Zussman J (1980) An Introduction to the Rock Forming Minerals, London, UK: Longman Group Limited.

20. Schwertmann U, Murad E (1983) Effect of pH on the formation of goethite and hematite from ferrihydrite. Clay Clay Miner 31: 277-284.

21. He TY, Traina SJ (2004) Transformation of magnetite to goethite during Cr(VI) reduction under alkaline pH conditions., 227th American Chemical Society Meeting. Anaheim, CA, 446-449.

22. Lowenstam HA (1981) Minerals formed by organisms. Science 211: 1126-1131.

23. Amemiya Y, Arakaki A, Staniland SS, Tanaka T, Matsunaga T (2007) Controlled formation of magnetite crystal by partial oxidation of ferrous hydroxide in the presence of re-combinant magnetotactic bacterial protein Mms6. Biomaterials 28: 5381-5389.

24. Arakaki A, Nakazawa H, Nemoto M, Mori T, Matsunaga T (2008) Formation of magnetite by bacteria and its application. J R Soc Interface 5: 977-999.

25. Cornell RM, Schwertmann U (2003) The Iron Oxides: Structure, Properties, Reactions, Occurrences and Uses, Wiley-VCH.

19

Newly-Isolated Laccase High Productivity *Streptomyces* Sp. Grown In Cedar Powder as the Sole Carbon Source

Akihisa Aoyama[*], **Kazuhiro Yamada, Yoshinobu Suzuki, Yuta Kato, Kazuo Nagai and Ryuichiro Kurane**

Department of Biological Chemistry, College of Bioscience and Biotechnology, Chubu University, Japan

[*]**Corresponding author:** Akihisa Aoyama, Department of Biological Chemistry, College of Bioscience and Biotechnology, Chubu University, 1200 Matsumoto-cho, Kasugai, Aichi 487-8501, Japan
E-mail: ap1s2000@gmail.com

Abstract

Microorganisms with greater potential to degrade lignin than well-known white-rot fungi were sought and identified, but the fungi were less frequently employed for slow growth and little enzyme productivity. They were subjected to enriched cultures in order to explore the bacteria instead from 300 soil samples with cedar powder as the sole carbon source were prepared. From these, a culture with actinomycetes which showed the most oxidation activity of 2,6-Dimetoxyphenol (2,6-DMP) known as laccase substrate was selected and labeled as KS1025A strain. Characteristics of the bacteria and behavior of the secreted enzymes were examined. As a result, it was identified as a strain of *Streptomyces* sp. from the 16S rDNA gene sequence homology. The optimum temperature and pH for laccase activity of the secreted enzyme of this strain are 50°C and 4.5, respectively. Since Mn^{2+} was not directly oxidized, it was assumed that it did not contain manganese peroxidase. However, when $MnSO_4$ was added during 2,6-DMP oxidation reaction, activity increased. After 120 hours of culture, 14 U/mL of laccase activity could be achieved by this strain, greatly exceeding known values by white rot fungus, namely 1.8U/mL after approximately 20 days. Furthermore, since reaction could continue without the addition of H_2O_2 during 2,6-DMP oxidation reaction, the culture solution is thought to contain free oxidizing agents. In addition, approximately 50% of 0.05% lignin sulfonic acid was decolorized by this strain in 5 days. The strain or the enzyme produced by it may be utilized for rapid biodegradation of lignin when adding hard (or soft) biomass containing lignin to produce bioethanol.

Keywords: Laccase; Streptomyces; Lignin; Bioethanol; Peroxidase; Lignosulfonate

Introduction

In recent years, effective use of wood-based (or grass-based) plant waste material (wood, bagasse, etc.) for biomass has been studied. For example, cellulose/hemicellulose within the plant material comprises approximately 50% of its dry weight, and by hydrolyzing, sugars such as glucose, etc. can be extracted [1-3]. The obtained sugars can be used to produce bioethanol, etc [4] through fermentation. However, plants generally contain very strong polymer fibers which are formed by bonding cellulose fibers or hemicellulose fibers with lignin to maintain their shape. Although, lignification by lignin is advantageous when these plants are used as construction material; when used as biomass, this is an obstacle. Since various problems arise from the standpoint of the environment and processing costs, such as the need for alkali or acidic degradation or mechanical grinding, the use of plant material is limited [5]. Therefore, reduction of processing costs is an important issue [6,7].

In addition, if lignin degradation is carried out chemically, phenol derivatives are generated, inhibiting the hydrolyzation of cellulose by enzymes [8] and subsequent fermentation of sugars gained by hydrolyzation [9].

On the other hand, plants also serve as important storage sites for organic matter in the biosphere. Some natural microorganisms can degrade and utilize such organic matter as a source of carbon and energy. In particular, white-rot fungi degrades lignin by extracellularly secreting enzymes [10,11] and are considered to play an important role in carbon circulation in the ecosystem. However, enzyme production requires approximately 20 days in culture [12,13] and is difficult to use. Hence, focusing on lignin degradation by extracellular enzymes from fast-growing microorganisms, microorganisms with a high ability to degrade lignin, excluding filamentous fungi, were sought.

In addition, among enzymes secreted extracellularly from white-rot fungi, lignin peroxidase (LiP) [14,15], manganese peroxidase (MnP) [16], and laccase [15,17-19] are known to be associated with lignin degradation. In this study, enzyme activity of newly-isolated actinomycete strains and that of white-rot fungi were comparatively considered.

Materials and Methods

Enriched culture

Two hundred milligrams of soil sample was placed in a microtube and suspended in 300 μL of added sterile saline solution. After standing for one-minute, 50 μL of the supernatant was added to 500 μL of screening medium (composition as below) with cedar powder as the sole carbon source. The medium was covered with a gas-permeable plate seal and cultured for 2 weeks at 150 spm. After 2 weeks, 50 μL of the cultured medium was transferred to a new medium of 500 μL and was further cultured. This process was repeated four times and the medium was subcultured four times.

After subculturing for the fourth time for 1 week, the medium was centrifuged at 4725 g for 10 minutes at room temperature. The supernatant was extracted as an extracellular crude enzyme liquid.

Approximately 300 soil samples were used as enriched cultures.

Screening medium compositions

The following composition was sterilized at 121°C for 15 minutes; 20 g/L cedar powder, 10 g/L $(NH_4)_2SO_4$, 10 g/L $NaNO_3$, 5 g/L KH_2PO_4, 1 g/L K_2HPO_4, 1 g/L $MgSO_4 \cdot 7H_2O$, 1 g/L NaCl, 1 g/L yeast extract, and 0.5 g/L $CaCl_2 \cdot 2H_2O$.

In addition to the above composition, 15 g/L of agar was added to the screening plate medium.

Lignosulfonate (a water soluble lignin derivative [20]) decolorization

For the lignosulfonate medium, 0.5 or 1 g/L of lignosulfonate was added instead of cedar powder. One loop was taken from the PD medium with the preserved strain, inoculated into 5 mL of ISP2 medium (composition g/L: 4 yeast extract, 10 malt extract, 4 glucose), and precultured at 28°C, 150 spm, for 48 hours.

One hundred microliter of preculture liquid was inoculated into 10 mL of lignosulfonate and cultured at 28°C, 150 spm, for 120 hours. Next, the culture was centrifuged for 10 minutes at 12000g, and A480nm was measured for the supernatant, which is the maximum absorption amount in the visible range of lignosulfonate.

Strain separation and preservation

The culture fluid was appropriately diluted, spread on the screening plate medium, and cultured statically at 28°C for up to 2 weeks. A single colony was transferred to the PDA medium as the preserved strain. PDA medium: potato dextrose agar medium (hereinafter referred to as "PDA medium").

Identification of the strain by 16S rDNA gene sequencing

A platinum loop was taken from the colony grown on the PDA medium and extracted by a DNA extraction kit, ultra clean microbial DNA Isolation kit (12224-50, MO BIO Laboratories, Carlsbad, CA, USA), as a template. As primers, 27f (5'-AGAGTTTGATCMTGGCTCAG-3') [21] and 519r (5'-CWATTACCGCGGCKGCTG-3') [22] were used. Sequencing was implemented from the template DNA, and the gene sequence was identified. The homology of gene sequence was searched in DNA Data Bank of Japan (DDBJ) [23]. In addition, the collected sequence data were multiple aligned by ClustalW in DDBJ.

The neighbor joining method was used to reconstruct the phylogenetic tree [24].

Enzyme production

The preserved strain was inoculated into 500 μL of screening medium, and cultured at 30°C, 110 spm for 120 hours, and the medium was centrifuged for 10 minutes at 4,725 g. The supernatant was extracted as the extracellular enzyme.

Measurement method of laccase activity

An enzyme liquid mixture containing 60 μL of enzyme liquid, 10 μL of 20 mM 2,6-dimetoxyphenol (2,6-DMP), 10 μL of 20 mM $MnSO_4$, 120 μL of 100 mM malonic Na buffer fluid (pH 4.5), and 10 μL of 2 mM H_2O_2 was added to a 96-well microplate and allowed to react for 10 minutes at 45°C, after which A469nm was measured [25,26].

Enzyme activity is shown in Units (U). One Unit is the enzyme level at which 1 μmol of substrate oxidizes in 1 minute.

2,6-DMP (ε_{469} = 49,600 $M^{-1}cm^{-1}$) [25]

Optimum pH measurement method

An enzyme liquid mixture containing 60 μL of enzyme liquid, 10 μL of 20 mM 2,6-Dimetoxyphenol (2,6-DMP), 10 μL of 20mM $MnSO_4$, 120 μL of 100 mM sodium malonate buffer (3.0, 3.5, 4.0, 4.5 5.0, 4.5, 5.0, 5.5 and 6.0), and 10 μL of 2mM H_2O_2 was added to a 96-well microplate and allowed to react for 10 minutes at 45°C, after which A469nm was measured.

Optimum temperature measurement method

An enzyme liquid mixture containing 60 μL of enzyme liquid, 10 μL of 20 mM 2,6-Dimetoxyphenol(2,6-DMP), 10 μL of 20 mM $MnSO_4$, 120 μL of 100 mM sodium malonate buffer (pH 4.5) was added to a 96-well microplate and pre-heated 10 minutes, and 10 μL of 2 mM H_2O_2 was added to a 96-well microplate and was allowed to react for 10 minutes various temperatures (30, 40, 50 and 60°C), after which A469nm was measured immediately.

Measurement method of lignin peroxidase activity

An assay was carried out at pH 3.0 using 0.1 M sodium tartrate buffer and 0.01 mM veratryl alcohol as the substrate. Reaction was initiated by adding 4 mM H_2O_2 to 100 mM sodium lactate buffer pH 4.5, 0.1 mM H_2O_2, and 0.5 ml culture supernatant and increased absorbance at 310nm could be observed [27].

Measurement method of Manganese (II) peroxidase activity

Manganese (II) peroxidase activity was identified by observing the formation of a Mn^{3+}-malonate complex at 275 nm [28]. The reaction mixture contained 0.5 ml of 10 mM Mn^{2+} in 100 mM sodium maloate buffer pH 4.5, 0.1 mM H_2O_2 and 0.5 ml culture supernatant.

Enzyme protein concentration measurement

Measurement of enzyme protein concentration was carried out by the Bradford method [29], using the Bradford Protein Assay kit (Bio-rad Laboratories, Atlanta, GA, USA) with BSA as the control.

Enzyme concentration

The enzyme was concentrated to 10 times using a centrifugal ultrafiltration filter, Vivaspin 20-3K (GE Healthcare Bio-Sciences AB, Uppsala, Sweden).

SDS-PAGE and Zymogram method

SDS-PAGE and Zymogram analysis was conducted with 8% acrylamide gel using a modified method of Díaz et al. [13]. The protein of the gel was embedded and denatured by SDS for 10 minutes at 45°C in the presence of a reducing agent.

After electrophoresis at 20 mA, the gel was placed in a cleaning solution (0.1% Triton X and 100 mM sodium acetate buffer, pH5.0). The cleaning solution was replaced after 15 minutes. Cleaning was carried out three times. The gel was transferred to a substrate solution (18 mM 2,6-DMP, 0.2 mM H_2O_2, 100 mM Acetate Na Buffer pH 5.0), allowed to react for 10 minutes at 45°C to color, and was then scanned. Next, the gel was placed to deionized water which was replaced after 15 minutes. The gel was cleaned three times with deionized water, stained by the CBB staining fluid, Gelcode Blue Stain Reagent (Thermo Fisher Scientific, Waltham, MA, USA), and then scanned.

Results

Screening

Three hundred types of soil samples were individually added to a screening medium to form an enriched culture. As a result, a high level of 2,6-DMP oxidation activity could be observed by enzymes in the enriched cultured with a soil sample collected from Aichi, Japan. Laccase activity of the sample could progress without the addition of H_2O_2.

Isolation of the KS1025A strain

The active culture fluid was appropriately diluted and spread on a screening plate medium, and an actinomycete-like colony could be subsequently obtained. This strain was designated as KS1025A. After it was preserved, the enzyme was measured, and a high level of laccase activity as illustrated in Figure 1 could be observed. The enzyme of this strain was active even without the addition of $MnSO_4$ and H_2O_2, but activity increased with such addition. The assay of the Mn^{3+}-malonic acid complex indicated oxidation activity from Mn^{2+} to Mn^{3+} (as manganase peroxidase activity). However, although veratryl alcohol oxidation (as lignin peroxidase activity) was examined, no activity was observed (data not shown). Hence, Laccase is thought to be the main enzyme group.

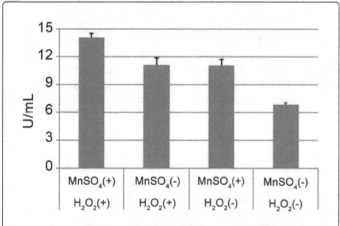

Figure 1: Laccase activity by KS1025A extracellular enzymes, Laccase activity in a pH4.5 sodium malonate buffer of KS1025A culture supernatant is shown. For $H_2O_2(+)$, 0.2mM H_2O_2 was added. The bars indicate the standard deviations of three samples.

After culturing the KS1025A strain in a PDA medium at 28°C for 7 days, a wrinkled, white to grey colony formed in 7 days. Mycelium was verified by microscope observation, but sporulation was not observed after a month of culture.

By identifying the 16S rDNA gene sequence of this strain, homology with *Streptomyces atratus* was determined to be 97.39% and a phylogenetic relationship could be shown (Figure 2).

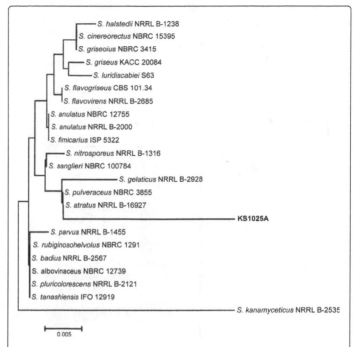

Figure 2: Phylogenic relationships of the isolated strains and some related species based on 16S rRNA gene-sequences. The neighbor joining method was used to generate branching pattern. The bar indicates 0.05 nucleotide substitutions per site. The sequence data were cited from DDBJ.

Enzymological properties of enzymes produced by KS1025A

The optimum temperature and pH for laccase activity are 50°C and pH 4.5 respectively, as shown in Figure 3.

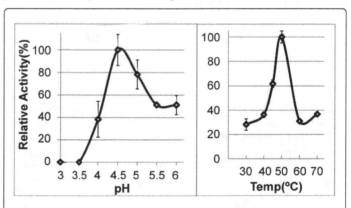

Figure 3: The optimum temperature and pH for laccase activity of KS1025A strain enzyme. Left: The relative activity when 2,6-DMP is oxidized in malonic acid buffer at various pHs, where the optimum pH is centered. Right: The relative activity when 2,6-DMP is oxidized in sodium malonate buffer at pH4.5 at various temperatures, where the optimum temperature is centered. The bars indicate the standard deviations of the two samples.

The zymogram image is shown in Figure 4A, where 2,6-DMP oxide turned yellow. From the culture supernatant of KS1025, two large bands could be verified approximately between MW>175 kDa and MW 90 kDa. The CBB stained image is shown in Figure 4B. A large band could be seen at approximately 42 kDa for the KS1025A enzyme, but a band could barely be seen in the area where a strong reaction could be verified by Zymogram. From these observations, KS1025A derived Laccase is thought to be an enzyme with highly specific activity.

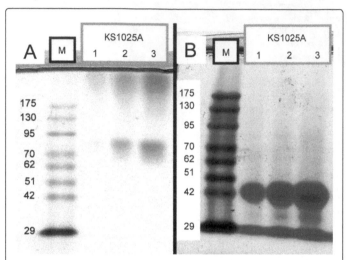

Figure 4: Zymogram image and CBB stained image of KS1025A secreted enzyme. Quantity of migration protein. KS1025A; Lane 1:1.25 μg, lane 2:2.5 μg, lane 3:5 μg. A) Zymogram image: After electrophorosis of protein in an 8% acrylamide gel, bands indicating activity with 18mM 2,6-DMP as a substrate were verified. B) CBB-stained image: After the verification of bands indicating activity by the Zymogram, the same gel was CBB-stained to verify a band for protein.

Lignosulfonate de-colorization

De-colorization rates of lignosulfonate at A480nm by KS1025A in 5 days were 52% for 0.05% lignosulfonate, and 22% for 0.1% lignosulfonate.

Discussion

In this study, actinomycete KS1025A with strong lignin-degradation was newly isolated and identified as *Streptomyces* sp., using a universal primer for PCR amplification and DNA-sequencing. The universal primer mentioned above contained V1-4 regions which could be used to determine the specific sequences of the microorganism [30]. In the phylogenetic tree, the isolated actinomycete was slightly distant to its most homologenic microorganism, *Streptomyces atratus* (97.39%). Thereby we named the identified microbe as *Streptomyces* sp. KS1025A.

Crude enzymes derived from KS1025A do not show manganese peroxidase or lignin peroxidase activity. Therefore, laccase activity is considered to be the main cause of the lignin degradation ability of KS1025A. There are no reports concerning laccase or peroxidase activity in *Streptomyces atratus* (97.39%), *Streptomyces pulveraceus* (97.38%) or *Streptomyces sanglieri* (97.38%).

Optimal pH and temperature of the crude enzyme(s) derived from KS1025A are 4.5 and 50°C, respectively. Acidic pH (4-5) has been reported to be optimal for most laccases, and pH 7-8 for others [31]. In the zymogram analysis, two types of laccases derived from KS1025A could be observed; low molecular weight laccases approximately 90 kDa and others greater than 175 kDa. However, most fungal laccases reported thus far are monomeric proteins with molecular weights between 50 and 80 kDa [17,32,33]. With the exception of the following, laccases from *Agaricus bisporus* [34] and *Trametes villosa* [35] are comprised of two subunits, and those derived from *Podospora anserine* [36] are comprised of four subunits. In addition, the molecular weight of laccases derived from *Streptomyces coelicolor* [31] and *Streptomyces ipomoea* [37] are approximately 32-33 kDa in monomeric form, and exist as dimers in their natural form with molecular weights of 67-69 kDa. The molecular weight of laccase derived from KS1025A is heavier than the above mentioned *Streptomyceses* laccase. Therefore, laccases from KS1025A may be comprised of more than two subunits, which may have been caused by the mild denaturing conditions for SDS-PAGE and zymogram analysis. This analysis was carried out at 45°C for 10 minutes to preserve enzyme activity, although SDS and reducing agents were added. Hence, subunits of laccases may not have completely separated.

Laccase activity and enzyme productivity of this strain was compared with white-rot fungus. *Pleurotus ostreatus* [13,38-41] is known as a powerful lignin-degrading enzyme producer. When laccase activity of *Pleurotus ostreatus* is compared, a maximum of 1.8 U/mL of enzyme is produced in 21 days of culture, provided that copper is added as a Laccase catalyst [41]. Meanwhile, KS1025A strain produced a maximum of 14 U/ml of 2,6-DMP oxidative enzyme in 5 days of culture. From this, culture time can be reduced by 1/4 if KS1025A is used for enzyme production, and enzyme yield per day would increase 33 times. Possible applications of this strain for lignin degradation include wood-based biomass process, rapid decomposition of lignin colored material, and the industrial production of laccase. To the best of our knowledge, this is the first report on isolating laccase high-productive Streptomyces sp. Application of this strain is potential for de-lignin in wood-based biomass and rapid decomposition of lignin colored substance. Moreover, laccase has wider applications for degrading environmental pollutants such as bisphenol A; endocrine-disrupting chemical.

Conclusions

In this study we have screened high de-lignin enzyme producing microorgasms, and *Streptomyces* sp. KS1025A strain was isolated. The strain produces laccase in a short period and at high activity than previously reported white rot fungi. Remarkably, the laccase from the KS1025 strain was not-needed in addition of oxidation agent such as H_2O_2. By adding lignosulfate to the medium of the strain, incubation of the A480 nm was decreased about 50%.

References

1. Sun Y, Cheng J (2002) Hydrolysis of lignocellulosic materials for ethanol production: a review. Bioresour Technol 83: 1-11.

2. Reddy N, Yang Y (2005) Biofibers from agricultural byproducts for industrial applications. Trends Biotechnol 23: 22-27.

3. Kumagai S, Yamada N, Sakaki T, Hayashi N (2007) Characteristics of Hydrothermal Decomposition and Saccharification of Various Lignocellulosic Biomass and Enzymatic Saccharification of the Obtained Hydrothermal-Residue. J Jpn Inst Energy 86: 712-717.

4. Sluiter JB, Ruiz RO, Scarlata CJ, Sluiter AD, Templeton DW (2010) Compositional analysis of lignocellulosic feedstocks. 1. Review and description of methods. J Agric Food Chem 58: 9043-9053.

5. Aoyama A, Kurane R, Nagai K (2013) Penicillium sp. strain that efficiently adsorbs lignosulfonate in the presence of sulfate ion. J Biosci Bioeng 115: 279-283.

6. Ragauskas AJ, Williams CK, Davison BH, Britovsek G, Cairney J, et al. (2006) The path forward for biofuels and biomaterials. Science 311: 484-489.

7. Sticklen MB (2008) Plant genetic engineering for biofuel production: towards affordable cellulosic ethanol. Nat Rev Genet 9: 433-443.

8. Ximenes E, Kim Y, Mosier N, Dien B, Ladisch M (2010) Inhibition of cellulases by phenols. Enzyme Microb Technol 46: 170-176.

9. Klinke HB, Thomsen AB, Ahring BK (2004) Inhibition of ethanol-producing yeast and bacteria by degradation products produced during pre-treatment of biomass. Appl Microbiol Biotechnol 66: 10-26.

10. Leonowicz A, Matuszewska A, Luterek J, Ziegenhagen D, WojtaÅ›-Wasilewska M, et al. (1999) Biodegradation of lignin by white rot fungi. Fungal Genet Biol 27: 175-185.

11. Higuchi T (2004) Microbial degradation of lignin: role of lignin peroxidase manganese peroxidase and laccase. Proc Jpn Acad Ser B Phys Biol Sci 80: 204-214.

12. Baldrian P, Gabriel J (2002) Copper and cadmium increase laccase activity in Pleurotus ostreatus. FEMS Microbiol Lett 206: 69-74.

13. Díaz R, Sánchez C, Bibbins-Martínez MD, Díaz-Godínez G (2011) Effect of medium pH on laccase zymogram patterns produced by Pleurotus ostreatus in submerged fermentation. Afr J Microbiol Res 5: 2720-2723.

14. Tien M, Kirk TK (1988) Lignin Peroxidase of Phanerochaete chrysosporium. Methods Enzymol 161: 238-249.

15. Higuchi T (1990) Lignin biochemistry: Biosynthesis and biodegradation. Wood Sci Technol, 24: 23-63.

16. Paice MG, Reid ID, Bourbonnais R, Archibald FS, Jurasek L (1993) Manganese Peroxidase, Produced by Trametes versicolor during Pulp Bleaching, Demethylates and Delignifies Kraft Pulp. Appl Environ Microbiol 59: 260-265.

17. Thurston C (1994) The structure and function of fungal laccases. Microbiology 140: 19-26.

18. Iimura Y, Katayama Y, Kawai S, Morohoshi N (1995) Degradation and Solubilization of 13C-, 14C-side Chain Labeled Synthetic Lignin (Dehydrogenative Polymerizate) by Laccase III of Coriolus versicolor. Biosci Biotechnol Biochem 59: 903-905.

19. Eggert C, Temp U, Eriksson KE (1997) Laccase is essential for lignin degradation by the white-rot fungus Pycnoporus cinnabarinus. FEBS Lett 407: 89-92.

20. Eggert C, Temp U, Eriksson KE (1996) The ligninolytic system of the white rot fungus Pycnoporus cinnabarinus: purification and characterization of the laccase. Appl Environ Microbiol 62: 1151-1158.

21. DeLong EF (1992) Archaea in coastal marine environments. Proc Natl Acad Sci U S A 89: 5685-5689.

22. Lane DJ (1991) 16S/23S rRNA sequencing. Nucleic acid techniques in bacterial systematics. Chichester United Kingdom. John Wiley & Sons 115–175.

23. Kaminuma E, Mashima J, Kodama Y, Gojobori T, Ogasawara O, et al. (2010) DDBJ launches a new archive database with analytical tools for next-generation sequence data. Nucleic Acids Res 38: D33-38.

24. Saitou N, Nei M (1987) The neighbor-joining method: a new method for reconstructing phylogenetic trees. Mol Biol Evol 4: 406-425.

25. Wariishi H, Valli K, Gold MH (1992) Manganese(II) oxidation by manganese peroxidase from the basidiomycete Phanerochaete chrysosporium. Kinetic mechanism and role of chelators. J Biol Chem 267: 23688-23695.

26. Mester T, de Jong E, Field JA (1995) Manganese regulation of veratryl alcohol in white rot fungi and its indirect effect on lignin peroxidase. Appl Environ Microbiol 61: 1881-1887.

27. Tien M, Kirk TK (1984) Lignin-degrading enzyme from Phanerochaete chrysosporium: Purification characterization and catalytic properties of a unique H2O2-requiring oxygenase. Proc Natl Acad Sci USA 81: 2280-2284.

28. Harazono K, Kondo R, Sakai K (1996) Bleaching of Hardwood Kraft Pulp with Manganese Peroxidase from Phanerochaete sordida YK-624 without Addition of MnSO(inf4). Appl Environ Microbiol 62: 913-917.

29. Bradford MM (1976) A rapid and sensitive method for the quantitation of microgram quantities of protein utilizing the principle of protein-dye binding. Anal Biochem 72: 248-254.

30. Ezaki T, Ohkusu K (2006) Identification of Intestinal Flora: DNA Probe and Primers. Journal of intestinal microbiology 20: 245–258.

31. Machczynski MC, Vijgenboom E, Samyn B, Canters GW (2004) Characterization of SLAC: a small laccase from Streptomyces coelicolor with unprecedented activity. Protein Sci 13: 2388-2397.

32. Bollag JM, Leonowicz A (1984) Comparative studies of extracellular fungal laccases. Appl Environ Microbiol 48: 849-854.

33. Yaropolov A, Skorobogatko OV, Vartanov SS, Varfolom-eyev SD (1994) Laccase-properties, catalytic mechanism, and applicability. Appl Biochem Biotechnol 49: 257–280.

34. Wood DA (1980) Production, purification and properties of extracellular laccase of Agaricus bisporus. J Gen Microbiol 117: 327–338.

35. Yaver DS, Xu F, Golightly EJ, Brown KM, Brown SH, et al. (1996) Purification, characterization, molecular cloning, and expression of two laccase genes from the white rot basidiomycete Trametes villosa. Appl Environ Microbiol 62: 834-841.

36. Durrens P (1981) The phenoloxidases of the ascomycete Podospora anserina: the three forms of the major laccase activity. Arch Microbiol 130: 121–124.

37. Molina-Guijarro JM, Pérez J, Muñoz-Dorado J, Guillén F, Moya R, et al. (2009) Detoxification of azo dyes by a novel pH-versatile, salt-resistant laccase from Streptomyces ipomoea. Int Microbiol 12: 13-21.

38. Giardina P, Aurilia V, Cannio R, Marzullo L, Amoresano A, et al. (1996) The gene, protein and glycan structures of laccase from Pleurotus ostreatus. Eur J Biochem 235: 508-515.

39. Giardina P, Palmieri G, Scaloni A, Fontanella B, Faraco V, et al. (1999) Protein and gene structure of a blue laccase from Pleurotus ostreatus1. Biochem J 341 : 655-663.

40. Palmieri G, Giardina P, Bianco C, Scaloni A, Capasso A, et al. (1997) A novel white laccase from Pleurotus ostreatus. J Biol Chem 272: 31301-31307.

41. Abo-State MAM, Khatab O, Abo-EL AN, Mahmoud B (2011) Factors affecting laccase production by Pleurotus ostreatus and Pleurotus sajor-caju. World Appl Sci J 14: 1607-1619.

20

Color Removal and COD Reduction of Dyeing Bath Wastewater by Fenton Reaction

Farouk KM Wali*

Chemical technology Department, The Prince Sultan Industrial College, Kingdom of Saudi Arabia

***Corresponding author:** Farouk K. M. Wali, Assistant professor, Chemical technology Department, The Prince Sultan Industrial College, KSA
E-mail: farook19702001@yahoo.com

Abstract

Wastewater produced from Al-Amel dyeing bathes was subjected to Fenton oxidation and three commercial disperse dyes were selected for this study. The selected dyes were, Disperse Yellow 23, Disperse red 167 and Disperse Blue 2BLN, which used for dying cellulose fibers. At first, the optimum conditions for removing dyes from their aqueous solutions were determined and found to be 3 g/l H_2O_2, 120 mg/l ferrous sulfate hepta hydrate, pH 3 and retention time of about 100 minutes; these conditions achieve color removal of dyes reach 94% from their aqueous solution.

For the treated wastewater, it's found that color removal for Disperse Yellow 23, Disperse red 167 and Disperse Blue 2BLN was 84.66%, 77.19% and 79.63% respectively after retention time 160 minutes. Chemical oxygen demand (COD) measurements indicate that Fenton reaction shows a very good reduction of COD, this was 75.81%, 78.03% and 78.14% for Disperse Yellow 23, Disperse red 167 and Disperse Blue 2BLN respectively. These results strengthen the using of Fenton reaction as a preliminary treatment prior to biological treatment for this wastewater.

Keywords: Fenton reaction; Color removal; COD reduction

Introduction

Al-Amel dyeing unit located at Mansoura city and concerned with dyeing fibers especially cellulose fiber, they use categories of commercial disperse dyes which cover a variety of colors. Wastewater produced from dyeing bathes drained directly to Sewerage network which increase organic pollutants in wastewater. In this study, Fenton's oxidation used for the reduction of COD and colors of the remaining disperse dyes in wastewater.

The problem of colored effluent has been a major challenge and an integral part of textile effluent treatment as a result of stricter environmental regulations. This is not only unsightly but dyes in the effluent may have a serious inhibitory effect on aquatic ecosystems [1-4].

Main pollution in textile wastewater came from dyeing and finishing processes. These processes require the input of a wide range of chemicals and dyestuffs, which generally are organic compounds of complex structure. Major pollutants in textile wastewaters are high suspended solids, oxygen consuming matter, heat, color, acidity and other soluble substances [5,6].

The effluent from the biological treatment still contains significant amount of colored compounds, microorganisms, recalcitrant organic compounds and suspended solids. Also, chemical oxygen demand (COD) cannot be removed effectively by biological treatment. Hence, advanced treatment is necessary to improve wastewater discharge quality and to reuse wastewater as process water [7].

Fenton process is one of advanced oxidation technologies which used for the degradation of organic compounds to simple products which is biodegradable. Fenton's reagent is a mixture of H_2O_2 and ferrous iron, which generates hydroxyl radicals according to the following reaction [6-11]:

$$Fe^{++} + H_2O_2 \rightarrow Fe^{+++} + HO^{\bullet} + {}^{-}OH \qquad (1)$$

In the presence of substrate, such as a target contaminant, the hydroxyl radicals generated are capable of detoxifying the contaminants via oxidation. Due to the formation of Fe^{+++} during the reaction, the Fenton's reaction is normally accompanied by the precipitation of $Fe(OH)_3$ which is a coagulant helpful in removing suspended solids [10,11].

Clarke and Knowles [12] indicated that produced hydroxyl radicals may attack organic molecules by abstracting a hydrogen atom from the molecule. Carey [13] described a common pathway for the degradation of organics by the hydroxyl radicals as follows [12-15]:

$$HO^{\bullet} + RH \rightarrow H_2O + R^{\bullet} \qquad (2)$$

$$R^{\bullet} + H_2O_2 \rightarrow ROH + OH^{\bullet} \qquad (3)$$

$$R^{\bullet} + O_2 \rightarrow ROO^{\bullet} \qquad (4)$$

$$ROO^{\bullet} + RH \rightarrow ROOH + R^{\bullet} \qquad (5)$$

Elham K and Mina F [15] studied the decolorization and degradation of Basic Blue 3 and Disperse Blue 56 dyes using Fenton Process, they found that more than 99% dye removal and 88% COD removal were obtained respectively for initial 10-100 mg/l dye solution in addition to Fenton's reagent at ambient conditions and pH 3.

Materials and Methods

Materials

All chemicals used were of the highest purity, sulfuric acid, ammonium hydroxide, hydrogen peroxide (30 % H_2O_2 w/v), ferrous sulphate hepta hydrate ($FeSO_4.7H_2O$), were produced by El-Nasr Pharmaceutical Company (Egypt), silver sulfate, mercuric sulfate and potassium dichromate were produced by Aldrich Chemical Co.

Dye solutions and dye house wastewater

Optimum conditions for the removal of dyes was determined for their aqueous solutions of initial dye concentration 100 mg/l, and applied for dye house wastewater. All experiments were carried out at room temperature (25 ± 2ºC).

Color measurement

All color measurements were carried out by colorimetric method using a UV/Visible spectrophotometer [16,17]. The remaining hydrogen peroxide concentration was measured by using the iodometric titration method [18].

Chemical oxygen demand (COD) measurement

Chemical oxygen demand (COD) was measured by dichromate reflux method [18,19]. The remaining hydrogen peroxide after the process must be rejected or decompose before COD measurements this was achieved by adding Manganese dioxide Powder (MnO_2) followed by centrifugation and filtration on 0.45 micrometermillipore filter paper to remove excess MnO_2 powder [3,12-14].

Suspended solids (SS) and surfactants measurement

Suspended solids and surfactants were measured by Pastel UV ESCOMAM (France).

Results and Discussion

Optimum conditions for color removal of dyes from their aqucous solutions

Maximum wavelength for dyes were determined and listed in Table 1 and calibration curves for relationship between concentration and absorbance were established.

Dye (commercial name)	Color	Wavelength (nm)
Disperse Yellow 23	Yellow	380
Disperse red 167	Red	460
Disperse Blue 2BLN	Blue	570

Table 1: Color and maximum wavelength for the selected dyes

Fenton oxidation process was carried out at one liter glass beakers and we determine each optimum condition alone and all other additions were constant, i.e. for determining optimum pH we add constant dose from each of ferrous sulphate hepta hydrate and hydrogen peroxide and start the process. The whole mixture was stirred for 1 minute at 200 rpm followed by additional 15 minutes at

25-30 rpm and finally let the mixture for settling. After settling and precipitation, samples were withdrawn from supernatant for performing analysis at different retention time intervals.

Effect of pH: Fenton reaction is strongly affected by pH, since pH influences the generation of OH• radicals and affect the oxidation efficiency [3,12-15].

Figures 1-3 indicate that the remaining dye concentration was strongly affected by pH; results revealed that the optimum pH value for the maximum removal of the color of the tested dyes was pH 3. At pH values above 6 the degradation strongly decreases because, the ferrous catalyst may deactivate by the formation of ferric hydroxo complexes [11-15,20-23]. Also at higher pH values iron precipitates as hydroxide which reduces the transmission of light and consequently deactivates the Fenton oxidation process [11,20].

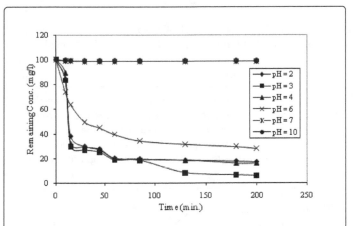

Figure 1: Effect of pH on removal of Disperse Yellow 23 from aqueous solutions at H_2O_2 dose=3 g/l and $FeSO_4.7H_2O$=120 mg/l.

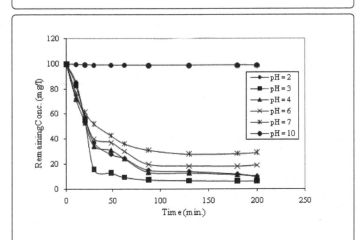

Figure 2: Effect of pH on removal of disperse red 167 from aqueous solutions at H_2O_2 dose=3 g/l and $FeSO_4.7H_2O$=120 mg/l.

Effect of hydrogen peroxide dose: During Fenton's process at the optimum pH, the limiting factor for the process is hydrogen peroxide dose [8,10,14,19]; A significant enhancement of removal was noticed when hydrogen peroxide dose increases.

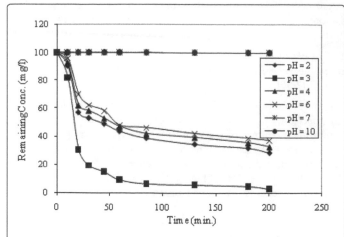

Figure 3: Effect of pH on removal of Disperse Blue 2BLN from aqueous solutions at H_2O_2 dose=3 g/l and $FeSO_4.7H_2O$=120 mg/l.

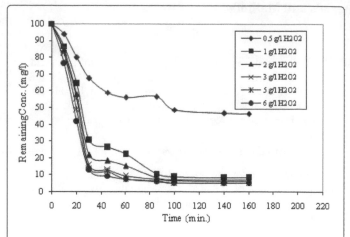

Figure 5: Effect of H_2O_2 dose on removal of disperse red 167 at pH 3 and $FeSO_4.7H_2O$=120 mg/l.

Results obtained in Figures 4-6 indicated that the color removal of dyes from their aqueous solutions increases by increasing hydrogen peroxide doses. It can be noticed from results that, the percent of removal was about 94% at a dose 3 g/l of H_2O_2 and after retention time 100 minutes. At higher doses than 3 g/l, there was a slight increase of dye removal.

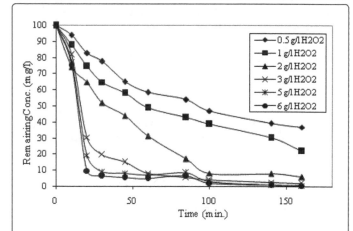

Figure 6: Effect of H_2O_2 dose on removal of dye Blue 2 BLN at pH 3 and $FeSO_4.7H_2O$=120 mg/l.

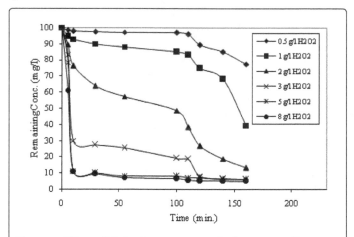

Figure 4: Effect of H_2O_2 dose on removal of Disperse yellow 23 at pH 3 and $FeSO_4.7H_2O$=120 mg/l.

The amount of H_2O_2 used can be minimized to 1 g/l if the process carried out for longer time, 180 minutes with a considerable color removal; this will be of great interest for economic reasons.

Effect of ferrous sulfate hepta hydrate dose: To obtain the optimal Fe (II) amounts, the processes were carried out with various amounts of iron salt under these conditions, pH=3, initial dye concentration 100 mg/l, and at hydrogen peroxide dose of 3 g/l.

Iron in its ferrous and ferric forms acts as a photocatalyst and requires a working pH below 4 to start the catalytic decomposition and enhance the removal of the tested dyes [20-24].

Results shown in Figures 7-9 indicate that the removal of the tested dyes in their aqueous solutions increases by increasing ferrous sulphate hepta hydrate dose to a certain limit at which the effect of ferrous salt is not significantly effective; this occur at doses higher than 120 mg/l.

Generally, the removal rate of colour increases as the dose of hydrogen peroxide increases until a critical value; after this critical value, the removal may decrease or not significantly increase [4,20,21]. This is in agreement with the fact that an excess amount of hydrogen peroxide in the solution will slightly retard the destruction and removal of dyes [20-22]. This behavior may be due to auto-decomposition of H_2O_2 to oxygen and water and the recombination of HO• radicals [20-22]. Since HO• radicals react with H_2O_2 itself and contributes to HO• scavenging capacity [6,11,20,22], so that H2O2 should be added at an optimal concentration to achieve the best degradation.

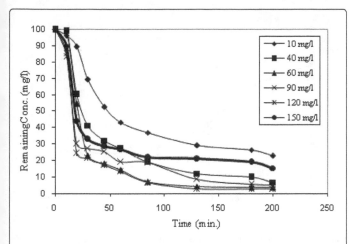

Figure 7: Effect of ferrous sulphate hepta hydrate doses on removal of Disperse Yellow 23 at pH=3 and H_2O_2=3 g/L.

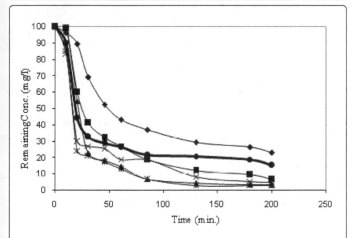

Figure 8: Effect of ferrous sulphate hepta hydrate doses on removal of disperse red 167 at pH=3 and H_2O_2=3 g/L.

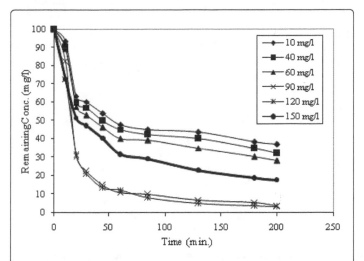

Figure 9: Effect of ferrous sulphate hepta hydrate doses on removal of Disperse Blue 2BLN at pH=3 and H_2O_2=3 g/L.

Parameter	Disperse Yellow 23		Disperse Red 167		Disperse Blue 2 BLN	
	Raw waste	Treated waste	Raw waste	Treated waste	Raw waste	Treated waste
pH	8.98	2.91	9.16	2.88	8.78	2.95
Remaining Dye concentration (ppm)	153.21	23.5	170.38	38.96	163.67	33.96
Total suspended solids (TSS, ppm)	800	46	948	52	1160	43
Chemical Oxygen Demand (COD, ppm)	2452.3	593.22	2274.62	499.73	2388.7	522.1
Surfactants (ppm)	62	36	158	18	84	12

Table 2: Characteristics of wastewater and treated waste from dyeing bathes.

It can be seen that the percent of removal was about 94% at 120 mg/l ferrous sulphate hepta hydrate after reaction time about 100 minutes. In case of higher doses of ferrous sulphate hepta hydrate the efficiency of color removal decreases. This may be due to the increase of brown turbidity of Fe^{3+} ions that hinders the absorption of light required for the Fenton process [11,20-24], so that ferrous sulphate hepta hydrate must be added in a critical dose, as well as it is necessary to use the smallest amount of iron in order to avoid the problems associated with its elimination.

Treatment of wastewater of dyeing bath by Fenton

Some characteristics of raw wastewater discharged from dyeing bathes and treated wastewater were shown in Table 2. These results show a high concentration of dyes, high COD and high suspended solids in raw wastewater.

Wastewater was subjected to Fenton's reagent at the determined optimum conditions. Results shown in Figure 10, indicate that removal percent for Disperse Yellow 23, Disperse red 167 and Disperse Blue 2BLN were 84.67%, 77.19% and 79.63% respectively after a retention time of 160 minutes. This reveals the destruction of dye molecules and chromophore groups by free radical reactions produced from Fenton's [6-10,13].

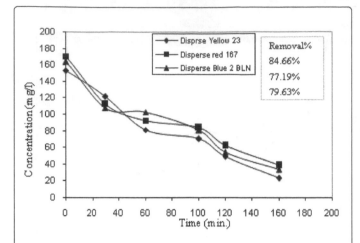

Figure 10: Remaining concentration of dyes in wastewater during Fenton process.

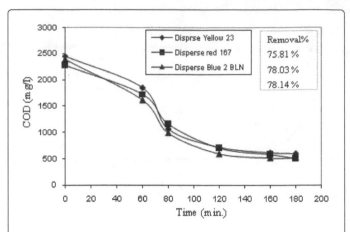

Figure 11: COD reduction for wastewater of dyes during Fenton reaction.

Removal percent in case of treated wastewater was lower than that in case of aqueous solutions at the same optimum conditions except retention time; this may due to in case of real wastewater some chemicals e.g. sodium chloride and surfactants were added during the dyeing process which inhibit the Fenton reaction or lower the formation of free radicals which oxidize the dye molecules and require a higher retention time. Similar results were reported by Stanisław and Lucyna [24]; they reported that inhibition effect of NaCl presence in textile wastewater on discoloration has been found: the higher content of NaCl the poorer is discoloration degree. The emulsification effect of surfactants present in textile wastewater causes a decrease of discoloration rate [24].

Evaluation of COD removal from wastewater of dyes

Fenton reaction has the advantage of both oxidation and coagulation processes due to the formation of Fe^{+++} ions. During the process, organic substances are oxidized by Fenton's reaction [6-12,24-31]; this leads to the degradation of dye molecules and reduce COD content.

COD measurements for treated waste water by Fenton reaction were shown in Figure 11, results indicate that COD reduction for Disperse Yellow 23, Disperse red 167 and Disperse Blue 2BLN were 75.81%, 78.03% and 78.14% respectively. These results in agreement with the degradation of these dyes by Fenton oxidation process.

Fenton oxidation occur at acidic pH close to 3, this must be followed by raising pH again to neutral conditions to help in coagulation and precipitation of ferric ions (Fe^{+++}); this can be achieved by using calcium hydroxide.

After Fenton process suspended solids greatly reduced, this due to the coagulant effect of Ferric iron (Fe^{+++}) which is a good coagulant helpful in removing suspended solids after precipitation. This will adapt wastewater to biological oxidation; also the remaining hydrogen peroxide when decompose increasing dissolved oxygen [6,9,23-31].

We can theoretically suggest a pathway for using Fenton's process as a preliminary treatment to biological oxidation as shown in the following; this can be investigated experimentally in a future studies.

Flow diagram for biological treatment of dye house wastewater is shown in Figure 12.

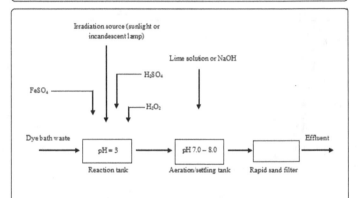

Figure 12: Diagram for adapting dye house wastewater for biological treatment.

Conclusions

The optimum conditions for removing dyes from their aqueous solutions were determined and found to be 3 g/l H_2O_2, 120 mg/l ferrous sulfatehepta hydrate, pH 3 and retention time of about 100 minutes; these conditions achieve color removal of dyes reach 94% from their aqueous solution.

The color removal for the treated wastewater for Disperse Yellow 23, Disperse red 167 and Disperse Blue 2BLN were 84.66%, 77.19% and 79.63% respectively. This reinforces the ability of using Fenton reaction for removing colors of these dyes from their wastewater.

Reduction percent in COD for dyes from their wastewater were 75.81%, 78.03% and 78.14% for Disperse Yellow 23, Disperse red 167 and Disperse Blue 2BLN respectively. It's necessary to inhibit the effect of remaining H2O2 by adding manganese oxide and filtration before measuring COD.

Fenton reaction not only leads to color removal of the dyes from wastewater but also leads to a significant COD reduction; this adapts the treated wastewater prior to biological treatment.

References

1. McMullan G, Meehan C, Conneely A, Kirby N, Robinson T, Nigam P, Banat IM, Marchant R, Smyth WF (2001) Microbial decolourisation and degradation of textile dyes. Appl Microbiol Technol 56: 81–87.

2. Little LW, Lamb JC, Chilling worth MA, Durkin WB (1974) Acute toxicity of selected commercial dyes to the fathead minnow and evaluation of biological treatment for reduction of toxicity. In: Proc. 29th Ind. Waste Conf., Purdue University, Lafayette, IN, USA, pp524–534.

3. Nigam P, Armour G, Banat IM, Singh D, Marchant R (2000) Physical removal of textile dyes and solid state fermentation of dye-adsorbed agricultural residues. Bioresour Technol 72: 219–226.

4. Azbar N, Yonar T, Kestioglu K (2004) Comparison of various advanced oxidation processes and chemical treatment methods for COD and color removal from a polyester and acetate fiber dyeing effluent. Chemosphere 55: 35–43.

5. Dae-Hee A, Won-Seok C, Tai-Il Y (1999) Biochemistry 34: 429–439.

6. Glaze WH, Kang JW, Chapin DH (1997) ozone. Sci Eng 9: 335-352.

7. Umran TU, Seher T, Emre O, Ulker BO (2015) Treatment of Tissue Paper Wastewater: Application of Electro-Fenton Method. International Journal of Environmental Science and Development 6: 415-418.

8. J. Chen, PhD Thesis (1997) Advanced Oxidation Technologies: Photocatalytic treatment of wastewater. Department of Environmental Tech of Wageningen, Agricultural University, Netherlands.

9. Fenton HJH (1894) J Chem Soc 65: 899-910, as cited in reference 8.

10. Kitis M, Adams CD, Daicger GT (1999) The effects of Fenton's reagent pretreatment on the biodegradability of nonionic surfactants. Water Res 3: 2561-2568.

11. Wali FKM (2005) Chemical studies on some new direct dyes and application of advanced oxidation technology for removal of their residuals in wastewater. PhD Thesis, Faculty of Science (Damietta), Mansoura University, Egypt.

12. Clarke N, Knowles G (1982) High purity water using H2O2 and UV radiation. Effluent Water Treat J 23: 335–341.

13. Carey JH (1990) An introduction to advanced oxidation processes (AOP) for destruction of organics in wastewater. In: A Symposium on Advanced Oxidation Process for Contaminated Water and Air Proceedings, Toronto, Canada.

14. Kuo WG (1992) Decolorizing dye wastewater with Fenton's reagent. Water Res 26: 881-886.

15. Keshmirizadeh E, Farajikhajehghiasi M (2014) Decolorization and Degradation of Basic Blue 3 and Disperse Blue 56 Dyes Using Fenton Process. Journal of Applied Chemical Research 8: 81-90.

16. Daniel CH (1995) Quantitative chemical analysis (4th edn), Michelson laboratory. China Lake, California USA.

17. Jeffery GH, Bassett J, Mendham J, Denney RC (1989) Vogel's Textbook of Quantitative Chemical Analysis. Longman Scientific and Technical, UK.

18. APHA, AWWA and WPCF (1995) Standard Methods for the Examination of water and Wastewater. Washington DC, USA.

19. Zollinger H (1987) Colour Chemistry-Synthesis, Properties of Organic Dyes and Pigments. VCH Publishers, New York, USA.

20. Tony MA (2005) Use of Photo-Fenton processes for wastewater treatment. MSc Thesis, Faculty of Engineering, El-Minia University, Egypt.

21. Rodriguez M (2003) PhD Thesis, Barcelona University, Spain.

22. Ghaly MY, Hartel G, Mayer R, Haseneder R (2001) Photochemical oxidation of p-chlorophenol by UV/H2O2 and photo-Fenton process. A comparative study. Waste Management 21: 41-47.

23. Pingatello JJ (1992) Dark and photoassisted iron(3+)-catalyzed degradation of chlorophenoxy herbicides by hydrogen peroxide. Environ Sci Technol 26: 944-951.

24. Perez M, Torrades F, Jose A, Domenech X, Peral J (2002) Removal of organic contaminants in paper pulp treatment effluents under Fenton and photo-Fenton conditions. Applied Catal B: Environmental 36: 63-74.

25. Stanisław L, Lucyna B (2012) Application of Fenton reagent in the textile wastewater treatment under industrial conditions. Proceedings of ECOpole 6.

26. USAEC (1997) CR-97013, 99-100.

27. Chung KT, Stevens SEJ, Serniglia CE (1992) The Reduction of Azo Dyes by the Intestinal Microflora. Crit Rev Microbiol 18: 175-197.

28. Panswad T, Luangdilok W (2000) Decolorization of reactive dyes with different molecular structures under different environmental conditions. Water Res 34: 4177-4184.

29. Kumar DR, Song BJ, Kim JG (2007) Electrochemical degradation of Reactive Blue 19 in chloride medium for the treatment of textile dyeing wastewater with identification of intermediate compounds. Dyes and pigments 72: 1-7.

30. Tony MA, Bedri Z (2014) Experimental Design of Photo-Fenton Reactions for the Treatment of Car Wash Wastewater Effluents by Response Surface Methodological Analysis. Advances in Environmental Chemistry, Volume 2014, Article ID 958134.

31. Gupta VK, Khamparia S, Tyagi I, Jaspal D, Malviya A (2015) Decolorization of mixture of dyes: A critical review. Global J Environ Sci Manage 1: 71-94.

21

Kinetic Study for Compost Production by Isolated Fungal Strains

Nassereldeen Kabbashi*, **Optakun Suraj, Md Z Alam and Elwathig MSM**

Bio-Environmental Research Centre, Department of Biotechnology Engineering; International Islamic University Malaysia, Malaysia

***Corresponding author:** Nassereldeen Kabbashi, Bio-Environmental Research Centre, Department of Biotechnology Engineering; International Islamic University Malaysia (IIUM), Jalan Gombak, P.O. Box 10, Malaysia-50728
E-mail: nasreldin@iium.edu.my

Abstract

Organic wastes, food wastes and trimming yard (FW and YT) were composted using selected fungal strains (*Phanerochaete chrysosporium* (PC), *Lentinus tigrinus* (LT), *Aspergilus niger* (ASP) and *Penicillium Spp* (PEN)) in a solid state bioconversion process. Results obtained at $P \leq 0.05$ after ten harvests indicated the minimum value of germination index (GI) in the open system was 43 ± 105% while in the closed system it was 46 ± 132% respectively. The simplest zero and first order kinetic models described the microbial mineralization of carbon to nitrogen (C/N) relatively (R^2 range of 0.87-0.99), but the second order model explained the observed kinetics of the solid state bioconversion (SSB) better with R^2 range of 0.87–0.98 and a positive decay coefficient (k). The decay coefficient which indicates if all the components of the biomass decomposed at the same rate increases from -0.0584 to 2×10^{-4} for *Phanerochaete chrysosporium* stream & -0.0578 to 2×10^{-4} for *Lentinus tigrinus* stream in the open system across the zero, first and second order.

Keywords: Composting; Organic wastes; Fungal strains; Germination index (GI); Reaction rate

Introduction

Asia produces the largest amount of Urban Food Waste (UFW), which is expected to increase from 251 to 418 million tonnes (45% to 53% of total world UFW) from 1995 to 2025. Currently, the 17000 tonnes of waste generated per day in Kuala Lumpur [1] comprises of 57% food wastes, 17% mixed papers, 4.7% yard trimmings and others constitute the municipal solid wastes (MSW) generated and disposed [2,3].

Unlike submerged fermentation, growth generally occurs on the surface of water-insoluble substrates in the absence of free water or at reduced water levels [4]. Traditionally SSB, has been applied to composting of agricultural wastes for mushroom cultivation and production of organic acids. SSB has also been found to be an efficient process for enzyme production [5]. Although often used for soil studies, community level physiological profiles (CLPPs) have been rarely applied to compost, probably for the lack of standardized methodology. Recently, however, CLPPs have been proposed as a tool to assess the degree of maturity of compost. One of the major problems is that the rate of colour development is a non-linear process related to both time and inoculum density. The aim of author's work was to investigate the suitability of data interpretation based on the kinetics of colour formation [6]. In another research carried by [7] for the degradation process was monitored, along with temperature, pH, total organic carbon, for the production of volatile fatty acids (VFAs) during the composting process of compost heaps in two different bioreactors (open and closed) at three different depth. Significant correlations were found between individual VFAs, as well as between VFA concentrations and organic carbon contents. Oxidizable carbon and mono- and oligosaccharides. Compost from vegetable residues is usually used as an organic amendment to soil; however, their thermal degradation characteristics show that it could be used as raw material in air gasification facilities. According to the obtained data by [8], hydrogen production is positively affected by composting, increasing hydrogen concentration. Using nth-order kinetic equations to describe component degradations, they have calculated a set of kinetic parameters which do not differ of the reported for other lignocellulosic materials.

In order to facilitate the operation and control of the composting process, however, there is a need for simple kinetic order of reaction that can accurately describe the dynamics of system. The rate of microbiological reaction comes from studies of the kinetics of microbial reduction of evaluation parameters such as C/N ratio, total organic matters (TOM) and others [9]. This paper presents results of an attempt undertaken to identify and understand the cycling of C/N ratio, GI and degree of degradation in the composting systems, there is need to describe the decomposition kinetics and the biodegradability of the substrate involved.

However, Malaysia waste treatment data revealed that 50% of these wastes are openly dumped, 30% land filled, 5% incinerated and only 10% composted [10]. This open dump of organically rich wastes could contributes significantly to the formation of leachate quality and quantity aside from the spread of disease vectors, odor, aesthetics and other environmental damages [11]. Leachates constitute major threat to underground water and the eco-system due to the presence of heavy metals. This organic content is beneficial for composting projects and not favorable for combustion or thermal technology as presently practiced [10]. Meanwhile, 89% of the entire waste generated are disposed while only 1% are converted to compost despite the suitability of the country climate for commercial compost production [12]. Therefore, composting of food waste and the institution yard waste with a bulking agent will solve the waste problem.

Materials and Methods

Experimental materials

The study is in two low technology adopted designs (open and close systems) in solid state bioconversion process experiment as indicated in Figure 1. The solid waste generated was sorted to remove the non-food components of the waste which includes plastics, papers and other non-organic components of the waste stream, while the yard trimming was collected separately as yard/lawn trimmings. The physicochemical properties of the comingled waste extract were analyzed and compared to the extract of major institutional wastes components food waste (FW), yard trimmings (YT) and soiled papers). All the substrates (food wastes, yard trimmings and sawdust) were characterized individually likewise the mixture, to determine the total organic carbon (TOC) content, total Kjeldahl nitrogen (TKN) contents, pH, moisture contents, ash content, hemicellulose, cellulose and lignin content among other parameters.

The fresh food waste collected was weighed and then dried in oven (MEMMERT GmbH Co. KG Germany) at a temperature of 105C. The dried substrates were then milled, grinded and sieved into smaller and uniform sizes of 1 to 2mm respectively to ensure homogeneity and faster degradation [13,14]. Yard trimming and flowers grown on the lawn across the land were regularly trimmed. These were collected and dried in the oven at 105C to remove the moisture content and ensure preservation. About 6-7 kilogrammes of fresh trimmings were raked on a 40m^2 area of the lawn. The dried samples were grinded using a Philip Twist home appliance (Model HR 1701, China), after which the electronic sieve (Model: AS 300, manufactured by Retsch GmbH Germany) was used to obtain 1-2mm sizes respectively as indicated for food wastes. The YT properties are determined while, the moisture content of the fresh trimmings before pretreatment was determined to be 69.88%. The sawdust (SD) used in the entire span of the experiment was collected from Forest Research Institute of Malaysia (FRIM) which are from chemically untreated log of woods (without preservative chemicals). The SD was dried as described above and sieve like the dried food and trimmings to obtain both 1mm and 2mm particle sizes.

These sources separated wastes were pretreated as substrates (food waste and yard trimmings) and composted with sawdust (SD) as the bulking agent after due characterization. Different ligninolytic fungi, *Phanerochaete chrysosporium, Lentinus tigrinus, Aspergilus niger* and *Penicillium spp.* were added at varying time interval to evaluate their effect using degradative indicators such as C/N, degradation degree (DD), and germination index (GI) coupled with the kinetics of the system.

Preparation of substrate mixture

All the substrates (food wastes, yard trimmings and sawdust) were characterized individually likewise the mixture, to determine the total organic carbon (TOC) content, total Kjeldahl nitrogen (TKN) contents, pH, moisture contents, ash content, hemicellulose, cellulose and lignin content among other parameters. The two main substrates (FW & YT) were mixed along with SD for the SSB process. The ratio of 1:1:0.5 (W/W) was used to mix FW (69.10% moisture, 48.85% TOC and 3.18% TKN) YT (69.88% moisture, 50.82% TOC and 1.46% TKN) and SD (19.32% moisture, 53.39% TOC and 0.27% TKN). The substrate to water ratio of 30:70 was used to maintain the optimum moisture content that is peculiar to SSB within the range of 50 – 60%

[15-18]. Nutrients (gkg^{-1}) of K2HPO4 (0.3), NaCl (0.3) and MgSO$_4$.7H$_2$O (0.3) were added to substrate as starter dose for activation of inoculated (*P. chrysosporium*) fungi Sawdust was added as a cheap source of carbon. In every container 400 grammes of substrates mixtures were used for the experiment. All substrates were autoclaved at 121C for 50 minutes after the addition of nutrients. Thereafter, the containers were inoculated with 6% fungal spores/mycelia out of the entire 70% water constituent (59% distilled water, 6% inoculum and 5% minerals) of each opened and closed system. The inoculum sizes used were 2.5 x 10^7 and 5.5 x 10^7 spores per ml for *P. chrysosporium* and *L. tigrinus* respectively; 84 x 10^6 and 92 x 10^6 CFU/g air dried inoculums for *A. niger* and *Penicillium spp.* respectively. Composting plastic containers bins were kept in the laboratory at room temperature with the open system uncovered and the holes created on the lids of the closed system were covered tightly with cotton wool as shown in Figure 2.

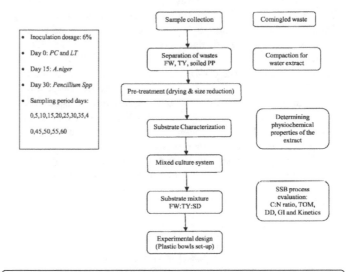

Figure 2: Closed (a) and Opened (b) Composting Systems.

The main characteristics of the mixture are thus : pH=5.68, Total Kjeldahl Nitrogen (TKN)=2.50%, Total Organic Carbon (TOC)=51.15%, Electrical Conductivity (EC)=5.62mS/dm (W.t), Salinity=5.62 0:00 (W.t) and Total Dissolved Solid (TDS)=2.77g/l. Likewise, the hemicellulose, cellulose, lignin and water soluble carbon content are 32.30%, 21.60%, 16.00% and 30.10% respectively.

Design of experiment

Two basic systems were deployed as Closed and Open Systems. Each of these systems comprises of the control (CR) and two treatment streams (as shown in Figure 2) to determine the optimal operating conditions, likewise, to compost the substrate mixture using

Phanerochaete chrysosporium (PC); *Lentinus tigrinus* (LT); *Aspergilus niger* (ASP) and *Penicillium spp* (PEN).

Results and Discussion

Isolation and identification of microorganism

Composting of the organic components of the institutional wastes (FW and YT) is a sustainable recycling process whereby organic matter are decomposed to shorter molecular chains, more stable, hygienic, humic rich and agriculturally useful product [19]. The selected fungi used in this study includes: *Phanerochaete chrysosporium* (PC); *Lentinus tigrinus* (LT); *Aspergilus niger* (ASP) and *Penicillium spp* (PEN). These selections were based on the physicochemical properties of the wastes or substrates involved. The weight is related to 69.1% and 69.88% moisture content of fresh FW and Yard Trimmings (YT) while sawdust has the lowest moisture content of 19.32%. Although all the substrates used for this experiment were dried at 105C for 24hrs to remove or reduce the moisture content. This is to ensure adequate preservation of the substrates throughout the experimentation process, thus most of the results are reported on dry matter basis (DMB). The initial pH of the substrate mixture is 5.68 a condition relatively good for aerobic composting using fungi, thus the pH of the system used was not adjusted. The TOC determined on a dry weight basis (DWB) was found to be 48.85% of FW (Table 1). Out of these, the water extractable carbon is 48.81% which justifies its total dissolved solids (TDS) concentration of 3.06 g/l. The Hemicellulose content of FW is 43.37% while the cellulose and lignin constitute 11.3% and 1.87% respectively (Table 1).

Substrates	FW	YT	SD	Mixture
pH	5.39	5.97	4.6	5.68
*TKN	3.18	1.46	0.27	2.5
*TOC	48.85	50.82	53.39	51.15
EC (mS/dm) w.t	3.04	4.13	0.32	5.62
Salinity(0/00)w.t	1.6	1.8	0.1	2.77
TDS (g/l) w.t	3.06	2.06	0.151	2.72
*Hemicellulose (%)	43.37	31.9	18.07	32.3
*Cellulose(%)	11.3	25.73	30.5	21.6
*Lignin(%)	1.87	21.53	45.93	16
**WSC (%)	43.46	20.84	5.5	30.1

Table 1: Characterization of substrates (FW=Food Waste; YT=Yard Trimmings; SD=Sawdust; *Dry weight basis; Wet basis (w.t); **Water Soluble Components).

Solid state bioconversion (SSB)

Since solid-state bioconversion (SSB) is considered as a hopeful novel, low- cost degradation of organic contaminants approach to control growth of microorganisms. The organic waste classification using simple waste separation technique shows that the organic component is 75% by weight and 18% by volume of the entire waste stream. This weight was related to 69.1% and 69.88% moisture content of fresh FW and Yard Trimmings (YT) while sawdust has the lowest

moisture content of 19.32%. Since the systems are microbially induced, occurrence of rapid decomposition in the thin liquid films on the surface of the organic particles releases CO_2 and H_2O as end products [20]. Thus, lower moisture content (< 30%) inhibits microbial activities while the higher one (> 70%) results in slow decomposition, odor formation and nutrient leaching. The open system moisture contents are evenly distributed and fluctuate due to activity of the microbes as shown in Figure 3. In the closed system, beside the ambiguous drop (probably due to trans - evaporation and initial heat generated by the reactor) in the value of the control sample, waste reactor A closed (WRAc) and waste reactor B closed (WRBc) turned out to be relatively higher. This can be traced to the covering which disallowed the evaporation process to take place within the system. Meanwhile, the water holding capacity of *P. chrysosporium* and *A. niger* was higher in this system compared *to L. triganus* and *A. niger*. Subsequently after the inoculation of *Penicillium spp.* the moisture content of WRBc increased and decreased with a greater rate, thereafter stabilized at day 60. This could be related to anti pathogenic tendency of Penicillium spp. Statistically, the open and close system are significant ($P<0.05$), meanwhile, the system with *P. chrysosporium* (WRAc) and *L. triganus* (WRBc) indicates a Least Square Difference (LSD) values of 0.001 with respect to control (CRc) while *P. chrysosporium* (WRAo) *and L. triganus* (WRBo) are slightly significant with $P=0.047$. This implied that at 95% confidence it can be proved that the moisture content of the close system is significantly different. In every container 400 grams of substrates mixtures were used for the experiment.

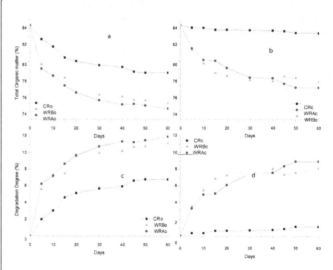

Figure 3: Degradation Degree (%) of (a) Open and (b) Close Systems.

Effect of Germination index (GI) on compost

GI of tomato seeds generally decreases gradually and significantly reaches the minimum at day 7 and 15 in the open and day 10 and 15 in the closed system respectively. This decrease might be due to the release of high concentration of ammonia and low molecular weight of organic acid [21]. The minimum value of GI in the open system was 53% and 55% while in the closed system it was 49 and 56% respectively. In all situations, the values of GI increased to about 90% above which correspond with the suggestion of Alberquerque (2006)

[22] that seed GI value above 50 is suitable for agriculture utilization while GI value above 80% indicates that the compost is mature [23,24]. The significant GI increase at day 15 through day 40 during the SSB process could be due to relief phytotoxins especially the ammonia volatilization, reduction of unstable organic acids and probably the anti-microbial strength of *Aspergilus* and *Penicillium spp*. At harvest the highest level of GI was 105 and 132 in open and closed system respectively. GI trend shows further increasing index even beyond the 60 day harvest period as shown in Table 2. This is a strong indication for the effectiveness of the anti-pathogenicity of *Penicilliium spp.* and the fitness of the product for use. Generally, the open system is significantly different at $P \leq 0.05$ level with value $[F(2, 27)=3.478, P=0.045]$ with post–hoc Least Square Difference (LSD) values 0.047 and 0.023 for Waste Reactor Open with PC (WRAo) and Waste Reactor Open with LT (WRBo) respectively. While, the closed system is not significant ($P \geq 0.05$) with value $[F (2, 27) = 1.305, P=0.288]$ and the post–hoc Least Square Difference (LSD) of 0.376 and 0.120 for Waste Reactor Closed with PC (WRAc) and Waste Reactor Closed with LT (WRBc) when compared with Close Control sample (CRc).

Test Parameters	Germany	Austria	USA	Produced Biofertilizer
C/N (%)	-	-	< 25	16.5-20
Salt ions (EC) mS/dm	2.5	2	2	2.72 ± 2.89
Germination Index (%)	> 90	80-90	< 80	105-132

Table 2: Characteristics of compost / biofertilizer produced with standards across countries.

Degree of degradation (DD) and ash content

The Degree of Degradation (DD) is inversely proportional to the TOC and the TOM (Figure 4a and 4b). *L. triganus* (WRBc) had a degradation degree of 3.02% on day 5 followed by *P. chrysosporium* (WRAo) with 2.80% while the control was 0.29% in the closed system. However in the open system degradation degree accelerated to be the highest for WRAo (5.29%) followed by WRBo (4.75%) and CRo (1.69%).

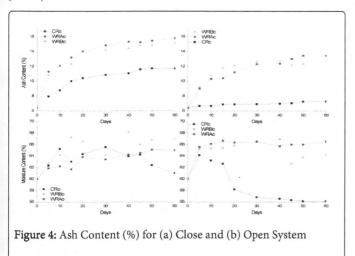

Figure 4: Ash Content (%) for (a) Close and (b) Open System

Generally the degradation degree of the open system were 10.12%, 9.44% and 5.76% for WRAo, WRBo and CRo respectively, while the

closed system relative performance were 7.57%, 6.42% and 0.94% for WRAc, WRBc and CRc respectively. This can be related to high microbial activity, peak ammonia loss and / or reduced organic matter content during the composting process. The ash content of the waste mixture was 6% in day 0, it then increases from 6% to 11% and approximately 15% in the control and treated reactors (WRAo and WRBo) of the open system. In the closed system, ash content increases from 6% to 12% and 13% respectively for WRAc and WRBc, while CRc was almost null as shown in Figure 3. In the post-hoc test wherein WRAc and WRBc are significant to each other with equal value of P=0.001, while WRAo and WRBo are also significantly different with lower P values of 0.015 and0.017 respectively all at 95% confidence interval. This is a reflection of the diversity in microorganisms, a phenomenon that is common when two or more organic wastes are combined for composting. This trend indicates that inorganic ash content is released when the organic matter in the compost material was actively decomposed by microorganisms [25].

Kinetic study of compost

The rates of microbial metabolism are related to observed (or estimated) properties such as reactant concentration, microbial biomass and the thermodynamic favorability of the reaction. Since biochemical process is never the result of a single elementary reaction, rather a multistep process involving one or more enzymes. Fortunately, in many cases, it is possible to use elementary rate expressions to describe the dependence of an overall multistep reaction on the concentration of one or a few reactions that control the rate of the overall reaction. Thus, kinetic expressions for rates of microbial reaction are macroscopic descriptions of overall reactions which are derived by considering the carbon to nitrogen ratio concentration [C/N].

The rates of microbial reaction, often considered as macroscopic descriptions of overall reactions for the decomposition of organic matter was evaluated based on the reaction order that best fit the experimental degradation data. The zero and first order reactions have been widely fitted for biodegradation of many elemental compounds during composting [26-28]. Zero-order kinetics commonly describe homogeneous chemical reactions in which the concentration of a catalyst controls the rate of reaction as expressed in Equation (1) below and has been suitable in describing reaction rates in experiments where the period of observation is relatively low. The first-order kinetics indicates the dependency of the reaction rate to the concentration of the reactant in direct proportion as expressed in Equation (2).

The rate of microbiological reaction comes from studies of the kinetics of microbial reduction of evaluation parameters such as C/N ratio, TOM and others [9]. Thus, to understand the cycling of C/N ratio in the composting systems, there is need to describe the decomposition kinetics and the biodegradability of the substrate involved. The simplest zero and first order kinetic models described the microbial mineralization that is done through the decay coefficient (k) which indicates if all the components of the biomass decomposed as explained in the following equations

$$r_A = k[C/N]^n$$

$$r_A = \frac{d[C/N]}{dt} = k_0[C/N]^n$$

Since n =0, for zero order

$$\frac{d[OM]}{dt} = k_0$$

$$d[C/N] = k_0 dt$$

Integrating both sides, when

$$[C/N]_o \, at \, time(t) = 0, and [C/N]_t \, at \, time(t) = 0$$

$$\int_{[OM]_0}^{[OM]_t} d[C/N] = \int_0^t k_0 dt$$

$$[C/N]\{(C/N_t) - (C/N_0)\} = k_0 t(t-0)$$

$$[C/N]_t - [C/N]_0 = k_0 t$$

$$[C/N]_t = k_0 t + [C/N]_0 \dots\dots\dots\dots(1)$$

First-order Reaction Derivation

$$r_A = k[C/N]^n$$

That is

$$r_A = \frac{d[OM]}{dt} = k_1 [C/N]^n$$

Since n=1 for first order

$$\frac{d[C/N]}{dt} = k_1 [C/N]$$

Therefore,

$$\frac{d[C/N]}{[C/N]} = k_1 dt$$

Integrating both sides, when [C/N]o at time (t) =0, and [C/N]t at time (t) =0

$$\int_{[OM]_0}^{[OM]_t} \frac{d[C/N]}{[C/N]} = \int_0^t k_1 dt$$

Thus,

$$ln[C/N]\{(C/N_t) - (C/N_0)\} = k_0 t(t-0)$$

$$ln[C/N]_t - [C/N]_0 = k_1 t$$

$$ln[C/N]_t = k_1 t + ln[C/N]_0 \dots\dots\dots\dots(2)$$

The microbial effect can be determined based on the significant percentage decreased values of C/N ratios 5.37%, 5.23% and 7.60% in the WRAo stream coupled with 6.68%, 4.73% and 5.58% in WRBo stream on days 15, 40 and 60 respectively in the open system. Similarly, 2.32%, 1.74% and 4.54% of the WRAc stream and 5.01%, 5.32% and 3.05% of WRBc for days 15, 40 and 60 of the close system reflect that LT activity is higher in the two (close and open) streams or systems than PC. Moreover, the DD is higher in open system (11.92%) compared to closed system (8.93%) which indicates that the effect of other microbes in the SSB process is almost not significant. Similarly, the significance of the SSB reaction order indicates R^2 valuesof 0.984 and 0.981 for open systems while closed systems were 0.865 and 0.965 for WRA and WRB respectively. However, between the closed systems LT performance was relatively better than PC, which suggests why LT had the lowest OM, TOC and C/N values in the closed system at day 15. Consequently, the germination index of the open system was considerably low compared to those of the closed system. This could be as a result of the higher microbial activities in the open system compare to the close system where intrusions of microbes are restricted. The toxicity strength as expressed by the GI indicated that open system (43 ± 67%) produced a significantly toxic biomass compared to the close system (55 ± 80%) especially during the most active degradation period (day 15). Similarly, TOM is significant (P=0.001) in both open and close system while C/N ratio of the closed system is significant only between systems. Table 2 provides the summary of the biofertiizer properties compare with some countries standards.

The simplest zero and first order kinetic models described the microbial mineralization of C/N relatively (R^2 range of 0.87-0.99), but the second order model explained the observed kinetics of the SSB better with R^2 range of 0.87–0.98 and a positive decay coefficient. The decay coefficient (k) which indicates if all the components of the biomass decomposed at the same rate increases from -0.0584 to 2×10^{-4} for PC stream and -0.0578 to 2×10^{-4} for LT stream in the open system across the zero, first and second order. Likewise, the closed system follows the same trend with k values of -0.0232 to 6×10^{-5} for PC and -0.0448 to 1×10^{-4} for LT. The positive values of the second order justify its fitness for the degradation order (Table 3). Comparatively, LT stream (R^2=0.984) performed narrowly better than PC stream (R^2=0.9813) within the open system. The same trend was indicated in the closed system with R^2 values of 0.9646 and 0.8652 for LT and PC stream respectively.

Samples	Zero Order			First Order			Second Order		
	K_0	[Conc]$_0$	R^2	K_1	[Conc]$_0$	R^2	k_2	[Conc]$_0$	R^2
WCAo[C/N]	-0.058	20.39	0.99	0	20.44	0.99	2×10^{-4}	20.49	0.98
WCBo[C/N]	-0.058	20.33	0.98	0	20.36	0.98	2×10^{-4}	20.41	0.98
WCAc[C/N]	-0.058	20.3	0.87	0	20.36	0.87	6×10^{-5}	20.33	0.87
WCBc[C/N]	-0.048	20.13	0.96	0	20.15	0.96	1×10^{-4}	20.16	0.97

Table 3: Zero, first and second order kinetics.

Decay of the biomass decomposed on first & second-order kinetics

The decay coefficient (k) which indicates if all the components of the biomass decomposed at the same rate increases from -0.0584 to 2×10^{-4} for PC stream and -0.0578 to 2×10^{-4} for LT stream in the open system across the zero, first and second order. Likewise, the closed system follows the same trend with k values of -0.0232 to 6×10^{-5} for PC and -0.0448 to 1×10^{-4} for LT. The positive values of the second order justify its fitness for the degradation order (Table 3). Comparatively, LT stream (R^2=0.984) performed narrowly better than PC stream (R^2=0.9813) within the open system. The same trend was indicated in the closed system with R^2 values of 0.9646 and 0.8652 for LT and PC stream respectively. Generally between open and closed systems especially with respect to the R^2 values as expressed in the Table 3, open system performance is better but the closed system is better in terms of decay coefficient.

Conclusion

The feasibility and efficacy of SSB process for biodegradability of composted source separated FW and YT wastes were evaluated by examining twelve parameters at (ten harvests) 0, 5, 10, 15, 20, 30, 40, 45, 50 and 60 days after inoculation of the organic wastes. Considering basic composting evaluation indicators, open system performance is better than the closed systems. This is predicated in the significant percentage decrease in the C/N ratio (18.20% and 16.99% in PC and LT streams respectively) of the open system compared to the close system with 8.60% and 13.38% of PC and LT streams. Similarly, a reflection of the biomass mineralization as shown by DD results indicated that open systems (10.12% WRAo and 9.44% WRBo) performed better than the close systems (7.57% WRAc and 6.42% WRBc).

This submission was further proved by the positive correlation of the ash contents of open and close systems maintained by DD. Statistically the close systems are significantly different at $P \leq 0.01$ while the open systems 95% degree of confidence at $P \leq 0.05$ is also acceptable.

References

1. Saeed M, Hassan M, Mujeebu M (2009) Assessment of municipal solid waste generation and recyclable materials potential in Kuala Lumpur, Malaysia. Waste Management 29: 2209-2213.

2. MH & LG (2008) National strategic plan for solid waste management. Unpublished Draft Strategic Plan, unpublished. Malaysia Local Government Department, Ministry of Housing and Local Government.

3. Nasir (2007) Institutionalizing solid waste management in Malaysia: Department of National Waste Management.

4. Cannel E, Moo-Young M (1980) Solid-state fermentation systems. Process Biochemistry 4: 2–7.

5. Krishna C (1999) Production of bacterial cellulases by solid state bio-processing of banana waste. Bioresource Technology 69: 231–239.

6. Mondini C, Insam H (2003) Community level physiological profiling as a tool to evaluate compost maturity: a kinetic approach. European Journal of Soil Biology 39: 141-148.

7. Plachá D, Helena Raclavská, Kucerová M, Kucharová J (2013) Volatile fatty acid evolution in biomass mixture composts prepared in open and closed bioreactors. Waste Management 33: 1104–1112.

8. Barneto AG, Carmona JA, Ferrer JAC, Blanco MJD (2010) Kinetic study on the thermal degradation of a biomass and its compost: Composting effect on hydrogen production. Fuel 89: 462–473.

9. Roden EE (2003) Fe (III) oxide reactivity toward biological versus chemical reduction. Environ. Sci Technol 37: 1319-1324.

10. World Bank (2008) Report on municipal solid waste treatment technologies and carbon Finance in East Asia.

11. Adhikari BK, Barrington SF, Martinez J (2009) Urban Food Waste generation: challenges and opportunities. International Journal of Environment and Waste Management 3: 4-21.

12. Periathamby A, Hamid F, Khidzir K (2009) Evolution of solid waste management in Malaysia: impacts and implications of the solid waste bill, 2007. Journal of Material Cycles and Waste Management 11: 96-103.

13. Adhikari BK, Barrington S, Martinez J, King S (2008) Characterization of food waste and bulking agents for composting. Waste Management 28: 795-804.

14. Diaz M, Madejon E, Lopez F, Lopez R, Cabrera F (2002) Optimization of the rate vinasse/grape marc for co-composting process. Process Biochemistry 37: 1143-1150.

15. Haug RT (1993) The Practical Handbook of Compost Engineering; BocaRaton, CRC Publishers Ltd, Florida, USA.

16. Morin S, Lemmay S, Barrington SF (2003) An urban composting system. [Written for presentation at the CSAE/SCGR 2003 meeting].

17. Pace MG, Miller BE, Farrell-Poe KL (1995) The Composting Process. Utah State University Extension.

18. Sherman R (2005) Large-scale organic materials composting. Retrieved October23, 2010,fromNorth Carolina Cooperative Extension Service, USA.

19. Sequi P (1996) The role of composting in sustainable agriculture in: The Science of Composting. In Bertoldi PSM, Lemmens B and Papi T (Ed.). The Science of Composting. Blackie Academic & Professional, London, UK (p23–29).

20. Betton CI (1992) Lubricants and their environmental impact. In:Chemistry and Technology of lubricants. Chemistry and Technology of lubricants. Blackies Academic & Professional Press, London, UK (p282-298).

21. Fang M, Wong JWC, Ma KK, Wong MH (1999) Co-composting of sewage sludge and coal fly ash: nutrient transformations. Bioresource Technology 67: 19-24.

22. Alberquerque JA, Gonzalvez J, Garcia D, Cegara J (2006) Effect of bulking agent on composting of "alperujo" the solid by-product of the two-phase centrifugation method for olive oil extraction. Process Biochemistry 41: 127–132.

23. Tiquia SM (2003) Evaluation of organic matter and nutrient composition of partially decomposed and composted spent piglitter. Environmental technology 24: 97-107.

24. Zucconi F,Pera A,Forte M, De Bertoldi M (1981) Evaluating toxicity of immature compost. BioCycle (USA).

25. Haddadin MS, Haddadin J, Arabiyat OI, Hattar B (2009) Biological conversion of olive pomace into compost by using Trichoderma harzianum and Phanerochaete chrysosporium. Bioresource Technology 100: 4773-4782.

26. Brezonik P (1994) Kinetics of biochemical reactions and microbial processes in natural waters. Chemical Kinetics and Process Dynamics in Aquatic Systems. Lewis Publishers, BocaRaton, Florida, USA.

27. Hunter K, Wang Y, Van Cappellen P (1998) Kinetic modeling of microbially- driven redox chemistry of subsurface environments: coupling transport, microbial metabolism and geochemistry. Journal of Hydrology 209: 53-80.

28. Roden EE (2008) Microbiological controls on geochemical kinetics1: Fundamentals and case study on microbial Fe (III) oxide reduction. Kinetics of Water-Rock Interaction 335-415.

22

The Influence of Gamma Irradiation on Repeated Recycling and Accelerated Acrylonitrile Butadiene Styrene Terpolymer Aging

Sandra Tostar[1]*, Erik Stenvall[2], Antal Boldizar[2] and Mark R StJ Foreman[1]

[1]*Department of Industrial Materials Recycling, Chalmers University of Technology, 412 96 Gothenburg, Sweden*

[2]*Department of Material and Manufacturing Technology, Chalmers University of Technology, 412 96 Gothenburg, Sweden*

*Corresponding author: Sandra Tostar, Department of Industrial Materials Recycling, Chalmers University of Technology, 412 96 Gothenburg, Sweden
E-mail: sandra.tostar@chalmers.se

Abstract

Electronic waste also referred to as waste electrical and electronic equipment (WEEE) is the fastest growing waste stream in Europe today. The fast exchange pace of mobile phones, television sets and computers create an important need to develop recycling areas to take care of, reuse and recycle all the materials they contain. One common plastic in WEEE is acrylonitrile butadiene styrene terpolymer (ABS). Repeated recycling of plastic causes it to chemically degrade and one of the unwanted effects is the shortening of the polymer chains. Gamma irradiation is known to be able to crosslink polymers and thus reverse the chain shortening. Therefore, the hypothesis of this study is that gamma irradiation of ABS would have a beneficial effect when recycling plastic. Comparative experiments of gamma irradiation has been done according to two methods: A single gamma irradiation (40 kGy) before the extrusion and aging cycles, and the effect of four 10 kGy doses delivered before each of the re-extrusion steps were completed. The results show that gamma irradiation has an impact on the mechanical and rheological properties of ABS. The yield stress increased with irradiation doses of 0, 10, 50 and 400 kGy. The viscosity also increased in the test samples with irradiation doses of 0, 10, 100 and 200 kGy. In the multi-recycling and accelerated aging test, there was a significant reduction in stiffness for the gamma irradiated samples after the second out of four cycles which cannot be fully explained.

Keywords: Recycling; ABS; Accelerated aging; Degradation; Gamma irradiation

Introduction

Recycling of plastics from electronic waste has become of increased importance since the WEEE directive (Directive 2012/19/EU, Annex V) requires increased recycling levels. This directive will be difficult to comply with if plastics are not included, since plastics can make up to 30 % of electronic products [1-3].

Acrylonitrile butadiene styrene terpolymer (ABS) is one of the main constituents in electronic plastics [1,4-6] since it is highly impact resistant, due to its rubber content, and its moldability [7].

Investigations on multiple recycling and accelerated aging have been reported by several researchers [8-10] but the degradation behavior, to the best knowledge of the authors, has never been compared with gamma irradiated ABS.

It is known that the degradation of plastic is the total amount of weathering it is exposed to during the products' lifetime. Therefore, to be able to simulate repeated recycling of the plastic, service life at room temperature has to be accounted [10]. This can be done by accelerated aging by thermo oxidative degradation.

The aging rate is doubled with every 10°C increase. A service time of one year at room temperature should thus be 72 h at 90°C [7,9,11].

Today's short working life of electronic products, such as mobile phones and computers, was the benchmark for the accelerated aging, simulating one year use of the product.

In this work a comparison between gamma irradiated and unirradiated ABS has been done to compare the degradation behavior of both materials after recycling. If the mechanical properties are enhanced by gamma irradiation, the same material can be recycled numerous times.

Experimental

Material

The ABS used was an electronic equipment grade, named Terluran® GP-35 produced by BASF, which does not contain any flame retardants. It was purchased at ALBIS, Sweden.

Sample preparation and material processing

Three series of sample preparation were performed: unirradiated ABS, ABS irradiated once at the start of the test at 40 kGy, and ABS irradiated at 10 kGy between each recycling cycle, which gives a total dose of 40 kGy.

Depending on the stage of the recycling cycle, the granulate and plastic flakes were irradiated in a Gammacell 220 (Atomic Energy of Canada). The dose rate to water in the irradiation chamber of the unit was on average 14 kGy·h^{-1}.

All recycling cycles had the same setup: drying, extrusion of strips, grinding and aging. Although, the drying step was excluded for cycle 1-4. The extrusions were performed by a Collin extruder 3250- 09-88 (screw length 694 mm, screw diameter 25 mm, and the rotational velocity was kept at 50 rpm for all extrusions). The extruder had five heating zones, three along the cylinder, one at the adapter, and one at the die, with the temperatures 200, 200, 200, 210 and 210°C respectively. The die was of slit type with an opening of 50 to 0.9 mm and gave strips that were 1.0-1.1 mm thick and 32-33 mm wide. Before extrusion, the material was dried at 80°C for 4 h in air using a Heraeus oven type UT 5042. The aging was performed in the same oven at 90°C for 144 h with constant air flow of 72 L·min⁻¹. At halftime, the ABS flakes were taken out and agitated to ensure that the oxygen was evenly distributed around them.

The strips were cut using a granulator of model SG 10 Ni from Dreher. The produced flakes had a size of circa 3.5-4.5 to 32-33 mm. Samples were collected after each extrusion occasion. Test specimens, in the shape of dog bones, were manually punched out from the extruded strips using equipment with the ISO standard: ISO-527-5A. For comparison, test specimens were also produced in an injection moulding machine, model Arburg - Allrounder 221 M 250-55 (maximum clamp force of 250 kN), to compare possible deviations of the test results within each test. The test specimens were conditioned at 25(±2)°C at 50(±5)% relative humidity for a minimum of 24 h before testing. The standard deviation is calculated on five test specimens per irradiation dose and shown in the figures with ± σ.

Gamma irradiation source

The gamma irradiation was performed in a Gammacell 220 (Atomic Energy of Canada, now trading as Norion). The dose rate to water in the irradiation chamber of this unit was 14 kGy·h⁻¹ on average. It was a little lower at the edges of the chamber (13 kGy·h-1) and higher at the centre (15 kGy·h⁻¹), and the dose rate was measured with the ferrous-cupric sulphate dosimeter test (19/04/2012). The temperature within the chamber was approximately 50°C.

Rheological tests

Viscosity

The viscosity measurements were performed on a Göttfert rheograph 2002 together with the WinRHEO software. Three repetitions were performed on each capillary length: 10, 20 and 30 mm. The diameter of the capillary was 1 mm and the machine held a temperature of 190°C for all tests. The pressure transmitter is rated at 2000 bar and the shear rate went from 20-1500 s⁻¹ in 10 steps. The two first measurements were excluded due to instability. All tests were subjected to both Bagley- and Rabinowitsch corrections.

Mechanical tests

Tensile test

The tensile test was performed on a Zwick 4031 with an Instron load cell (500 N) together with the testXpert® software. The equipment was used with the tensile speed of 2.8 mm·min⁻¹, which is approximately 10% of the length of the dog bone's waist. An additional test was also performed with different tensile speeds, 5.6, 28, 280 and 560 mm·min⁻¹, to investigate if the material became more brittle with increased speed.

Results and Discussion

Effect of the gamma irradiation dose on ABS

Excluding the 10 kGy sample which was not following the pattern, a decrease in stiffness with increasing irradiation dose was observed and illustrated in Figure 1, which is also supported by Figure 2. This is contradictory to the results Perraud et al. found. They saw an increase in stiffness for the hydrogenated nitrile butadiene rubber (HNBR), an elastomer commonly used in the automotive and printing industries, which includes two components of ABS, A- and B-, when it was irradiated with an electron beam [12]. The elongation at break decreased with increasing irradiation dose, which was the same behaviour as we noticed in our test (Figure 3). In addition, the tensile strength decreased in their test whereas it remained similar for our different irradiation doses.

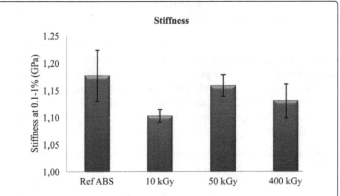

Figure 1: The effect of gamma irradiation (unirradiated, 10, 50 and 400 kGy) on ABS stiffness.

The acrylonitrile content influence how sensitive different plastics are to gamma irradiation. Cardona et al. analysed acrylonitrile/butadiene rubber and saw an increased effect on the rubber with increasing acrylonitrile content when it was irradiated.

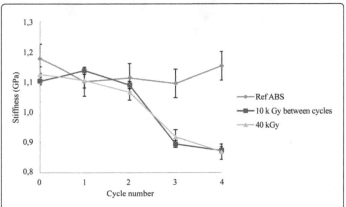

Figure 2: Stiffness comparisons (GPa) between the Reference ABS, 10 kGy between each cycle, and 40 kGy gamma irradiated materials.

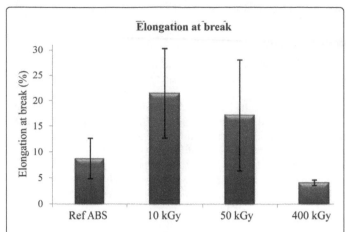

Figure 3: The effect of gamma irradiation (unirradiated, 10, 50 and 400 kGy) on ABS elongation at break.

Secondly, the acrylonitrile part was more sensitive to gamma irradiation than the butadiene part, and thirdly, the reactions observed were the consumption of double bonds and crosslinking, no chains scissoring occurred. This behaviour can explain the increase in viscosity noticed in our comparison of unirradiated and gamma irradiated ABS [13].

A trend towards tougher material with an increasing irradiation dose can be noticed for the yield stress in Figure 4.

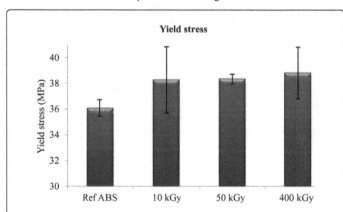

Figure 4: The effect of gamma irradiation (unirradiated, 10, 50 and 400 kGy) on ABS yield stress.

Degradation test of ABS with different gamma irradiation doses and accelerated aging

Elongation at break:

The overall hypothesis was that gamma irradiation would counteract the expected accelerated aging, thus degrading the ABS, and withstand or even increase the mechanical properties. Previous results presented by Boldizar and Möller indicated that the elongation at break increased for each cycle in the extrusion steps up until cycle 6 [9]. Since the deviation of the elongation at break was very high, it was difficult to draw any conclusions about any change during the recycling cycles and aging.

Yield stress:

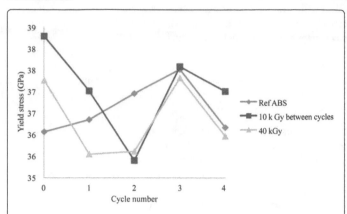

Figure 5: Yield stress comparison (GPa) between the Reference ABS, 10 kGy between each cycle, and 40 kGy gamma irradiated materials.

Stiffness:

The standard deviation is calculated on five test specimens per material and shown in Figures (1,2,5 and 6) with ± σ.

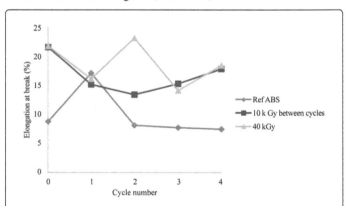

Figure 6: Elongation at break (%) comparison between Reference ABS, 10 kGy between each cycle, and 40 kGy gamma irradiated material.

Ductile to brittle transition test:

A trend to increasing yield stress illustrated in Figure 7 is valid for all tensile speeds except the last, at 560 mm/min, which is a very high tensile speed.

The standard deviation is calculated for five test specimens per tensile speed and shown in Figure 7 with ± σ.

Viscosity comparison of unirradiated and gamma irradiated ABS:

The effect of the gamma irradiation could be observed in the ABS viscosity illustrated in Table 1 and Figure 8, and it increased with increasing irradiation dose. The most likely cause is that the ABS crosslinked, as has been seen in other studies made by Cardona et al and Perraud et al. [12,13].

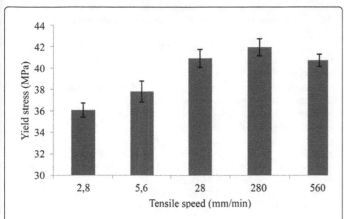

Figure 7: The effect of tensile speed (mm/min) on yield stress (MPa) for ABS.

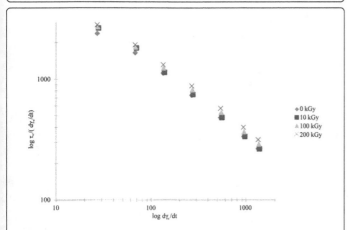

Figure 8: Summary of the viscosities for unirradiated ABS, 10, 100 and 200 kGy gamma irradiated ABS.

Shear rate (s^{-1})	0 kGy Viscosity (Pas)	10 kGy Viscosity (Pas)	100 kGy Viscosity (Pas)	200 kGy Viscosity (Pas)
20	2400	2600	2700	2800
50	1600	1800	1800	1900
100	1100	1100	1200	1300
200	740	750	820	880
400	480	480	530	580
700	340	340	370	400
1000	270	270	290	320
1500	200	200	220	240

Table 1: Viscosity data of ABS treated with different gamma-irradiation doses.

Combining the viscosity curves in the same graph shows that the viscosity increased with increasing gamma irradiation dose; this can be observed in Figure 8. 0 and 10 kGy have similar values, especially for higher shear rates, while the 100 kGy lines is between the 10 and 200 kGy lines.

Conclusions

The mechanical properties of multi-recycled, accelerated aged and gamma irradiated ABS test, a difference could be seen in the stiffness results between the different cycles, as well as between the different irradiation doses. The unirradiated ABS sample was stable during all four recycling cycles and did not degrade during the multi-recycling and accelerated aging. Although, the 10 kGy (irradiated between each cycle) sample and the 40 kGy sample, showed a decrease in stiffness after two cycles which cannot yet be explained. Due to the very high standard deviation for the elongation at break, no relevant conclusions could be drawn from the test. For the yield stress test no trend could be distinguished.

The 10 kGy sample irradiated between each cycle, and the 40 kGy sample behaved similarly for all the tested properties and no conclusion could be drawn on whether the material would degrade less for a repeated small dose of gamma irradiation between each recycling cycle than for a higher dose prior to the repeated recycling.

The gamma irradiation had an influence on the mechanical properties of ABS. The ABS seems to crosslink with an increasing irradiation dose, which could be seen as the increase in viscosity and higher yield stress compared to the unirradiated ABS. The high variation in EB results may be explained by the butadiene rubber part of the ABS's structure. It was almost a randomized event if the ABS sample breaks or elongates further. The punching of dog bones from the samples has not caused any major surface damage that can explain the high variation in the EB test results. The comparison with the injection moulded dog bone test specimens supported the theory, since the variation in EB was similar for those test specimens. A trend could be noticed towards higher yield stress with an increasing gamma irradiation dose, and the opposite behaviour was noticed for the stiffness, a decrease in stiffness with higher irradiation dose. However, the 10 kGy sample did not fit the trend line very well, but it may be explained by the high variation within the sample.

No ductile to brittle transition was seen when increasing the tensile speed. The samples showed a plastic deformation (necking) before breakage that did not appear to have any correlation with the tested strain rates. The absence of a brittle to ductile transition with respect to strain rate is possibly due to the multi-component ABS system with the rubber component, which is relatively insensitive to the high strain rates used in this test.

Gamma irradiation also had an influence on the rheological properties; the viscosity increased with increasing irradiation dose.

Acknowledgement

This work was sponsored by Materials Science: a Chalmers Area of Advance.

References

1. Martinho G, Pires A, Saraiva L, Ribeiro R (2012) Composition of plastics from waste electrical and electronic equipment (WEEE) by direct sampling. Waste Manag 32: 1213-1217.

2. Menad N, Björkman Bo, Allain EG (1998) Combustion of Plastics Contained in Electric and Electronic Scrap. Resources Conservation and Recycling 24: 65-85.

3. Taurino R, Pozzi P, Zanasi T (2010) Facile characterization of polymer fractions from waste electrical and electronic equipment (WEEE) for mechanical recycling. Waste Manag 30: 2601-2607.

4. Dimitrakakis E, Janz A, Bilitewski B, Gidarakos E (2009) Small WEEE: determining recyclables and hazardous substances in plastics. J Hazard Mater 161: 913-919.

5. Schlummer M, Gruber L, Mäurer A, Wolz G, van Eldik R (2007) Characterisation of polymer fractions from waste electrical and electronic equipment (WEEE) and implications for waste management. Chemosphere 67: 1866-1876.

6. Stenvall E, Tostar S, Boldizar A, Foreman MR, Möller K (2013) An analysis of the composition and metal contamination of plastics from waste electrical and electronic equipment (WEEE). Waste Manag 33: 915-922.

7. Shimada J, Ando, Kabuki K (1972) The Mechanism of Photo-Oxidative Degradation and Stabilization of Abs Resin. Rev Electr Commun Lab 20: 553-563.

8. Bai X, Isaac DH, Smith K (2007) Reprocessing Acrylonitrile-Butadiene-Styrene Plastics: Structure-Property Relationships. Polymer Engineering and Science 47: 120-130.

9. Boldizar A, Möller K (2003) Degradation of ABS during repeated processing and accelerated ageing. Polymer Degradation and Stability 81: 359-366.

10. Pérez JM, Vilas JL, Laza JM, Arnáiz S, Mijangos F, et al. (2010) Effect of Reprocessing and Accelerated Weathering on ABS Properties. Journal of Polymers and the Environment 18: 71-78.

11. Rapoport N, Livanova N, Balogh L, Kelen T (1993) Simulation of the Durability and Approach to the Stabilization of Polyolefins Undergoing Oxidative Degradation under Mechanical Stress. International Journal of Polymeric Materials 19: 101-108.

12. Perraud S, Vallat MF, David MO, Kuczynski J (2010) Network Characteristics of Hydrogenated Nitrile Butadiene Rubber Networks Obtained by Radiation Crosslinking by Electron Beam. Polymer Degradation and Stability 95: 1495-1501.

13. Fatt MSH, Ouyang X (2008) Three-Dimensional Constitutive Equations for Styrene Butadiene Rubber at High Strain Rates. Mechanics of Materials 40: 1-16.

14. Cardona F, Hill DJT, Pomery PJ, Whittaker AK (1999) A Comparative Study of the Effects of Uv- and γ- Radiation on Copolymers of Acrylonitrile/Butadiene. Polymer International 48: 985-992.

23

Towards Sustainable Resource and Waste Management in Developing Countries: The Role of Commercial and Food Waste in Malaysia

Effie Papargyropoulou*, **Rory Padfield, Parveen Fatemeh Rupani and Zuriati Zakaria**

University Technology Malaysia (UTM), Kuala Lumpur, Malaysia

*Corresponding author: Effie Papargyropoulou, University Technology Malaysia (UTM), Kuala Lumpur, Malaysia
E-mail: effie@ic.utm.my

Abstract

Rising commercial waste generation poses a significant environmental and public health issue, especially in rapidly expanding urban centres in developing countries. A commercial district in Malaysia was selected to explore the challenges and opportunities for minimisation of commercial waste. This research provides empirical data on commercial and food waste generation rates, the problems faced by waste producers, and the priorities for improvement. It is argued that whilst commercial waste offers opportunities for waste minimisation, current challenges related to amenity and public health such as pests, odour and littering, can be addressed by the provision of additional bins, grease traps and improved public areas cleansing. The study concludes that food waste plays a key role in the progression towards a more sustainable waste management system in a developing country such as Malaysia, due to its high generation rates, its contribution to public health and amenity problems, and its high potential for resource recovery.

Keywords: Commercial waste; Food waste; Sustainable waste management; Waste generation; Waste minimization; Recycling; Composting; Malaysia

Introduction

The management of solid wastes continues to be a major challenge in urban areas throughout the world, especially in rapidly growing cities and towns in developing countries (United Nations World Commission on Environment and Development). The Population Reference Bureau [1], estimated the world population to have reached seven billion in 2011, with more than half residing in urban areas (Global Health Observatory 2013). In 2010, 11 billion tons of solid waste were collected worldwide and an even larger but unknown quantity generated [2]. As urbanization continues to take place, the management of solid waste is becoming a significant environmental and public health issue. Various technical, financial, institutional, economic, and social factors are responsible for constraining the development of effective solid waste management systems in developing countries [3].

Even with the increased interest in solid waste management, one waste stream that has not received sufficient attention is that generated from commercial premises [4]. The term commercial waste covers a wide range of commercial businesses and in some cases includes wastes from industrial processes and the construction industry [4-6]. In line with the scope of this research, commercial waste in this study was defined as waste from premises used for the purpose of business or trade and it does not include industrial, construction and demolition waste. Strategies to collect, reduce, recycle or otherwise manage commercial waste are often challenging and complex [4,7]. However, there is a clear consensus that commercial waste offers attractive opportunities for resource efficiency, waste prevention, re-use and recycling [6-10]. Malaysia is selected as the country of study as it represents a developing country in the rapidly growing region of Southeast Asia that has achieved noteworthy economic growth in the

past few decades [11]. During this period, an increase in urbanization has been observed leading to a number of waste management challenges [12]. In turn, the Malaysian governments have developed a number of policies and initiatives to address this issue. The Solid Waste Management and Public Cleansing Bill forms the basis of policy to support the government's targets of 20 per cent recycling, 100 per cent separation at source and closure of all historic, unsanitary disposal sites by 2020 (Ministry of Housing and Local Government Malaysia 2005). Currently, official recycling is estimated to be 3-5 per cent (Ministry of Housing and Local Government Malaysia 2012), although this figure is almost certainly higher due to the contribution of the informal waste management sector [13]. In terms of waste disposal, landfill dominates with 95-97 per cent of the waste collected being disposed in one of the 112 landfills [13]. According to the Ministry of Housing and Local Government Malaysia, the majority of landfills is at full capacity and operates with minimal leachate and landfill gas control. Collection costs make up 83 per cent of the total waste management budget [14] which constrains current efforts for upstream sustainable waste management activities such as minimisation, reuse, recycling, waste treatment and energy from waste [11]. Furthermore, despite efforts by the Ministry of Housing and Local Government Malaysia, public awareness of the environment and waste remains low.

The change in lifestyle and consumption patterns linked to the rapid economic development and population growth in Malaysia has contributed to an increase in waste generation [13]. Waste generation has increased from 0.5 kg per capita per day in the late 1980s, to more than 1.3 kg per capita per day in 2009 [13]. In urban centres, such as Kuala Lumpur and Petaling Jaya, the generation of waste has increased to 1.5–2.5 kg per capita per day [13,15,16]. To date, annual waste generation has reached 11 million tonnes with approximately half comprising of food waste (Japan International Cooperation Agency 2006, Ministry of Housing and Local Government Malaysia 2012) [17]. The increase in the volume of waste is also contributing towards Malaysia's rapidly growing carbon footprint [18].

This study provides new empirical data in the less studied field of commercial waste generation in urban areas of the developing world, by presenting a case study in Malaysia. The research explores the generation, challenges and opportunities for sustainable solid waste management of commercial waste. There is a focus on food waste generation, as the food service sector (i.e. restaurants and cafes) is the main waste producer in this study. Policy recommendations for improvement and the relevance of this research are also discussed.

Methodology

Description of the study area

A commercial district in urban Malaysia was selected as the study area to investigate the generation of commercial waste, the challenge it poses, and the opportunities it offers for sustainable solid waste management. The commercial district of SS2 Petaling Jaya (3.0833° N, 101.6500° E) was selected as it represents a typical commercial area in urban Malaysia, comprising a variety of local small to medium commercial businesses (e.g. restaurants, cafes, convenience shops, health and beauty salons, electronics, clothing and furniture retail) and servicing a thriving residential area, in this case the neighborhood of Petaling Jaya. As with most other commercial districts in Malaysia, SS2 faces challenges related to waste management, public cleaning and amenity [19]. These issues are expected to intensify as the surrounding residential area of Petaling Jaya continues to grow.

Research design

The research design was organized into two main sections. Firstly, a background data assessment on status of the commercial and waste management sectors in Malaysia was conducted. Secondly, a focus group was undertaken with local environmental Non-Governmental Organizations (NGOs), academics and members of the public, to guide the design of a questionnaire (questionnaire template provided in the supplementary material).

Qualitative and quantitative data was collected with the use of an interviewer-administered questionnaire, translated in the three main languages spoken in Malaysia, Bahasa Malaysia (Malay), Mandarin Chinese and English. The team of researchers spoke all three languages which facilitated the research process. All 285 registered commercial businesses in the SS2 area of Petaling Jaya, Malaysia were approached to participate in the survey. In total, 63 questionnaires were completed.

The questionnaire was organized in four thematic sections: background information of the business, waste generation, waste collection, and feedback on the overall waste management service currently provided. The data was further analyzed in quantitative and qualitative ways, in order to add depth to the data analysis and draw fuller picture of the current practices, awareness level, expectations and perceptions related to the management of commercial waste. Statistical analysis was carried out using the PSAW software (version 18). The Analysis of variance (ANOVA) test was applied to determine the statistical significance amongst the results, and assess the correlation between the different variables.

Results and Discussion

The results are presented and discussed under the four thematic sections described in the methodology section which are i) profile of participating businesses, ii) waste generation, iii) waste collection and iv) feedback on overall waste management service provided.

Profile of participating businesses

The commercial premises were grouped according to the nature of their business. As illustrated in Table 1, 'Food service' is the largest group including restaurants, cafes and other food service establishments, followed by 'Health and wellbeing' such as pharmacists, gyms, medical doctors, spas, nutrition and healthy eating specialists.

Type of commercial business	Number	% of business type in relation to total sample
Food service	24	38%
Health and wellbeing	13	21%
House furnishing	5	8%
Convenience shop	4	6%
Electric and electronic equipment	4	6%
Clothes retailer	4	6%
Car repair garage	4	6%
Miscellaneous (2 optical centres, 1 pet shop, 1 florist and 1 internet cafe)	5	8%
Total	63	100%

Table 1: Type of participating businesses

The size of the business was measured by the number of staff employed and customers served per day. It was established that all the businesses are relatively small in size (i.e. Small to Medium Enterprises, according to Secretariat to National SME Development Council, 2005) with most of them employing less than 10 staff and serving less than 50 customers per day, as illustrated in Table 2. The largest businesses, both in terms of staff employed and customers served, were from the food service sector.

From the background information collected, it was anticipated that businesses from the food service sector produce most of the waste, as they represented the majority of businesses in the area, they employed more staff and they served more customers. This was confirmed by the data collected about waste generation presented in the section below.

Number of staff	Percentage	Number of customers per day	Percentage
<10	80%	<50	46%
20-Nov	18%	51-100	21%
>20	2%	>100	33%

Table 2: Size of participating businesses

Waste generation

A total of 58 respondents reported on their daily waste generation and the results are presented in the form of a frequency histogram. As illustrated in Figure 1, 26 businesses generated less than 25 kg of waste per day, 8 businesses generated between 25 and 50 kg, and 5 businesses between 50 and 75 kg per day. The average waste generation was 51 kg per day, although half of the businesses generated less than 25 kg per day, reflecting the fact that most businesses generate relatively small quantities of waste on a daily basis.

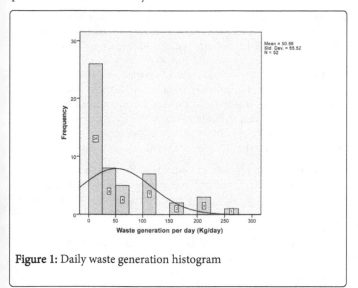

Figure 1: Daily waste generation histogram

Analyzing the waste generation trends in relation to the type of business, revealed the food service group as the largest producer; Table 3 shows that the food service group generated an average of 98 kg per day. This is followed by the house furnishing businesses with 26.6 kg per day, car repair garages and convenience shops with 26 kg per day, health and wellbeing businesses at 25 kg per day, and finally electric, electronic equipment and clothes retailers that generated on average 16 kg per day. The statistical analysis of the waste generation for the different business groups, revealed that the mean difference between the groups was statistically significant at $p<0.05$.

Type of business	N	Minimum	Maximum	Mean	Std. Deviation
Food service	19	1	265	98.184	80.1076
Health and wellbeing	12	0.5	106	25.083	38.5008
House furnishing	5	1	106	26.6	44.6016
Convenience shop	4	0.5	100	26.125	49.2551
Electric and electronic equipment	4	0.5	39	16.625	19.0848
Car repair garage	4	5	53	26.375	21.8532
Clothes retail	4	1	53	17.125	24.5505

Table 3: Average daily waste generation by business type

The average daily waste generation per customer is presented in Table 4. This table allows comparisons between the different types of businesses. The statistical analysis of the waste generation per costumer for the different business groups, revealed that the mean difference between the groups was statistically significant at $p<0.05$. The car repair garages had the highest average daily waste generation per customer with 3.3 kg/customer/day, followed by the food service businesses with 2.2 kg/ customer/day and health and wellbeing with

1.5 kg/ customer/day. The relatively high daily waste generation per customer by the car repair garages businesses was attributed to the relatively high weight of the waste generated by this type of business. The same applied for food service businesses producing primarily food waste which is relatively heavy due to its high water content and density. Convenience shops and electric and electronic equipment retailers had a lower daily waste generation per customer as explained by the production of relatively lighter wastes such as packaging cardboard and paper.

Type of business	N	Minimum	Maximum	Mean	Std. Deviation
Car repair garage	2	0.02	6.63	3.325	4.67398
Food service	17	0.18	21.2	2.2129	5.03502
Health and wellbeing	6	0.1	4.24	1.5533	1.82843
Electric and electronic equipment	3	0.01	0.65	0.2733	0.33471
Convenience shop	2	0	0.33	0.165	0.23335
House furnishing	4	0.05	0.42	0.155	0.17711

Table 4: Average daily waste generation per customer

Businesses were asked to describe the type of waste they produce in terms of composition. Paper (including newspapers and cardboard) and plastic waste (including plastic packaging, plastic bottles and Styrofoam containers) were the most commonly produced wastes, followed by food waste, aluminum, glass and medical waste. Having food waste as the third most common waste produced was attributed to the high percentage of food outlets amongst the participating businesses.

Comparing the results from this study with other published data revealed several points of discussion. Firstly, the term commercial waste covers a wide range of commercial businesses and in some cases includes wastes from industrial processes. The diverse nature of commercial waste producers led to a wide range of waste generation rates, making it difficult to draw conclusions for the whole waste stream. In addition, possibly due to the difficulties in defining commercial waste, only limited data is available on waste generation and composition. Though this does not allow for extensive comparison between the values derived from this study and other sources, it supports the argument that more empirical data is required and highlights the importance of the present study.

Although methodologies, system boundaries, the social and the material context might differ between studies, a preliminary comparison with other sources reporting on commercial waste generation rates from the food service sector was possible (Table 5). The Californian Department of Resources Recycling and Recovery reported an average waste generation of 8 kg/employee/day for commercial premises from the food service sector [20]. The average from this study was higher, at 12 kg/employee/day.

A study across Massachusetts reported an average food waste generation rate at restaurants of approximately 40 tons/establishment/ year (Massachusetts Department of Environmental Protection 2002), whereas the Sustainable Restaurant Association (SRA) in the UK reported an average of 24 tons/establishment/year (Sustainable

Restaurant Association 2010). Both figures are lower than the average from this study at 51 tons/establishment/year [21].

The study by the SRA in the UK also reported an average food waste generation per customer of 0.5 kg/customer/day. In the current study, the average waste generation from restaurants was 2.2 kg/customer/day. Assuming that 53 per cent of the total waste in restaurants was food waste (California Integrated Waste Management Board) [20], it was derived that 1.1 kg/customer/day of food waste was generated by the restaurants in this area. This average figure is significantly higher than the one reported in the UK by the SRA.

Waste generation	Waste type	Country	Comparison with this study
8 kg/employee/day	total waste	USA	12 kg/employee/day
40 tons/restaurant/year	food waste	USA	51 tons/restaurant/year
24 tons/restaurant/year	food waste	UK	51 tons/restaurant/year
0.5 kg/customer/day	food waste	UK	1.1 kg/customer/day

Table 5: Waste generation trends from the food service sector. Sources: (California Integrated Waste Management Board 2006; Sustainable Restaurant Association [20]; Massachusetts Department of Environmental Protection [21])

These findings raise some concerns considering that the USA and the UK report some of the highest waste generation rates from developed countries. Whilst it is problematic to draw satisfactory conclusions as to why food waste generation rates are high in Malaysia, the relatively low cost of food and the heavily subsidized economy, may provide some clues to the wasteful practices in relation to food.

The authors argue that the high food waste generation rates can be attributed to the relatively low cost of food in Malaysia and the consumers' wasteful behaviors observed in this fast growing economy. Although food prices have increased in recent years, Malaysia's abundance of natural resources and fuel subsidies maintain the relatively low food prices, allowing for wasteful practices in relation to food surplus and waste.

Waste collection

The participants were asked to give details on the type of waste receptacles they use, the frequency and cost of waste collection. Sixty eight per cent of the businesses had their own collection bin, whereas 29 per cent stated that they simply place their waste in plastic bags at the front of their shop or at the back lane because they do not have bins. No participants admitted that they dispose their liquid waste directly into the drains by the road, although this is a common practice amongst the food outlets observed during the survey. Only half of the food service sector businesses stated they have grease traps installed.

The majority of the respondents (89 per cent) receive waste collection services, 8 per cent stated they do not receive any waste collection, while 3 per cent did not respond. The respondents who stated they do not receive any waste collection services highlighted the lack of waste storage facilities as the main reason for not receiving waste collection. In addition, twenty per cent of the respondents did not know who is responsible for their waste collection.

When asked how frequently waste is collected, the majority responded daily, 6 per cent stated they did not know, and the rest gave a variation of answers ranging from once every 2 days, to once a week. The variation in responses was attributed to the lack of interest and / or awareness regarding waste management by the waste producers.

The majority of the participants (73 per cent) did not provide an answer when asked how much they pay for waste collection. Following this up further and discussing in a more informal setting, it was revealed that some participants did not know how much they pay, others did not want to disclose this type of information, and some did not pay at all. Similar results were obtained when the researchers inquired regarding the method of payment for the waste collection services received. The majority did not provide an answer, nearly a third admitted they did not know, and only 26 per cent stated they pay through their rent or directly to a private waste collection company.

The study revealed a number of issues related to waste collection, similar to the ones faced by other developing countries especially in Southeast Asia [22,23]. Firstly, there appeared to be a lack of waste receptacles, both for individual businesses and for public use. The lack of waste bins contributed to littering, and other amenity issues. The provision of additional bins ranked as the top recommendation by the business owners, as explained in the following section.

The second key finding related to the low interest and / or awareness regarding waste management and collection by the business owners. This was revealed by the fact that a large number of businesses were unclear as to who is providing the service, how much it costs and how this fee is recovered, or do not pay at all for waste collection. This issue highlighted the need for a more straightforward and systematic waste collection system, with a simplified cost recovery mechanism accompanied by a communication campaign to improve awareness amongst the waste producers.

In terms of food waste, the limited number of grease traps installed and the common practice observed amongst food outlets to dispose liquid wastes, such as oils and fats, directly into the open drains is another issue requiring attention. This practice not only contributes to pollution of surface water bodies but also leads to amenity issues such as odour and pests.

Feedback on the overall waste management service

Participants were asked to offer their feedback on the overall waste collection and management service provided, by indicating how satisfied or dissatisfied they are on a Likert style scale. As Figure 2 illustrates, most participants indicated they were 'Moderately satisfied' (35), 18 were 'Satisfied', 3 'Very satisfied', 3 'Very dissatisfied' and 2 'Dissatisfied'.

The participants that responded as 'Moderately satisfied', 'Dissatisfied' and 'Very dissatisfied' were then asked to comment on the issues that contribute to their dissatisfaction. The presence of pests, the odour from the drains and litter on the streets were amongst the most commonly cited problems. These problems were also confirmed by the researchers' observations. Infrequent waste collection appeared to effect only four participants and no one quoted the cost of waste collection as the reason for being dissatisfied with the service received.

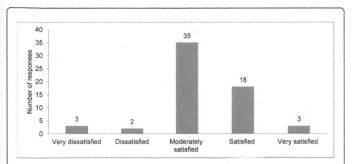

Figure 2: Level of satisfaction with the current waste management system

Priority	Recommendations
1	Provide public bins
2	Improve waste collection services
3	Improve law enforcement
4	Increase fines for polluters
5	Reduce service cost
6	Carry out regular public awareness campaigns
7	Offer financial and technical assistance

Table 6: Prioritized list of recommendations for improving waste management in the area.

The participants' suggestions for improving waste management in the area are presented in Table 6 in the form of a prioritized list of recommendations. In line with findings from the previous sections of this survey, providing more bins was the highest priority, and improvement of waste collection the second most important recommendation. Better law enforcement and higher financial penalties for polluters were the third and four most important suggestions, whereas reducing cost, carrying out public awareness campaigns and offering financial or technical assistance appeared less important interventions.

Finally, when asked whether they would use recycling bins if they were provided without additional cost, 86 per cent responded in a positive way, whereas 9 per cent gave a negative answer. In addition, the majority of the food and drinks businesses (75 per cent) suggested that they would participate in a recycling scheme for used cooked oil, if they could sell it.

The points emerging from this section offer insight on the concerns, expectations and suggestions by the commercial waste producers. The participants' views provided a guide to the existing problems and can inform the direction of future interventions.

In more detail, the feedback indicated a moderately satisfied sample with the waste management service provided. The presence of pests, odorous drains and litter on the streets are the main reasons for dissatisfaction, whereas the waste collection frequency and cost are not. These concerns are related to amenity and public health, rather than commercial waste management per se. However, they can be attributed to the disposal of liquid wastes, such as oils and fats from

food outlets, directly into drains. This practice significantly contributed to the problem of pests and odour. Resolving the uncontrolled disposal of liquid wastes would improve these amenity issues, as would the reduction in littering.

The issue of insufficient waste receptacles (both individual for each shop and for public use) was highlighted again in this section, and provision of more bins was suggested as the top priority recommendation for improvement. Finally, businesses in the study area suggested that they would participate in recycling, as well as a used oil collection scheme, if such options were available. The positive response to recycling indicated the waste producers are ready and indeed willing to recycle.

Limitations

The collected sample (63) represented approximately 23 per cent of the total surveyed population (285) of the study area. Replicating the survey in other commercial districts across Malaysia would further validate the results of this study. As employed by other waste generation studies (Massachusetts Department of Environmental Protection [21]; California Integrated Waste Management Board 2006; Sustainable Restaurant Association [20]) this research relied on waste producers to report on their waste generation rates. Therefore the margin of error related to the waste generation rates is unknown.

Conclusions

This paper presented a study of commercial and food waste in Malaysia. The study highlighted a number of issues related to the management of commercial waste, such as amenity and public health problems (pests, odour and littering) caused primarily by irresponsible disposal practices of food and liquid wastes, low awareness and interest in waste management, and inefficient provision of waste receptacles. In addition, the food service sector emerged as the main waste producer with food waste generation rates exceeding the ones reported from the UK and the US. These characteristics are typical of developing countries and not unique only to Malaysia [24-27]. However, these same amenity and public health problems can indeed act as drivers for change to a more advanced waste management system. The financial capital to support this transition and maintain the new waste management system can be provided by a simpler and more efficient cost recovery mechanism.

This research has wider implications to waste management policy. It is argued that commercial and food waste has the potential to act as drivers for sustainable resource and waste management. The study suggests that commercial waste in a middle income developing country such as Malaysia provides opportunities for resource efficiency and minimisation through recycling. The willingness of the waste producers to participate in recycling can act as the foundation of business and industry driven changes in waste management. Before capitalizing on this opportunity, problems related to amenity and public health can be resolved by simple infrastructure adjustments such as the provision of bins, grease traps and improved public areas cleansing.

In terms of food waste, the high generation rates reported by this study suggest that food waste should be a priority waste stream in waste management policy prevention and minimisation, in the case of Malaysia. This point is potentially relevant to other countries with similar developmental status as Malaysia, where food waste forms a significant portion of the overall waste stream and is the main contributor to public health and amenity problems. Food waste offers an opportunity to formulate waste management strategies appropriate to the specific needs of Malaysia, rather than just adopt ones produced by developed countries for the needs of developed countries.

Acknowledgements

The researchers would like to express their gratitude to all the research support staff that contributed during the data collection and analysis process; without which this study would not have materialized.

Funding

This work was supported by the Research Management Centre of the University Technology Malaysia (UTM) [research grant V21000].

References

1. Population Reference Bureau (2011) The world at seven billion world population data sheet.

2. Renner M (2012) Making the Green Economy Work for Everybody. State of the World 2012.

3. Ogawa H (2000) Sustainable solid waste management in developing countries. 7th ISWA International Congress and Exhibition. Kuala Lumpur Malaysia. World health Organization.

4. Iman A, Chak C, Alan C (2011) Commercial solid waste management for New York City. New York.

5. Salhofer S, Isaac NA (2002) Importance of public relations in recycling strategies: principles and case studies. Environ Manage 30: 68-76.

6. Purcell M, Magette W (2009) Prediction of household and commercial BMW generation according to socio-economic and other factors for the Dublin region. Waste management 29: 1237-1250.

7. WRAP (2011) Co-collection of household and commercial waste and recyclables. Banbury.

8. Papargyropoulou E, Padfield R, Harrison O, Preece C (2012) The rise of sustainability services for the built environment in Malaysia. Sustainable Cities and Society 5: 44–51.

9. Helftewes M, Flamme S, Nelles M (2012) Greenhouse gas emissions of different waste treatment options for sector-specific commercial and industrial waste in Germany. Waste Manag Res 30: 421-431.

10. Yi S, Yoo KY, Hanaki K (2011) Characteristics of MSW and heat energy recovery between residential and commercial areas in Seoul. Waste Manag 31: 595-602.

11. Manaf LA, Samah MA, Zukki NI (2009) Municipal solid waste management in Malaysia: practices and challenges. Waste Manag 29: 2902-2906.

12. Hezri A, Nordin Hasan M (2006) Towards sustainable development? The evolution of environmental policy in Malaysia. Natural Resources Forum 30: 37-50.

13. Agamuthu P, Fauziah HS, Khidzir K (2009) Evolution of solid waste management in Malaysia: impacts and implications of the solid waste bill 2007. Journal of material cycles and waste management 11: 96-103.

14. Isa MH, Asaari FA, Ramli NA, Ahmad S, Siew TS (2005) Solid waste collection and recycling in Nibong Tebal, Penang, Malaysia: a case study. Waste Manag Res 23: 565-570.

15. Kathirvale S, Muhd Yunus MN, Sopian K (2003) Energy potential from municipal solid waste in Malaysia. Renewable Energy 29: 559-567.

16. EPU (2007) Ninth Malaysia Plan 2006 – 2010.

17. Fauziah S, Agamuthu P (2009) Recycling of household organic waste in Malaysia: The challenges Proc of the International Symposium of Environmental Science and Technology. China: Shanghai Science Press USA: 2234-2240.

18. Padfield R, Papargyropoulou E, Preece C (2012) A preliminary assessment of greenhouse gas emission trends in the production and consumption of food in Malaysia. International Journal of Technology 3: 56-66.

19. Murad MW, Siwar C (2007) Waste management and recycling practices of the urban poor: a case study in Kuala lumpur city, Malaysia. Waste Manag Res 25: 3-13.

20. California Integrated Waste Management Board (2006) Targeted Statewide Waste Characterization Study: Waste Disposal and Diversion Findings for Selected Industry Groups. California.

21. Massachusetts Department of Environmental Protection (2002) Identification Characterization and Mapping of Food Waste and Food Waste Generators in Massachusetts. Massachusetts.

22. Seng B, Kaneko H, Hirayama K, Katayama-Hirayama K (2011) Municipal solid waste management in Phnom Penh, capital city of Cambodia. Waste management & research 29: 491–500.

23. Heisler T (2004) Lessons from Experience: A Comparative Look at Solid Waste Management Policies In Cambodia India The Philippines and Sri Lanka.

24. Guerrero LA, Maas G, Hogland W (2013) Solid waste management challenges for cities in developing countries. Waste Manag 33: 220-232.

25. Asase M, Yanful EK, Mensah M, Stanford J, Amponsah S (2009) Comparison of Municipal Solid Waste Management Systems in Canada and Ghana: a Case Study of the Cities of London Ontario and Kumasi Ghana. Waste management 29: 2779–2786.

26. Henry RK, Yongsheng Z, Jun D (2006) Municipal solid waste management challenges in developing countries--Kenyan case study. Waste Manag 26: 92-100.

27. Mrayyan B, Hamdi MR (2006) Management approaches to integrated solid waste in industrialized zones in Jordan: a case of Zarqa City. Waste Manag 26: 195-205.

24

Some Physical and Chemical Properties of Compost

El-Sayed G. Khater*

Agricultural Engineering Department, Faculty of Agriculture, Benha University, Egypt

***Corresponding author:** Farouk K. M. Wali, Assistant professor, Chemical technology Department, The Prince Sultan Industrial College, Saudi Arabia
E-mail: alsayed.khater@fagr.bu.edu.eg

Abstract

The main objective of this research was to study the physical and chemical properties of compost made of different row materials. These materials are cattle manure, herbal plants residues and sugar cane plants residues. These properties include: bulk density, moisture content, water holding capacity, porosity, pH, EC, total organic carbon, total organic matter, total nitrogen, total phosphorus, total potassium and C/N ratio. The bulk density value ranged from 420 to 655 kg m^{-3}. The moisture content values ranged from 23.50 to 32.10 %. The water holding capacity values ranged from 3.50 to 4.40 g water/g dry. The porosity values ranged from 60.69 to 72.47 % for different compost types. The pH value ranged from 6.3 to 7.8 and EC values ranged from 2.6 to 4.1 dS m-1 for different compost types. The total organic carbon values ranged from 16.6 to 23.89 %. The total organic matter values ranged from 28.60 to 41.20 %. The total nitrogen values ranged from 0.95 to 1.68 %. The total phosphorus and total potassium values ranged from 0.27 to 1.13 % and 0.27 to 2.11 %, respectively, for different compost types. The C/N ratio values ranged from 14.22:1 to 18.52:1.

Keywords: Physical properties; Chemical properties; Compost

Introduction

Composting is a technique which can be used to reduce the amount of organic waste through recycling and the production of soil fertilizers and conditioners. Compost is primarily used as a soil conditioner and not as much as a fertilizer because it contains a high organic content (90-95%) but generally low concentrations of nitrogen, phosphorus, potassium as well as macro and micro nutrients compared to commercial fertilizers. It is comparable to peat moss in its conditioning abilities. Areas where composting can be beneficial is in the recycling of the organic fraction of the municipal waste. It reduces as much as 30% of the volume, in the form of organic matter, entering our already overcrowded landfill sites. Furthermore the composting process, if performed correctly, transforms wet and odorous organic waste into an aesthetically, dryer, decomposed and reusable product [1].

Crop residues, unused bedding materials, silage, manures, and similar on-farm materials can be used as co-compost cover materials, along with many off-farm residues and wastes. Since a mortality compost pile cannot be turned until the bio-decomposition of the carcass body has been largely completed, the type and thickness of the cover and base layer materials play a key role in influencing the biodegradation of carcasses, and the development and retention of heat that is necessary for pathogen inactivation [2].

Quality control during compost production should ensure adequate chemical and physical properties [3], as well as an adequate degree of stability and maturity [4]. The beneficial effects on crop production and soil quality reported in literature [5,6] are directly related to the physical, chemical and biological properties of the composts [7].

The physical and chemical properties of organic wastes and the factors that affect their performance in composting require easily identifiable and reliable methods to control the process *in situ*, in order to make proper decisions about its performance [8].

Although the characteristics of yard waste will vary, depending upon the predominant vegetation in the area and the season of the year for its collection, composted green waste typically contains low levels of heavy metals, commonly present in sludge-based composts, which makes them more environmentally sound [9].

To produce a sound and a good quality compost, due to the lake of physical and chemical properties of the compost should be determined by the end of processing period, therefore, the main objective of this research was to study some physical and chemical properties of compost made of different row materials. These properties include: pH, EC, total organic carbon, total organic matter, total nitrogen, total phosphorus, total potassium, C/N ratio, bulk density, moisture content, water holding capacity and porosity.

Materials and Methods

The experiment was carried out at the Compost Unit at the Experimental Research Station, Faculty of Agriculture, Moshtohor, Benha University. Some of agricultural wastes are used for compost making, these wastes are cattle manure, herbal plants residues and sugar cane plants residues. The physical and chemical properties that used in the manufacturing the compost are listed in tables (Tables 1 and 2).

Properties	Raw materials		
	Cattle manure	Herbal plants residues	Sugar cane plants residues
Bulk density (kg m^{-3})	750.00	335.00	426.00
Moisture content (%)	58.30	16.20	36.20
Water holding capacity (g water/g dry sample)	3.00	3.50	3.30
Porosity (%)	41.57	80.62	69.96

Table 1: Physical properties of the raw materials used in compost making.

Properties	Raw materials		
	Cattle manure	Herbal plants residues	Sugar cane plants residues
pH	8.1	4.3	7.1
EC (dS m^{-1})	4.2	1.3	3.1
Total organic carbon (%)	18.16	9.4	20
Total organic matter (%)	31.3	43.1	61.3
Total nitrogen (%)	0.93	1.35	1.62
Total phosphorus (%)	0.21	0.36	1.12
Total potassium (%)	0.17	0.42	1.36
C/N ratio	19.53:1	6.97:1	12.35:1

Table 2: Chemical properties of the raw materials used in compost making.

Five different types of compost were obtained by mixing cattle manure with herbal plants residues and sugar cane plants residues at different ratios to form:

1. C_1: cattle manure (100:0)
2. C_2: cattle manure and herbal plants residues (50:50)
3. C_3: cattle manure and sugar cane plants residues (50:50)
4. C_4: herbal plants residues (100:0)
5. C_5: sugar cane plants residues (100:0)

The mixtures of wastes were composted in piles (1.5 m high, 3 m width and 80 m long). The piles were turned periodically to maintain adequate O_2 levels. The piles were turned weekly during the maturation phase in order to improve the O_2 level inside the pile. Pile moisture was controlled by adding enough water to keep the moisture content not less than 50%. Samples were taken at the end of the composting process to determine the chemical and physical properties.

Each sample was made by mixing five subsamples taken from five points in the pile. Samples were placed in polyethylene bags and transferred to the laboratory for analysis.

Physical properties

The physical properties include: bulk density, moisture content, water holding capacity and porosity.

Moisture contents (MC)

Moisture content (wet basis) throughout this study was measured by drying at 105°C for approximately 24 h or at constant weight [10].

Water holding capacity (WHC)

A wet sample of known initial moisture content was weighed (Wi) and placed in a beaker. After soaking in water for 1–2 days and draining excess water through Whatman #2filter paper, the saturated sample was weighed again (Ws). The amount of water retained by dry sample was calculated as the WHC. The water holding capacity (g water/g dry material) is calculated as [11]:

$$WHC = \frac{\{(Ws - Wi) + MC \times Wi\}}{\{(1 - MC) \times Wi\}} \tag{1}$$

Where:

Wi is the initial weight of sample (g)

Ws is the final weight of sample (g)

MC is the initial moisture content of sample (decimal)

Bulk density and porosity

Bulk density was measured using an approximately 10 liters volume container. The container was filled with material, and then the material was slightly compacted to ensure absence of large void spaces.

The bulk density was calculated by dividing the weight of the material by the volume of material in the container.

Compost porosity (ε_a) was determined using the known density of water (ρ_w; 1000 kg m^{-3}) and estimated densities of organic matter (ρ_{om}; 1600 kg m^{-3}), and ash (ρ_{ash}; 2500 kg m^{-3}), as well as the moisture content and bulk densities of the sample [12-14]. If the moisture content (MC), dry matter (DM), organic matter (OM), and wet bulk density (ρ_{wb}) of samples are known, the porosity can be calculated using the following equation:

$$\varepsilon_a = 1 - \rho_{wb}\left(\frac{MC}{\rho_w} + \frac{DM \cdot OM}{\rho_{om}} + \frac{DM \cdot (1 - OM)}{\rho_{ash}}\right) \times 100$$

Where:

ε_a is the porosity (%)

ρ_{wb} is the wet bulk density (kg m^{-3})

ρ_w is the density of water (kg m^{-3})

ρ_{om} is the density of organic matter (kg m^{-3})

ρ_{ash} is the density of ash (kg m^{-3})

MC is the moisture content (decimal)

DM is the dry matter (decimal)

OM is the organic matter (decimal)

Chemical properties

The chemical properties include: pH, EC, total organic carbon, total organic matter, total nitrogen, total phosphorus, total potassium and C/N ratio.

Electrical conductivity was measured using EC meter (Model ORION 105 – Range 0 - 199.99 dS m^{-1} ± 0.01, USA). pH was measured using pH meter (Model ORION 230A – Range -2 - 19.99 ± 0.01, USA). Total organic carbon (TOC) by the dry combustion method at 540C for 4 h according to [15]. Total organic matter was measured by combustion at 550C for 8 h according to [16] and total nitrogen (TN) by Kjeldahl digestion [17]. Potassium (K) was determined by atomic absorption and phosphorus (P) was determined colorimetrically following the [18] method.

Results and Discussion

Physical properties

Table 3 shows the physical properties (bulk density, moisture content, water holding capacity and porosity) of the different types of compost (cattle manure, cattle manure and herbal plants residues (50:50), cattle manure and sugar cane plants residues (50: 50), herbal plants residues and sugar cane plants residues).

The results indicate that the bulk density value ranged from 420 to 655 kg m^{-3} for different compost types. The highest value of bulk density (655 kg m^{-3}) was found for cattle manure compost and the lowest value of bulk density (420 kg m^{-3}) was found for sugar cane plants residues compost. [8] found that the bulk density values were between 447 and 502 kg m^{-3} for different compost types, as agreed with [19-22] results.

It could be seen that the bulk density of compost decreases with increasing the compost total organic matter. Figure 1 shows the

relationship between the bulk density and the total organic matter. It decreases from 655 to 420 kg m^{-3} when the total organic matter increased from 28.6 to 41.2%.

Properties	Compost types				
	C_1	C_2	C_3	C_4	C_5
Bulk density (kg m^{-3})	655	625	573	582	420
Moisture content (%)	25.6	23.5	30.1	31.2	32.1
Water holding capacity (g water/g dry sample)	3.5	3.7	4.1	3.9	4.4
Porosity (%)	60.69	62.67	63.52	66.56	72.47

Table 3: Physical properties of different compost types. C_1: cattle manure (100: 0); C_2: cattle manure and herbal plants residues (50: 50); C_3: cattle manure and sugar cane plants residues (50: 50); C_4: herbal plants residues (100: 0); C_5: sugar cane plants residues (100: 0).

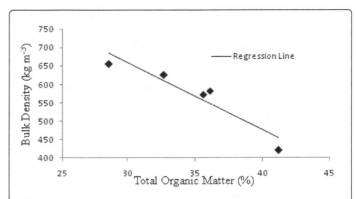

Figure 1: The relationship between the bulk density and the total organic matter.

The regression between the bulk density of compost and the compost total organic matter is show the following equation:

BD = -18.399 TOM + 1212 $R^2 = 0.89$ (3)

Where:

BD is the bulk density (kg m^{-3})

TOM is the total organic matter (%)

The moisture content values ranged from 23.50 to 32.10% for different compost types. The lowest value of moisture content (23.50%) was found for cattle manure and herbal plants residues (50: 50) compost and the highest value of moisture content (32.10%) was obtained for sugar cane plants residues compost.

Regarding the water holding capacity values ranged from 3.50 to 4.40 g water/g dry sample for different compost types. The lowest value of water holding capacity (3.50 g water/g dry sample) was found for cattle manure compost and the highest value of water holding capacity (4.40 g water/g dry sample) was found for sugar cane plants residues compost.

The porosity values ranged from 60.69 to 72.47% for different compost types. The lowest value of the porosity (60.69%) was found for cattle manure compost and the highest value of the porosity (72.47%) was found for sugar cane plants residues compost. The porosity depends on bulk density and moisture content of compost. The porosity decreased with increasing bulk density and moisture content. Figures 2 and 3 shows the relationship between the porosity and bulk density and the porosity and moisture content. The results indicate that the porosity of compost decreased from 72.47 to 60.69% when the bulk density increased from 420 to 655 kg m^{-3}. The results indicate that the porosity of compost decreased from 72.47 to 60.69% when the moisture content increased from 25.6 to 32.1%. These results agreed with those obtained by [11].

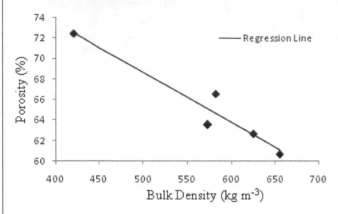

Figure 2: The relationship between the porosity and bulk density.

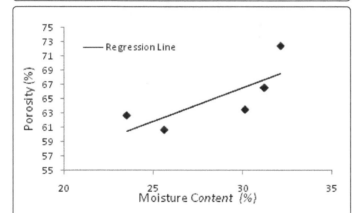

Figure 3: The relationship between the porosity and moisture content.

The regression between the porosity and the bulk density of compost and the porosity of compost and the compost total organic matter are shown in the following equations:

$$\varepsilon_a = -0.0487\ BD + 92.981 \qquad R^2 = 0.93 \qquad (4)$$

$$\varepsilon_a = 0.9407\ TOM + 38.373 \qquad R^2 = 0.60 \qquad (5)$$

Chemical properties

Table 4 shows the chemical properties (pH, EC, total organic carbon, total organic matter, total nitrogen, total phosphorus, total potassium and C/N ratio) of different types of compost (cattle manure, cattle manure and herbal plants residues (50:50), cattle manure and sugar cane plants residues (50:50), herbal plants residues and sugar cane plants residues). It could be seen that the pH value ranged from

6.3 to 7.8 for different compost types. The highest value of pH (7.8) was found for herbal plants residues compost, while, the lowest value of pH (6.3) was obtained for sugar cane plants residues compost. This pH range is in the optimum range for growing media as mentioned by [23] who stated that the optimal range is from 5.2 to 7.3.

The EC values ranged from 2.6 to 4.1 dS m^{-1} for different compost types. The highest value of EC (4.1 dS m^{-1}) was found for cattle manure compost and the lowest value of EC (2.6 dS m^{-1}) was determined for the herbal plants residues compost. This EC range is in the optimum range (2.0 to 4.0) for growing media as mentioned by [24].

Properties	Compost types				
	C$_1$	C$_2$	C$_3$	C$_4$	C$_5$
pH	7.3	7.6	7.2	7.8	6.3
EC (dS m^{-1})	4.1	3.1	3.6	2.6	3.2
Total organic carbon (%)	16.6	18.98	20.64	20.93	23.89
Total organic matter (%)	28.6	32.7	35.6	36.1	41.2
Total nitrogen (%)	0.95	1.26	1.32	1.13	1.68
Total phosphorus (%)	0.31	0.27	0.51	0.32	1.13
Total potassium (%)	0.27	0.35	0.62	0.51	2.11
C/N ratio	17.47	15.06	15.64	18.52	14.22

Table 4: Chemical properties of different compost types. C$_1$: cattle manure (100: 0) C$_2$: cattle manure and herbal plants residues (50: 50); C$_3$: cattle manure and sugar cane plants residues (50: 50); C$_4$: herbal plants residues (100: 0); C$_5$: sugar cane plants residues (100: 0).

Regarding the total organic carbon results it was found that it ranged from 16.6 to 23.89% for different compost types under study, where, the lowest value of total organic carbon (16.6%) was found for cattle manure compost and the highest value of total organic carbon (23.89%) was obtained for sugar cane plants residues compost. These results are in agreement with [25] who found that the optimum value of total organic matter higher than 10%.

The total organic matter values ranged from 28.60 to 41.20% for different compost types. The lowest value of total organic matter (28.60%) was found for cattle manure compost and the highest value of total organic matter (41.20%) was found for sugar cane plants residues compost. These results are in agreement with [9] whose found that the highest value of total organic matter about 44%.

The total nitrogen values ranged from 0.95 to 1.68% for different compost types under study. The lowest value of total organic carbon (0.95%) was found for cattle manure compost and the highest value of total organic carbon (1.68%) was found for sugar cane plants residues compost. These results are in agreement with those obtained by [9] whose found that the total nitrogen rate ranged from 0.99 to 2.01%.

The total phosphorus and total potassium values ranged from 0.27 to 1.13% and 0.27 to 2.11%, respectively, for different compost types. The lowest value of total phosphorus (0.27%) was found for herbal plants residues and cattle manure (50: 50) compost and the highest value of total phosphor (1.13%) was found for sugar cane plants residues compost. The lowest value of total potassium (0.27%) was found for cattle manure compost and the highest value of total potassium (2.11%) was found for sugar cane plants residues compost.

Regarding the C/N ratio, it ranged from 14.22:1 to 18.52:1 for different compost types. The lowest value of C/N ratio (14.22: 1) was found for sugar cane plant residues compost and the highest value of C/N ratio (18.52: 1) was found for herbal plant residues compost. These results are in agreement with the results obtained by [26] whose found that the C/N ratio ranged from 15:1 to 20:1 is ideal for ready-to-use compost.

Conclusions

An experimental study was carried out successively to determine some physical and chemical properties of different compost types. The obtained results indicate that the pH value ranged from 6.3 to 7.8 and EC values ranged from 2.6 to 4.1 dS m^{-1} for different compost types. The total organic carbon values ranged from 16.6 to 23.89%. The total organic matter values ranged from 28.60 to 41.20%. The total nitrogen values ranged from 0.95 to 1.68%. The total phosphorus and total potassium values ranged from 0.27 to 1.13% and 0.27 to 2.11%, respectively, for different compost types. The C/N ratio values ranged from 14.22:1 to 18.52:1. The bulk density value ranged from 420 to 655 kg m^{-3}. The moisture content values ranged from 23.50 to 32.10%. The water holding capacity values ranged from 3.50 to 4.40 g water/g dry. The porosity values ranged from 60.69 to 72.47% for different compost types.

References

1. Knight W (1997) Compost convective airflow, N and C conservation with passive and active aeration. M. Sc. Thesis, Agric. And Biosystems Eng. McGill University, Canada.

2. Fonstad TA, Meier DE, Ingram LJ, Leonard J (2003) Evaluation and demonstration of composting as an option for dead animal management in Saskatchewan. Canadian Biosystems Engineering 45: 19–25.

3. Inbar Y, Chen Y, Hoitink HA (1993) Properties for establishing standards for utilization of composts in container media. In: Hoitink, H.A.J., Keener, H.M. (Eds.), Science and Engineering of Composting: Design, Environmental. Microbiological and Utilization Aspects. Ohio State University, USA.

4. Benito M, Masaguer A, Moliner A, Arrigo N, Palma RM (2003) Chemical and microbiological parameters for the characterization of stability and maturity of pruning waste compost. Biol Fert Soils 37: 184–189.

5. Hoitink HA, Stone AG, Han DY (1997) Suppression of plant diseases by composts. HortScience 32: 184–187.

6. Atiyeh RM, Edwards CA, Subler S, Metzger JD (2001) Pig manure vermicompost as component of a horticultural bedding plant medium: effects on physicochemical properties and plant growth. Bioresour Technol 78: 11–20.

7. He X, Logan TJ, Traina SJ (1995) Physical and chemical characteristics of selected US Municipal solid waste composts. J Environ Qual 24: 543–552.

8. Hurerta-Pujol O, Soliva M, Martinez-Farre FX, Valero J, Lopez M (2010) Bulk density determination as a simple and complementary too in composting process control. Bioresour Technol 101: 995–1001.

9. Benito M, Masaguer A, Moliner A, De Antonio R (2005) Use of pruning waste compost as a component in soilless growing media. Bioresour Technol 97: 2071–2076.

10. Moisture measurement-meat and products. Adopted and published by ASAE 1998, St. Joseph, MI: ASAE. 45th Edition P552. ASAE Standards, Washington DC, USA.

11. Ahn HK, Richard TL, Glanville TD (2008) Laboratory determination of compost physical parameters for modeling of airflow characteristics. Waste Management 28: 660–670.

12. Raman S (1995) Food Properties Handbook. CRC Press, Boca Raton, Florida.

13. van Cingel CT, Raats PA, van Haneghem IA (1999) Bulk density and porosity distributions in a compost pile. Netherlands Journal of Agricultural Science 47: 105–121.

14. Richard TL, Hamelers HV, Veeken AH, Silva T (2002) Moisture relationships in composting processes. Compost Science and Utilization 10: 286–302.

15. Abad M, Noguera P, Puchades R, Maquieira A, Noguera V (2002) Physico-chemical and chemical properties of some coconut coir dusts for use as a peat substitute for containerised ornamental plants. Bioresour Technol 82: 241–245.

16. Test Methods for the Examination of Composting and Compost (TMECC) (2001) The Composting Council Research and Education Foundation, Bethesda, MD, USA.

17. Bremmer JM, Mulvaney CS (1982) Nitrogen-total. In: Page, A.L., Miller, R.H., Keeney, D.R. (Eds.), Methods of Soil Analysis, Part 2. Chemical and Microbiological Properties, second ed., Agronomy series No. 9 ASA, SSSA, Madison, WI.

18. Murphy J, Riley JP (1962) A modified single solution method for determination of phosphate in natural waters. Anal Chem Acta 27: 31–36.

19. Raviv M, Tarre S, Geler Z, Shelef G (1987) Changes in some physical and chemical-properties of fibrous solids from cow manure and digested cow manure during composting. Biol Waste 19: 309–318.

20. Larney FJ, Olson AF, Carcamo AA, Chang C (2000) Physical changes during active and passive composting of beef feedlot manure in winter and summer. Bioresour Technol 75: 139–148.

21. Mohee R, Mudhoo A (2005) Analysis of the physical properties of an in-vessel composting matrix. Powder Technol 155: 92–99.

22. Romeela M, Ackmez M, Unmar GD (2008) Windrow co-composting of shredded office paper and broiler litter. Int J Environ Waste Manage 2: 3–23.

23. Bunt AC (1988) Media and Mixes for Container-Grown Plants, 2nd edn. Unwin Hyman Ltd, London, UK.

24. Hanlon EA (2012) Soil pH and electrical conductivity: a count extension soil laboratory manual. IFAS Extension, University of Florida, FL, USA.

25. Batjes NH (1996) Total carbon and nitrogen in the soils of the world. Eur J Soil Sci 47: 151–163.

26. Rosen CJ, Halbach TR, Swanson BT (1993) Horticultural uses of municipal solid waste components. Hortic Technol 3: 167–173.

Waste Prevention in a "Leasing Society"

Susanne Fischer[1], Meghan O'Brien[1], Henning Wilts[1]*, Sören Steger[1], Philipp Schepelmann[1], Nino David Jordan[2] and Bettina Rademacher[1]

[1]Research Group on Material Flows and Resource Management, Wuppertal Institute for Climate, Environment and Energy, Germany

[2]Institute for Sustainable Resources, University College London, London, UK

*Corresponding author: Henning Wilts, Wuppertal Institute for Climate, Environment and Energy, Germany
E-mail: henning.wilts@wupperinst.org

Abstract

A future-oriented and sustainable "Leasing Society" is based on a combination of new and innovative service-oriented business models, changed product and material ownership structures, increased and improved eco-design efforts, and reverse logistic structures. Together these elements have the potential to change the relationship between producers and consumers, and thereby create a new incentive structure in the economy regarding the use and re-use of resources. While the consumer in a leasing society buys a service (instead of a product), the producer in a leasing society retains the ownership of the product (instead of selling it) and sells the service of using the product. This creates producer incentives to re-use, remanufacture, and recycle products and materials and could become a cornerstone of the circular economy, depending on how the leasing society is implemented. While a predominantly positive picture of the success of a leasing society model and related business cases emerges from the bigger part of the available literature, this paper argues that the resource efficiency of respective business cases is highly dependent on the specific business case design. This paper develops a more cautious and differentiated definition of the leasing society by discussing relevant mechanisms and success factors of leasing society business cases. The leasing society is discussed from a micro business-oriented and a macro environment-oriented perspective complemented by a discussion of conditions for successful business models that reduce environmental impacts and resource footprints.

Keywords: Leasing society; Eco-design; Re-use of resources; Business

Introduction

Facing the globally ever-increasing consumption of natural resources and thus the increasing generation of waste, the prevention of waste in the first place has been identified as a key strategy for an increased efficiency of resource use. But defining the prevention of waste as the top priority of the waste hierarchy—as confirmed by the revised European waste framework directive (WFD, Directive 2008/98/EC)—is much more than a simple amendment of ways how to deal with waste. It is nothing less than a fundamental change of the socio-technical system of waste infrastructures with all its economic, legal, social and even cultural elements [1,2] and requires a transition from end-of-pipe technologies towards an integrated management of resources [3]. In the public opinion large-scale systems based on municipal waste collection schemes and end-of-pipe technologies like waste incineration, shredding or other volume reducing waste treatment procedures seem to literally have minimized these sorrows —in most developed countries and especially in the EU waste seemed to be a "solved problem".

Only recently has this perception been contested and the idea of a circular economy raised, increasing interest in the public debate, e.g. in the European Commission's Communication on Zero Waste: *Since the industrial revolution, our economies have developed a 'take-make-consume and dispose' pattern of growth — a linear model based on the assumption that resources are abundant, available, easy to source and cheap to dispose of. It is increasingly being understood that this threatens the competitiveness of Europe. Moving towards a more circular economy is essential to deliver the resource efficiency agenda established under the Europe 2020 Strategy for smart, sustainable and inclusive growth"* [4]. Therefore waste prevention has to be put into the context of eco-innovations in production and consumption that potentially might reduce environmental impacts and at the same time save costs for the different actors alongside the value chain [5].

This paper will look at how the development of a "leasing society" may contribute to waste prevention. Specifically, it focuses on the future potential of the leasing society in the EU for the transition toward a resource efficient, circular economy. It defines what a leasing society is, examines successful case studies, highlights transition challenges and presents key barriers and drivers to the further uptake of the leasing society, as well as the policies needed to overcome these barriers. The purpose of this paper is to strengthen understanding of the potential of leasing society business models and of the kind of challenges, which need to be overcome in order for the leasing society to contribute to a "green transition" in a meaningful way.

Understanding the Leasing Society

The conceptual roots

The term "Leasing Society" (or "Lease Society") is rather new. However, it is related to other already existing models of a sustainable society such as the "Service Society" or "Functional Society" [6], the Performance Society [7,8], "Collaborative Consumption" [9], the Sharing Economy [10,11], the "Circular Economy" [12-14], or "Using instead of Owning" [15,16]. One common element of these models is based on the idea that customer needs can be met by changing the business value proposition towards a higher service-orientation and changing product ownership structures towards an increased producer responsibility. Both of these underlying concepts, increased service-orientation and changed product ownership, have been discussed in the scientific realm for decades. Especially the sustainability research and marketing literature [17,18] have paved the way for this new business model approach.

On the part of sustainability literature, the idea of the leasing society is strongly connected with the concept of "Product-Service Systems" (PSS). The discussion about PSS was spawned by Stahel and Reday in 1976, who called for a shift of activities from manufacturing to service, that would concentrate on long-term leasing, maintenance and reconditioning activities in order to support economic development while saving resources at the same time [17]. In 1999, the first paper on PSS was published by Goedkoop et al. defining PSS as a "marketable set of products and services, capable of jointly fulfilling a user's need" [20]. Since then, a number of academic papers have picked up and developed the term and the concept, paving the way for the present understanding of PSS that recent articles and studies are based on [19-24]. E.g. Baines et al. have defined it as a "market proposition that extends the traditional functionality of a product by incorporating additional services. Here the emphasis is on the 'sale of use' rather than the 'sale of product'. The customer pays for using an asset, rather than its purchase, and so benefits from a restructuring of the risks, responsibilities, and costs traditionally associated with ownership" [25]. PSS span a wide range of activities on the continuum between pure products and pure services unfolding different subcategories. A number of subcategories of PSS have been developed [26-28], Tukker et al. have identified the following sub-categories [29]:

1. *"Product-oriented strategies"* put the product into the focus of the business activity. The customer buys the product and retains ownership of it, but also extra after-sale services are provided. These services could be product-related (e.g. financing scheme, maintenance, repair or take-back agreement) or include training and advice in order to optimise the product application.

2. *"Use-oriented strategies"* change the ownership structure of the traditional selling and buying activities—use, not the product itself, is sold. Different forms of consumption (e.g. alone or shared with others) and payments (e.g. per time unit or service unit) are possible.

3. *"Result-oriented strategies"* meet the real customer needs in new ways. Instead of selling the product or selling the use of a product, the result of using a product is sold. For example, the customer may purchase an outsourced activity from a third party (e.g. cleaning) or may buy a predefined functional result (e.g. cooling). The producer remains the owner of the product used and the customer pays for the provision of the results.[1]

Parallel to the development of the PSS continuum in environmental sustainability research, the marketing literature developed more service-oriented and value-adding concepts, such as *"Full Service Contracts"* [31], *"Functional Sales"* [32], *"Functional Product"* [33], or *"Performance-Based Contracting"* [34], which are very similar to the PSS concept. They focus on how business can improve their value proposition by meeting customer needs in more service-oriented ways while reducing total costs of product functionality. Environmental impacts here are rather subordinate; however the call for more service-

orientation (e.g. through selling functionality or performance) implies a change of incentives on the producer side that could affect the environmental performance of production and consumption, indirectly.

Besides the sustainability and marketing literature, sector-specific discussions in the chemical and energy-consuming industry [17], have brought forward the idea of new business models focussing on the retention of product ownership and selling only the result of using a product, in order to reduce costs for resources and optimize customer satisfaction. Both of the sector-specific concepts, namely "Chemical Management Services" [35] and "Energy Service Contracting" [36], focus on the outsourcing of former in-house activities (varnishing cars, optimizing energy use)and the contractual arranged procurement of performance in order to save resources and thereby costs.

The idea

The leasing society stands for a society or an economy that is characterized by a new relationship between producers and customers connected with new incentives of how to use resources. Thereby, this new producer customer relation is offering the potential of reducing environmental impacts by diminishing raw material extraction, resource consumption, waste generation and associated environmental impacts—as such, the leasing society has the potential to fundamentally contribute to the societal challenge of increasing resource efficiency and preventing waste. The leasing society is based on two main pillars:

1. More innovative and service-oriented business models to fulfil customer needs, focusing on the provision of product use and result of product use, and

2. A product ownership staying in the realm of the producer, while the customer either uses the actual product or consumes the actual result of the product use.

As indicated by the term, the leasing society refers to the established concept of leasing. However, leasing in its original meaning is not a new concept. It comprises a special contract between the owner of an asset and the user of that asset, which gives the latter one the right to use that asset for a certain amount of time. During the contract period, the owner, not the user, is responsible for maintenance and repairs. After the contract has expired, the owner receives the asset back. The leasing society has been coined as such, because a core part of its characteristics, namely the service orientation and the changed product ownership structures, is related to what is traditionally understood as leasing.

These pillars set the incentives to use resources more efficiently. However, responding to this incentive requires further elements, which allow for keeping the used products in circular flows. These include a product design enabling later remanufacturing, respectively the process of rebuilding a product (including cleaning, inspection, disassembly, replacement of defective components, reassembly, testing and inspection of the new product) [37]. Moreover, it requires reverse logistic structures in order to collect and transport used products and thereby physically supporting the remanufacturing of the used products. In their entirety and interplay, the named pillars and elements can be labelled as "leasing society".

---

[1] What we want from these products is not ownership per se, but the service the product provides: transportation from our car, cold beer from the refrigerator, news or entertainment from our television [30].

In a leasing society, what business sells to customers (value proposition) is different. Products are still manufactured, but from the customer perspective they are complemented, if not even substituted, by services. Selling services instead of products changes ownership structures, responsibility for upkeep, maintenance and disposal is transferred from the customers to the producers. As the products are returned to the owner (the producer) at the end of their use instead of disposal, the producer has the incentive to design and produce its products according eco-design criteria, e.g. making them more resource-efficient, optimising their utilisation, extending their lifetime, enabling easier remanufacturing or recycling and thereby avoiding waste. As such, the leasing society re-orientates the value chains for physical goods towards more circularity. In the vision of a leasing society, the changed producer-customer relationships contribute to shifting conventional production and consumption models, largely based on linear supply chains, towards more circular value chains and a circular economy in the big picture.

An ideal business case

Taking a washing machine producer as a hypothetical example, demonstrates how a conventional business model (1. product-oriented PSS: selling washing machines with little additional services) could be transformed intoa conventional washing machine leasing (2. use-oriented PSS: selling the use of washing machines) or into an ideal leasing society business case by delivering fresh laundry (3. result-oriented PSS: selling "cleanliness of laundry" as a result of using washing machines), see also [13,23,26]. Furthermore, it explores how the different economic incentives and environmental implications change within the transformation of the established business model to the leasing society business models.

Like any other producers of consumer goods, also a washing machine producer (1st example referring to a product-oriented PSS) strives for maximising profits. Profits in this conventional business model are mainly generated by selling washing machines at a competitive price, which is higher than the production costs. As the business model is based on the product sale, it tends to create incentives for designing the machine in a way that it just covers the warranty period. This can lead to an artificial reduction of the product's lifetime, so-called built-in or planned obsolescence [19,38]. It also motivates business to stimulate changing trends by means of advertising, to motivate a maximum exchange of products by new ones. Producing as many machines as possible results in a high demand for resources. Accordingly, the economic incentives within a conventional, sales-based business model tend to steer the producer and consumer towards resource-intense behaviour.

In the case of a conventional leasing business model (2nd example referring to a use-oriented PSS) that is based on certain contract arrangements (like agreements regarding regularly maintenance), still requires the production of a washing machine, but the producer is also responsible for additional services, like the installation and repair of the machine. In order to avoid the costs of repair or replacement of the washing machine, the producer has an interest in creating a product with a long lifetime. After the contract expires, the machine is returned to the producer, who remanufactures it and either leases or sells it to another customer. The shift from selling to leasing washing machines

in a leasing society business model could lead to more durable products and to an increased re-use of machines, machine parts or built-in material. Environmental pressures could be decreased by a reduced number of manufactured—and increased numbers of remanufactured machines, resulting in less use of resources and less waste production.

In case of a performance-oriented business model (3rd example referring to a result-oriented PSS), the producer is neither selling nor leasing washing machines, but delivering a result or a performance in terms of fresh laundry. The producer, who is more a service provider now, operates the washing machine. The machine remains not only in the ownership but also in the possession of the producer (or third party). This has implications for the incentive structure. Due to his professional know-how, the producer is able to ensure the best possible machine utilisation, which reduces costs of use, maintenance and repair. Further, as operating costs for energy, water or detergent are now shifted from the customer to the service provider, the producer has an increased interest to design its washing machines with low energy, water or detergent requirements. In order to be environmentally beneficial, it is necessary that the consumer demand does not increase to avoid a rebound effect (e.g. having clothes washed more frequently because customers no longer have the work of doing laundry or because of the attractive rebate schemes). As the washing machine is designed to be cost-effective also during the use phase, it meets resource efficiency requirements. Due to capacity optimisation, the number of washing machines that need to be produced decreases. Furthermore, a decreased amount of resources is required for maintenance and repair because of the knowledge driven optimum operation of the machine.

Another conceivable constellation for the performance-oriented business model (3rd example) would consist in the inclusion of a third party (e.g. a laundrette) delivering the laundry performance. However, the above discussed example shows that it is of crucial importance to confront the original washing machine producer with the entire product lifecycle costs in order to change the producer's incentive structure towards a cost-effective and resource-efficient operating of machines. Respective incentives could be set by contractually arranged gain sharing mechanisms or agreements on the equipment with the most efficient machines combined with regularly maintenance arrangements.

Potentials of a Leasing Society

The micro perspective

While conventional leasing activities are well established in society and as such also well documented, the activities related to a leasing society do not correspond to common practices yet and are comparatively poorly and non-systematically documented. However, there are a number of studies that investigated exemplary business cases without labelling them as leasing society business practices but e.g. as entrepreneurial eco-innovation, green business models or sustainable PSS [15,20,26,38-47][2]. The various cases comprise different brands as for example chemical industry, waste disposal, office equipment, textiles, automotive, and machinery and equipment. They cover as well business-to-business (B2B), as business-to-consumer

---

[2]   Nonetheless, not all eco-innovation, green business models or sustainable PSS can serve as leasing society business cases. E.g. the introduction of a process-related eco-innovation, green business models that built on an innovative new product or a prolonged product guarantee in terms of a product-oriented PSS are not automatically leasing society business cases.

(B2C) as consumer-to-consumer (C2C) constellations. Furthermore, they include different types of PSS such as the use-oriented and results-oriented types.

From selling chemicals to Chemical Management Services (SAFECHEM example): The provision of CMS[3] instead of selling pure chemicals constitute one of the most well-known, and discussed examples of a possible leasing society (and related concepts) business model. In a conventional business model, a chemical producer or retailer sells chemicals. Incentives to help the customer using its chemicals more efficiently are limited. In contrast, the buyer is interested in a decrease of the chemicals' volume and costs. In this business model, the customer is still responsible for the chemicals' application. With the contractual agreement on a special result (like e.g. a coated car), the activity of using the chemicals can be outsourced to the chemicals supplier in a leasing society business model. In this constellation, the supplier still owns the chemicals and is responsible for an agreed performance and is paid on the basis of this performance (e.g. cost savings delivered). The costs of using chemicals are shifted to the supplier, who will seek for an efficient use of the chemicals by reducing the lifecycle costs of materials, labour and waste management. Geldermann et al. highlights the economic benefits for users and suppliers that often sum up to 20%. Often cited examples comprise SAFECHEM, Ashland, and Castrol [21,35,43,49].

From selling waste disposal to resource management (GM example):

In the conventional business model, the manufacturing company manages its resources on its own and charges another company with the waste disposal. As the waste disposing company is paid by the waste volume, potential efforts of the waste disposing company to support resource-efficient structures within the manufacturing company do not exist. In a business model, in which the manufacturing and the waste disposing company set up a contract and agree on a resource management performance payment (that rewards resource-efficiency), they form a strategic alliance. Together, they have the same economic incentives: Savings through resource-efficiency efforts. Thus, instead of rewarding waste-creating behaviour, resource-efficiency and waste avoidance are recompensed. In addition to the traditional waste disposal, activities of the contractor in the new business model include services over the whole value chain activity of the producer, like the design of products and processes, procurement and delivery, inventory and storage, use and recovery of resources, monitoring and reporting and training. The analysis of case studies e.g. in Leipzig or Berlin shows cost saving potentials of around 20-30% .The scheme of resource management contracting corresponds to energy performance contracting. Often cited examples comprise GM, Public Service Enterprise Group, and Innotec [39-50-52].

From selling jeans to leasing jeans (Mud Jeans example): Under the slogan "Using instead of Owning" the Dutch company Mud Jeans developed a leasing model for Jeans in 2013, which is based on a contract in which consumers pay a deposit and a monthly rate for the use of a pair of jeans, while they have the possibility to exchange it in case it breaks. After a year of leasing, the customer can either keep the jeans and pay four more months, or get new jeans and keep paying his monthly rates [53]. The jeans returned to the company are either sold in the shop again, or they are recycled and turned into new jeans or

other products like up cycled bags. This innovation serves as a best practice example for a circular economy, as it allows retailers to ensure that products are recycled in the best possible way. But also consumers are educated and their awareness of an interest in sustainability and environmental problems is raised. So far, there are eight Mud Jeans stores in the Netherlands and 19 in Germany. Mud Jeans already received different awards, among which the Sustainia100 Study, the NCD Change Award and the Circular Economy Award[54].

From selling electric cars to leasing batteries (Better Place example):
In 2007, the Israeli electric-car battery technology start-up. Better Place raised $ 200 million venture funding for a leasing society business model being based on giving its customers mobility guarantee[55]. The basic idea was to separate the ownership of the electric car (owned by the customer) from the battery (owned by Better place). The customer buys the electric car (from Better Place partner Renault-Nissan) but leases the electric car battery (from Better Place partnersA123 Systems and AESC), which allows to just switch depleted batteries (in a dense network of quick-swap battery stations) after having driven long distances and to drive on. This business idea targets the weak point of the e-mobility diffusion—the dependence of the driver on the battery, being connected with a limited power density and thus driving distance, and waiting time for charging the battery including limited flexibility and independence. Keeping the ownership of the battery by Better Place and partners would have created the incentive to produce effective, efficient and long-living batteries—thereby supporting resource efficiency targets. However, in 2013, better Place had to apply for insolvency. It is reported that the company failed due to missing customer acceptance and cooperation with the automotive industry [56].

The macro perspective

The various PSS subcategories identified by Tukker et al. [57] come along with different potential to reduce environmental impacts—only a few PSS have the potential for a considerable improvement of the ecological situation. While the product-related PSS have the potential for an incremental reduction of environmental impacts (< 20 % compared with a reference product) of environmental impacts (the traditional product lease might even worsen the situation), the use- and result-oriented PSS can be connected with a considerable reduction of environmental impacts (<50 %). Amongst the latter PSS type, especially the functional result delivering PSS (in contrast to a pay per unit use) is associated with a potential radical reduction of environmental impacts. Accordingly, the more the focus switches from the products to the service functions, the higher the potential for environmental savings. In general, there seems to be a prevalent assumption in literature and society that PSS solely have positive ecological effects and economic benefits, so-called win-win situations, e.g. [15,26,38,39,41]. However, empirical studies [44,58,59] demonstrate that significant environmental improvements can only be reached under certain conditions. Some of these conditions will be presented here in more detail.

Product lifetime optimization on the part of the leasing society producer needs to take into account intensified and more careless product use: Despite improved eco-design being part of a leasing society, some PSS constellations may rather shorten than extend the use-phase of products. For example, in cases where products are

---

[3]   Similar expressions that can be found in the literature are chemical product services, chemical leasing, shared savings contracts, service contracts, servicing, performance contracts, contracting, total care and total gas and chemical management [48].

rented or pooled, they may be used much more intensively. The joint and intensified use of a product itself is not an impaired situation compared with the normal business case situation—however, two factors could contribute to negative environmental effects: First, the offer of an always and easily available product may increase the customer's need to make use of that product. Second, customers might tend to show less care when they use a rented or leased product instead of a product that they own. Under these circumstances, the intensified and more careless product use could lead to a withdrawal of the product out of the use phase before its originally planned end and the production (and use of resources) of more products.

Product lifetime optimization on the part of the leasing society producer needs to take into account user behavior: Users of leased products often expect new equipment or machinery, e.g. regarding mobile phones, car sharing or photocopying machines. The product lifetime optimization in a leasing society must take this into account as resources would be used inefficiently if all products were solely designed with the aim being as durable as possible. Also, at least nowadays leasing contracts are often designed in a way (amongst others as it is prescribed by law) that the lease term does not exceed 75 % of the lifetime, with the product being detracted from the use phase before its potentially possible working life has finished. These regulatory conditions automatically lead to an inefficient use of resources and would need to be revised in order to further a resource-efficient leasing society.

Product lifetime optimization on the part of the leasing society producer should take into account dynamics of technological advance: Some of those products with an artificially reduced use phase are often sold as used products at the end of their leasing period. It is however not self-evident that selling leased products as used products leads to a reduced resource use (compared with buying a new product). This is especially true for products where environmental impacts are mostly incurring during the use phase and less in the production phase of the product. For consumer goods such as washing machines, where the environmental impacts are connected mainly to their use phase, it might even be useful to replace these products well before reaching the end of their lifecycle. For example, from a resource saving point of view it would make sense to replace consumer goods even rather quickly, because of the technical improvements in energy and water efficiency [58]. If the efficiency gains of the new products are combined with a remanufacturing of the used goods, it may even result in a net decrease of material and energy use. For part of the leased products it thus may make sense to reduce their use phase—thus designing products in a leasing society a priori as long-lasting products is not entirely true.

Remanufactured goods in a leasing society should not create additional markets: Furthermore, former leased and subsequently remanufactured goods that create an additional market instead of solely replacing new goods might lead to increased resource consumption—this is particularly problematic if a second or third use phase of goods and service takes place in regions that are not yet characterized by appropriate recycling facilities. While selling used and remanufactured goodsin those regions make sense from a business perspective, as new markets can be opened to sell their products in different price segments, the economy-wide environmental consequences are far less clear. Under certain circumstances, the global energy and material consumption of such PSS solutions would be higher than in the traditional purchasing model.

In a leasing society, a direct producer-customer relationship has some advantage instead of interposing a third party: Another point concerns the incentives for the producer to design and run its products more resource efficient. Those incentives are only true for those constellations, where the original equipment manufacturer offers the PSS. However, PSS are often offered by specialized companies that are interposed between the consumer and the producer (e.g. specialised leasing companies, independent remanufactures). In this case, the producers may lose interest in producing goods, which require less material and energy or in designing products with provisions for easy repair or recycling.

In Transition Towards a Leasing Society

Discussion of trends

The selected case studies present anecdotal evidence that specific leasing society business models have the potential to increase resource efficiency and improve the firm's competitive position. They also clearly highlight that waste prevention needs integrated approaches that go beyond technology-dominated end-of-pipe infrastructures, including production and consumption patterns. A circular economy aims at overcoming the division between waste production and waste treatment. PSS point out the potential of institutional solutions to radically reduce the waste intensity of consumption and production. However, the evidence presented in this paper suggests that not every PSS reduces environmental impacts. Depending on the type of PSS, leasing society business cases may even increase negative environmental impacts.

Three key trends can be observed:

1. Although the intense discussion about PSS began in the late 90s, *"the uptake of such ideas by industry appears limited"* [25].
2. PSS success-stories are repeatedly reported in specific market segments (such as chemical management or car-sharing).
3. So far, the effective use of PSS, and especially of the result-oriented product-services, seems to have found a stronger foothold in commercial B2B activities rather than B2C activities [60].

Drivers and barriers

The current dynamics of technology push and market pull are probably not sufficient to promote the transition toward a leasing society [54] (with regard to eco-innovations more generally see [55]). In order to realise the economic and ecological potentials of PSS, government intervention would be required. Transition management towards more sustainable patterns of production and consumption will require the identification of existing barriers and drivers that *"offer the best leverage for guiding change in a desirable direction"* [61-63].

Depending on the type of chosen PSS combination, the market and the producer-customer relationship, and the concrete sector, there is not only one leasing society business model, but diverse business strategies possible. Details regarding specific arrangements regarding e.g. maintenance, product take-back, gain sharing mechanisms etc. again multiply the number of possible business model options. The variety of possible business strategies, relate to different barriers and drivers that influence the implementation of PSS.

The following tables (Tables 1 and 2) sum up the identified drivers and barriers of leasing society business cases for a result-oriented PSS in a B2B relationship—each from the perspective of the producer (respectively service provider) and the customer.

Producer / Service provider	Customer
• Increased competition and declining margins in traditional markets • Maintain and gain new market shares, customers and profits • Diversification / increased range of services possible • Benefit from gain sharing mechanisms • Built-in material is not lost and residual material value can be retrieved • Technological advancement that enables new solutions • Business customer makes buying decisions rather rationally than emotionally	• Demand for more services • Discontinuation of ownership responsibility / risk • Reduced contract complexity • Flexible contract conditions (ability to purchase, renew, cancel focus on core competencies) • Possibility to upgrade and access to latest technology • Improved production process efficiency and reduced complexity • Reduced life cycle costs • Predictable costs • Advantage of tax benefits • Improved liquidity • Reduced environmental pressure • Advantages for health and safety

Table 1: Drivers of result-oriented B2B leasing society business model.

Producer / Service provider	Customer
• Investments into new infrastructure • Increased fixed and operating costs • Long-term relationship as a risk (coupled with success of customer) • Lack of capital • Diversity of regulations • Variable client requirements afford expert experience, knowledge and skills • Lack of (skilled) personnel • Lack of flexibility • Public procurement guidelines • Technological progress that benefits resource-inefficient production patterns • Dependency from other business model partners • Risk of underperformance	• "We can do it better" • Long-term relationship as a risk (changing supplier is more difficult) • Fear of loss of control • Need to include supplier in confidential processes • Lack of awareness and priority towards resource-efficiency • Unknown total costs of ownership • Uncertainty about saving potentials • Certain level of company size needed in order to be profitable

Table 2: Barriers of result-oriented B2B leasing society business model.

Drivers and barriers on the customer side take on a new dimension when it comes to the relationship between business and private customers (B2C). In comparison to the decision-making of commercial customers (B2B), decisions of private customers tend to be more influenced by emotions. This might promote but also hinder the distribution of PSS. For reasons of flexibility, safety, time, convenience, personal identity perception, status symbols and living standards, people tend to prefer owning the products they use. This affects cars as well as white ware, computers, toys, tools and other private equipment. However, as sharing products could become a lifestyle change of a new generation, PSS have high potentials in private consumption.

Leasing society policy measures

General policy measures can set the right background conditions for a leasing society and at the same time they can counteract rebound effects that may arise from the utilisation of PSS. PSS will require a change in political and economic framework conditions, as well as information campaigns and grants. Otherwise, product substituting business models run the risk of becoming costly and short-lived solutions for marginal niche markets. Changing the framework conditions and thereby indirectly supporting PSS is more likely to meet the requirements of a Circular Economy. However, direct measures aimed at supporting the uptake or up-scaling of PSS can have a secondary but nevertheless important effect: They can help to overcome resistance against changing framework conditions by raising producer interest in general framework conditions that are supportive for their own business models and thus amplify interests in favour of a Circular Economy. While active support for PSS in research and development is important, an exclusive focus on innovative business models instead of framework conditions runs the risk of providing insufficient incentives for innovation as there are insufficient or no cost to environmentally harmful activities [64]. Nevertheless, a number of specific recommendations can be given. They are divided into research-based, market-based, regulatory, information and participation policy instruments.

Research-based policy instruments:

• The transition towards a leasing society would require a better assessment of social, economic and environmental impacts. The different manifestations of PSS have completely different intended and unintended macro-economic effects. This calls for more systemic research, including on impact assessments and transition management towards a leasing society.

• Research and assessment on the impacts of a leasing society could be complemented with demonstration and pilot projects, monitoring and comparison of existing PSS as well as diffusion of best practice, and targeted experimental public procurement initiatives. Eventually, public-private partnerships of relevant actors along the innovation cycle could be initiated.

Market-based policy instruments:

- An ecological tax reform could shift taxation from labour towards resource consumption to give the right incentives for a transition towards less resource-intensive products, lifestyles and more labour intensive maintenance and repair of more durable products and innovative PSS.
- Reduced VAT rates for maintenance, repair and remanufacturing could give the right incentives for a longer lifetime of products and leasing business models e.g. for electronic and household equipment. In addition, a reduced VAT rate could be granted to producers who offer an extended warranty going beyond the typical time period.
- Landfill and incineration charges could give incentives for re-use, remanufacturing and recycling, including supporting product service supply.
- Longer depreciation periods could contribute to extending the average use phase of a product.
- Public procurement could create niche markets for developing a leasing society.

Regulatory policy instruments:

- The Eco-Design regulations could be extended towards resource savings and efficiency including requirements for materials.
- Producer responsibility could be strengthened including deposit refund schemes in areas such as end-of-life vehicles or electric and electronic equipment.
- Minimum warranties for products could be further expanded.
- The introduction of communal laundry and car sharing facilities for housing complexes exceeding a certain number of housing units should be tested.
- The legal framework for PSS for standardised and harmonised contracting could be developed in economic areas such as the EU Single Market.

Information policy instruments:

- In general, economic and ecological impacts over the entire life cycle of conventional products in comparison to PSS are insufficiently explored. Better assessment procedures and their results need to be shared among producers and consumers.
- Voluntary labelling for leasing-solutions could be encouraged with tax credits for assessment and auditing expenses to facilitate consumer choices and public procurement.
- Knowledge on life cycle costs advantages of PSS including public assessments of PSS and products should be integrated in government procurement procedures.
- Research, pilot projects, education and the dissemination of information on PSS can be supported in the framework of policy programs.

Participation policy instruments:

- Business, civil society, policy-makers and scientists should be consulted for improving a shared understand of possible opportunities and risks connected to the leasing society.

Discussion

This article has raised a number of key issues to be considered in assessment of a leasing society, with the overall aim of contributing to a transition toward a resource efficient, circular economy. Achieving this aim depends on how the leasing society is implemented. Business is already developing innovative PSS strategies and business models. The challenge for policy makers is to act in a timely manner to establish the framework conditions and support mechanisms for shaping these activities so that they contribute in an effective way to smart, sustainable and inclusive growth. In conclusion, four challenges for further research and policy may be distinguished:

- Taking into account lock-ins: Complex products with long lifetimes combined with multiple remanufacturing steps may tend toward more incremental innovation. Especially in "use-oriented" PSS small incremental steps could "lock-in" opportunities for disruptive innovation. Research on the risk of potential lock-ins for different markets would help to better understand such dynamics to shape innovation.
- Expanding beyond niche markets: Leasing society business models seem to be well established in certain areas (like in chemical markets) whereas there is little evidence of leasing in other markets. Research to better understand the barriers and assess the suitability of leasing business models for other markets could help to provide more targeted policy support for the wider diffusion of successful leasing business models (in particular as regards B2C relationships).
- Addressing value systems and rebounds: Better understanding of consumer behaviour in relation to new leasing business models would help to anticipate rebounds (e.g. associated with intensified and careless product use), overcome barriers (e.g. related to value systems concerning ownership) and more effectively engage citizens in the transition process. Limits toa leasing society related to customers' preferences of ownership might be analysed, too.
- Quantifying environmental effects: The environmental performance is highly dependent on the design of the individual business case. The mechanisms described in this paper need further case-related validations. From a macro perspective especially those case studies with high economic saving potentials show relevant resource efficiency potentials due to an expectable market uptake – nevertheless these inter-linkages between eco-innovation patterns, changed economic incentive structures and resource savings will require further research.

Acknowledgements

This paper is based on the study "Leasing Society" the authors have prepared for the European Parliament in the year 2012.

References

1. Wilts H, Rademacher B (2014) Potentials and Evaluation of Preventive Measures - A Case Study for Germany. Int J Waste Resour 4: 1–7.

2. Wilts H, Nachhaltige (2014) Innovations prozesse in der kommunalen Abfallwirtschaftspolitik – eine vergleichende Analyse zum Transition Management städtischer Infrastrukturen in deutschen Metropolregionen.

3. ISWA Working Group on Recycling and Waste Minimization (2011) ISWA Key Issue Paper on Waste Prevention Waste Minimization and Resource Management.

4. Towards a circular economy: A zero waste programme for Europe (2014) European Commission.

5. Berkhout F, Smith A, Stirling A (2003) Socio-technological regimes and transition contexts. SPRU – Science & Technology Policy Research, University of Sussex, UK.

6. Stahel WR (2003) Perspectives on Industrial Ecology (eds Bourg D & Erkmann S) Greenleaf Publishing, Sheffield, UK. p264-282.

7. Giarini O, Stahel WR (2000) Die Performance-Gesellschaft: Chancen und Risiken beim Übergang zur Service Economy. Metropolis-Verlag Publishers, Germany.

8. Stahel WR (1997) The service economy: 'wealth without resource consumption'?. Philos. Trans R Soc Lond Ser Math Phys Eng Sci 355: 1309-1319.

9. Botsman R, Rogers R (2011) What's Mine Is Yours: How Collaborative Consumption is Changing the Way We Live. HarperCollins Publishers, UK.

10. Heinrichs H, Grunenberg H (2012) Sharing Eocnomy, Auf dem Weg in eine neue Konsumkultur?. Centre for Sustainability Management, Leuphana Universität Lüneburg, Germany.

11. The sharing economy - All eyes on the sharing economy (2013) The Economist.

12. Resilience in the Rond - Seizing the Growth Opportunities of a Circular Economy (2012) Aldersgate Group.

13. Economic and business rationale for an accelerated transition (2012) Towards the circular economy. Ellen MacArthur Foundation, Cowes, United Kingdom.

14. Braungart, M (1991) Umweltschutz als Wettbewerbsvorteil für Unternehmen - Intelligente Lösungen für Umweltprobleme durch bessere Produkte, das Leasingkonzept für Gebrauchsgüter 109–121.

15. Leismann K, Schmitt M, Rohn H, Baedeker C (2012) Nutzen statt Besitzen, Auf dem Weg zu einer ressourcenschonenden Konsumkultur. Im Auftrag und herausgegeben von der Heinrich-Böll-Stiftung. Zusammenarbeit mit dem Naturschutzbund Deutschland.

16. Scholl G, Schulz L, Süßbauer E, Otto S (2010) Nutzen statt besitzen - Perspektiven für ressourceneffizienten Konsum durch innovative Dienstleistungen. Institut für ökologische Wirtschaftsforschung, Wuppertal Institut für Klima, Umwelt, Energie, Germany.

17. Lay G, Schroeter M, Biege S (2009) Service-based business concepts: A typology for business-to-business markets. Eur Manag J 27: 442–455.

18. Stahel WR, Reday G (1976) The potential for substituting manpower for energy. Report to the Commission of the European Communities.

19. Hockerts, Kai (2008) Property Rights as a Predictor for the Eco-Efficiency of Product-Service Systems. CBS Center for Corporate Social Responsibility Porceleænshaven 18BDK - 2000 Frederiksberg, Germany.

20. Goedkoop MJ, van Halen CJG, te Riele HRM, Rommens PJM (1999) Product Service Systems, Ecological and Economic Basics.

21. Mont O (2000) Product-Service Systems. The International Institute of Industrial Environmental Eocnomics, Lund University, Sweden.

22. Meijkamp R (2000) Changing consumer behaviour through eco-efficient services: an empirical study on car sharing in the Netherlands. Delft University of Technology, Netherlands.

23. Manzini E, Vezzoli C (2000) Product-Service System and Sustainability: Opportunities for sustainable solutions. United Nations Environment Programme, Division of Technology Industry and Economics, Production and Consumption Branch, Kenya.

24. Brezet JC, Bijma AS, Ehrenfeld J, Silvester S (2001) The design of eco-efficient services: Method, tools and review of the case study based 'Designing Eco-efficient Services' project. Delft University of Technology, Netherlands.

25. Baines TS et al. (2007) State-of-the-art in product-service systems. Proc Inst Mech Eng Part B-J Eng Manuf 221: 1543–1552.

26. Cooper T, Evans S (2000) Products to services. The Centre for Sustainable Consumption, Sheffield Hallam University, England.

27. Hockerts K, Petmecky A, Hauch S, Seuring, S (1994) Kreislaufwirtschaft statt Abfallwirtschaft: Optimierte Nutzung und EInsparung von Ressourcen durch Öko-Leasing und Servicekonzepte. Universitätsverlag Ulm, Germany.

28. White AL, Stoughton M, Feng L (1999) Servicizing: The Quiet Transition to Extended Product Responsibility. Office of Solid Waste, U.S. Environmental Protection Agency, USA.

29. Tukker A, Berg C, van den, Tischner U (2006) Product-services: a specific value proposition. New Bus Old Eur Prod-Serv Dev Compet Sustain. Greenleaf Publications, Sheffield, UK. p22-34.

30. Hawken P (1993) The Ecology of Commerce. HarperCollins Publishers, UK.

31. Stremersch S, Wuyts S, Frambach RT (2001) The Purchasing of Full-Service Contracts: Ind Mark Manag 30: 1–12.

32. Ölundh G (2003) Environmental and Developmental Perspectives of Functional Sales. Department of Machine Design, Royal Institute of Technology, Sweden.

33. Markeset T, Kumar U (2005) Product support strategy: conventional versus functional products. J Qual Maint Eng 11: 53–67.

34. Kim SH, Cohen MA, Netessine S (2007) Performance Contracting in After-Sales Service Supply Chains. Manag Sci 53: 1843–1858.

35. Reiskin ED, White AL, Johnson JK, Votta TJ (2000) Servicizing the Chemical Supply Chain. J Ind Ecol 3: 19–31.

36. Sorrell S (2007) The economics of energy service contracts. Energy Policy 35: 507–521.

37. Sundin E, Bras B (2005) Making functional sales environmentally and economically beneficial through product remanufacturing. J Clean Prod 13: 913–925.

38. Fishbein BK, McGarry LS, Dillon PS (2000) Leasing: A Step Toward Producer Responsibility. INFORM Inc, NY, USA.

39. COWI. Promoting Innovative Business Models with Environmental Benefits, Final Report.(2008).

40. 'Green Servicizing' for a More Sustainable US Economy: Key concepts, tools and analyses to inform policy engagement (2009). Office of Resource Conservation and Recovery, US Environmental Protection Agency, USA.

41. Green business models in the Nordic Region: A key to promote sustainable growth (2010) Danish Enterprise and Construction Authority, Denmark.

42. Provisional collection of Nordic cases and expert interviews - 'Green Business Models in the Nordic Region' (2010) Danish Enterprise and Construction Authority, Denmark.

43. Holliday CO, Schmidheiny S, Watts P (2002) Walking the talk⊠: the business case for sustainable development. World Business Council for Sustainable Development. Greenleaf Publishers, UK.

44. Intlekofer K (2010) Environmental implications of leasing.

45. Tukker A, Tischner U (2006) Product-services as a research field: past, present and future. Reflections from a decade of research. J Clean Prod 14: 1552–1556.

46. Wimmer R, Kang MJ, Verkuijl UTM, Fresner J, Möller M (2007) Erfolgsstrategien für Produkt-Dienstleistungssysteme. Bundesministerium für Verkehr, Innovation und Technologie.

47. Zaring O, Bartolomeo M, Eder P, Hopkinson P, Groenewegen P, et al. (2001) Creating eco-efficient producer services. Gothenburg Research Institute, Sweden.

48. Kortman J, Theodori D, van Ewijk H, Verspeek F, Uitzinger J, et al. (2006) Chemical product services in the European Union. European Commission, Spain.

49. Saecker, S. The SAFECHEM Business Model - sustainable solutions for industrial cleaning. (2011).

50. Advancing Resource Management Contracting in Massachusetts: Reinventing Waste Contracts and Services (2002) Tellus Institute, Boston, USA.

51. Resource Management: Strategic Partnerships for Resource Efficiency. Office of Solid Waste, US Environmental Protection Agency, USA.

52. Abfallmanagement (2012) Innotec, Hungary.

53. Mud Jeans. Lease How It Works | Lease A Jeans von nur 5,95 im Monat. (2014).

54. Mud Jeans (2014) Awards.

55. Better Place: Charging into the Future? (2010) ERB Institute & William Davidson Institute. University of Michigan, USA.

56. Elektroautoprojekt Better Place gibt auf. Manager Magazin.

57. Tukker A, Tischner U, Verkuijl M (2006) Product-services and sustainability. New Bus Old Eur Prod-Serv Dev Compet Sustain. Greenleaf Publications, Sheffield, UK.

58. Intlekofer K, Bras B, Ferguson M (2010) Energy Implications of Product Leasing. Environ- Sci Technol 44: 4409–4415.

59. Agrawal VV, Ferguson M, Toktay LB, Thomas VM (2012) Is Leasing Greener Than Selling? Manag Sci 58: 523–533.

60. Tukker A, Berg C, van den (2006) Product-services and competitiveness. New Bus Old Eur Prod-Serv Dev Compet Sustain. Greenleaf Publications, Sheffield, UK.

61. Cook MB, Bhamra TA, Lemon M (2006) The transfer and application of Product Service Systems: from academia to UK manufacturing firms. J Clean Prod 14: 1455–1465.

62. Hemmelskamp J (2000) Innovation Oriented Environmental Regulation. Physica-Verlag.

63. Smith A, Stirling A, Berkhout F (2005) The governance of sustainable socio-technical transitions. Res Policy 34: 1491–1510.

64. Taxation, Innovation and the Environment (2010). OECD Publishing, Paris, France.

65. Geldermann J, Daub A, Hesse M (2009) Chemical Leasing as a Model for Sustainable Development. Research Papers University Göttingen, Germany.

66. Wilts H, Palzkill A (2014) Suffizienz als Geschäftsmodell: Contracting. PolRess-Kurzanalyse.

26

Sustainable Water Resources Management in Arid Environment: The Case of Arabian Gulf

Elnazir Ramadan*

Department of Geography, Sultan Qaboos University, Muscat, Oman

*Corresponding author: Elnazir Ramadan, Department of Geography, Sultan Qaboos University, Muscat, Oman
E-mail: alnazir@squ.edu.om

Abstract

Over the past decades the Arabian Gulf region has witnessed a great economic development and social transformation. The region is facing a water insufficiency problem that is one of the biggest in the world. The level of available renewable water in the region is one fifth of what the rest of the world enjoys on a per capita basis. The population of the region is growing 55% more quickly than the population at the rest of the world. By 2020, water needs is predicted to be around 341 million imperial gallons per day. The threat of water scarcity ensures that investment in developing freshwater supply, along with the recycling and reuse of waste and sea water is an urgent priority across the region moreover strategies for sustainable water use have to be adopted. This means that if strategies for rational water use that entail educational component is not established the water storage is going to run out. This paper reviews the existing situation identifies the gap and proposes an appropriate institutional framework which involves assignment of responsibilities among various levels. Ensures stakeholders participation, accommodates adaptive change and remain self sustainable.

Keywords: Arabian Gulf ; Water scarcity; Sustainable management; Arid environment

Introduction

The availability of adequate freshwater has become a limiting factor of the quality of life, worldwide. More than availability, the problem is often the rational use of water than ensure its continuality. In the semi-arid and arid regions water scarcity was always a dominant problem moreover, the interference with the natural hydrologic cycle as a result of overexploitation of both surface and ground waters and of changes in land usage resulted not only in the reduction of the available water amounts but also in the deterioration of the water quality due to pollution from urban, industrial and agricultural practices and salinity build-up in soil and water. Such a situation is in all cases a pretext for discord and assignment of blame on those supposedly responsible for the deterioration of the water accessibility , especially in view of the possible high cost and technical complexity of the measures that need to be taken for remedying and alleviating the situation. Arabian Gulf countries are located in an arid area with limited water resources [1]. Hydrological investigations point to limited resources of underground water. The best choice for providing fresh water in the Arabian Gulf countries is through seawater desalination with ground water as a backup. About 65% of desalination plants that are in operation worldwide are located in the Arabian Gulf countries, most of which are the dual-purpose multistage flash (MSF) plants, producing power and water. Integrated water resource management (IWRM) is the need to take a holistic approach to ensure the socio-cultural, technical, economic and environmental factors are taken into account in the equitable development and management of water resources.

Sustainable and Integrated Water Resources Management

The relationship between "Sustainable" and "Integrated" Water Resources Management is essentially that sustainability is the general goal whereas Integrated Water Resources Management (IWRM) is a strategy for pursuing this goal. Water is a resource under considerable pressure. Effective and sustainable management of water resources is vital for ensuring sustainable development [2]. However, efforts of water resource management seem to demonstrate inappropriate practices, especially when compared to water consumption trends in Arabian Gulf states. Being a major and vital ingredient to human kind, water resources influence all sectors. However, there have been increased problems over time that subject water resources to a number of crisis and pressures. Poor water resources management have stimulated and sustained a number of problems related to health, socio-economic and environment, which need to be solved. Integrated water resources management (IWRM), is a process, a change, and an approach that mainstream water resource use and management into the national economic in an equitable manner without compromising the sustainability of vital ecosystems. Successful community involvement demonstrates the importance of IWRM. Public/community involvement is crucial for a successful and sustainable water resource management. It has been emphasized that natural resources management related policies including water requires the use of knowledge, experience and opinions of local communities who are the key stakeholders in resource conservation.

Water Resources of the Region

In the area under discussion the arid climate extending over the Gulf States is characterized by large geographic and temporal disparities of the precipitation distribution, with relatively more annual amounts of precipitation in winter season, a relatively variable regime, with alternating drought and excessive rain periods further acerbates the situation. Scarcity of water resources in the region is evident some studies earlier this year said that the imminent shortage of water resources in the region has been compounded by the real estate boom, with new construction projects taking a larger share of resources. This is alarming since this region is already the driest in the world. The Gulf remains the largest market for water desalination in the world and local municipalities are seriously examining ways to double existing capacity to meet regional demand [3].

The arid climate of the region where rainfall is sparse, limits conventional water resources and recent studies show that temperatures are increasing. In August 2009, a temperature in some areas of the region has exceeded 47.3°C. The available water resources in the Arabian Gulf countries are mainly the ground water which is not enough to satisfy the water demand due to the rapid urban and industrial developments. Due to the improper planning and mismanagement of the water abstraction from the wells, the ground water is very much reduced and in many countries the withdrawal of water is now done from the non renewable layers. Abstracting water from the non-renewable layer causes a serious problem because this layer was formed during very long decades and abstraction from this layer will definitely deteriorate the ground water storage. Desalination of seawater is one of the main alternatives for the substitution of water shortage in the Arabian Gulf countries and other countries. Although desalinating the seawater is costly, it is still an important option for compensating for the water shortage. Most of Gulf countries built power and desalination plants for water and power production. We should be aware of the fact that the effluent discharges from the plant back to the sea may have a negative impact on environment.

The Gulf remains the largest market for water desalination in the world and local municipalities are seriously examining ways of looking to double existing capacity to meet regional demand. However, it is North Africa that will experience the greatest growth in desalination, Algeria and Libya experiencing a 300 per cent growth. Water desalination, specifically membrane processes based on reverse osmosis, is becoming more efficient as more technologically advanced methods are being developed. The price of desalinating one cubic meter of water has come down more than a third of what it used to be 10 years ago.

environmental concern for the region. water deficit in the Arab Gulf countries is estimated to be 16 billion cubic meters, this would have an impact on the quality of ground water. Consumption per capita, which is currently at 1,100 cubic meters annually, is projected to fall to 550 cubic meters a year by 2050. While Gulf countries spend around $133bn in water and wastewater management yearly, more effective strategies are needed, such as convincing companies and even residents to set up their own sustainable infrastructure.

The region has the lowest availability of actual renewable water resources per capita in the world. It is thus imperative that gulf governments invest in technologies that can ensure adequate supply for the region's vibrant industries and its more than 300 million inhabitants [4].

Building and industrial projects alone in the region will consume more than 112 billion liters of drinkable water from 2008 to 2009 this part of the world hosts 5% of the global population and yet possesses only 1% of the world's renewable fresh water. Desalination is likely to become one of the world's biggest industries. Growing communities and new industries must have dependable water supplies in order to prosper. Desalting systems have long proven effective in Kuwait, Bahrain, Qatar, the United Arab Emirates, Oman, and Saudi Arabia. Where once there were bleak villages on barren deserts there are now bright modern cities with tree lined streets. There are homes with lush gardens. In the countryside there are productive farms. The big desalting plant at Jubail, Saudi Arabia, is a model for the world. A pipeline carries a river of freshwater 200 miles inland to the capital city of Riyadh, and desalted seawater has given a large region an entirely new future filled with opportunities. There are more than 7,000 desalination plants, mostly small ones, in operation worldwide. About two thirds are located in the Middle East, and others are scattered across islands in the Caribbean and elsewhere. Aruba's high-tech water plant has for many years met the needs of a thriving tourist industry.

The largest plant in the United States is the pioneering $158 million project of the Tampa Bay Water agency. The project was left to contract in 1999 and after overcoming some technical problems in its early years is now performing well and causing no significant environmental problems. But no U.S. water agency has yet undertaken a really big project comparable to those found along the Arabian Gulf.

	2001	2002	2004	2007	2010
UAE	165.1	182.4	204.9	229.1	229.1
Bahrain	27.9	29.8	30.4	32.1	33.87
KAS	293.14	214.98	233.99	270.46	275.14
Oman	20.79	21.54	22.72	23.98	26.19
Qatar	32.3	33.5	34.8	35.6	35.6
Kuwait	82.134	107.27	90.89	95.17	97.87
Total	621.364	589.49	617.7	686.41	697.77

Table 1: Water production in Gulf States.

	2002	2003	2004	2007	2010
UAE	142.8	149.4	196.9	212.5	229.1
Bahrain	27.9	29.8	30.4	32.2	33.82
KAS	------	252.7	279.52	318.24	207.28
Oman	20.69	21.5	22.85	23.95	26.14
Qatar	------	33.5	34.8	35.6	35.6
Kuwait	82.11	102.06	91.05	94.99	97.88
Total	273.5	588.96	655.52	717.48	629.82

Table 2: Water consumption in Gulf States.

Water Resources Deficiency

The problem of water shortages in the Gulf States is quite evident despite the fact that these countries adopting advanced technology. Water scarcity in the region is rapidly becoming part of a widespread

Towards a Strategy for Water Management in the Region

Certainly, water conservation programs should come first as a strategy for regions facing water problems. Many jurisdictions are already imposing water use limits. Other communities try drilling wells deeper and deeper until their aquifer is maxed out, or they propose to pipe water from distant streams. But such shortsighted strategies can do incalculable damage to the environment

Water conservation programs should be carried out by the agency responsible for water resources management. The execution of such plans for the various sectors (such as municipal, agricultural, or industrial) should be co-ordinated by the relevant government body in each sector. Close co-ordination and partnership should be institutionalized between the agencies responsible for water supply, demand management, and education, media, and awareness. Unfortunately, the ministries of education, Awqaf, and Islamic Affairs rarely participate in water conservation programs. in the region, although this is essential for effective awareness activities.

Equally, the public's involvement and its co-operation in designing and implementing conservation measures are essential to the success of water conservation programs. The public includes consumers, service providers, managers, and planners as well as policy-makers. Raising public awareness using Islamic conservation concepts should always be integrated with the use of other communication tools and channels.

To achieve greater co-operation and involvement, the public must understand the water supply situation, including the cost of delivery, the overall water resources situation, and the need to conserve water resources and to maintain them for future generations. This increased understanding is the first step in any successful public awareness activity. However, the credibility of this information is essential. Because honesty is a core principle in Islam, the public expects the truth from imams and other Islamic sources. Most water conservation activities require changes of behavior and attitudes, which is usually a slow process. Therefore, ad hoc and occasional public awareness activities are not effective. Water authorities should plan continuous, long-term activities in close collaboration and coordination with ministries of Education and Islamic Affairs. Some water conservation activities involve costs that must be paid by the public, such as fixing water taps, upgrading irrigation systems, or modifying industrial production lines. These costs of water conservation programs must be offset by some incentives. In addition to the physical incentives, the spiritual incentives offered by Islam can be of value.

Water conservation activities and awareness campaigns typically focus solely on domestic users. This is shortsighted, and the focus should be on all water users. Mosques are ideal places for awareness campaigns, since all kinds of people meet there at least weekly. However, imams should be aware of the need to address all sections of the population. In the GCC countries, a shift in societal values from a development-oriented to a conservation-oriented view of water resources is occurring. It is believed in the GCC countries that conservation of natural resources in general, and water resources in particular, are a principal component of Islamic teachings. It is also believed that the most important and effective way to make the public aware of conservation from an Islamic perspective is through the media and the educational system [5]. Islamic messages are being used in the preparation of posters and video clippings for these campaigns. On the occasion of World Water Day 1998, and upon request of ministries of Islamic Affairs, imams were requested to devote their Friday speeches to the theme of Islam and water conservation [6]. But such occasional public awareness water conservation campaigns need to be integrated through a comprehensive and long-term plan of action that targets behavioral change [7], otherwise their effect will

limited water treatment and reuse is now one of the top priorities for governments across the GCC as populations increase and the demand on existing infrastructure and potable water generating capacity is stretched to the limits. More than funds will be invested in the wastewater and water reuse sector over the coming 20 years as Gulf states embrace the concept of full sewerage coverage and the need for conservation of scarce water resources. The investment drive in nearly every GCC state will be carried out by the private sector either in tandem with the government in public-private partnerships (PPPs) or through fully-fledged privatization.

Technology will continue to play an important role. Most states are still reluctant to utilize the more recent membrane bioreactor (MBR) technology, preferring to stick to conventional wastewater treatment applications, the exception being some of the smaller real estate players in Qatar and UAE. However, MBR technology is expected to take hold in the region. Sharjah is building a small-scale pilot STP for Ashghal using MBR and reverse osmosis technology. There are no plans, at the moment at least, for states to connect the treated wastewater network to potable water networks. 'It's a psychological fear,' said Salah al-Mutawa, manager of the treated sewage effluent department at Bahrain's Works & Housing Ministry. 'People aren't generally aware of the treatment process' [8,9].

Education and increasing public awareness are now firmly on the agenda. Both Dubai and Doha are thinking about public visitor centres to teach people about the benefits of water reuse and the processes involved in treatment. Cultural and religious factors will remain an obstacle to the implementation of a full-water reuse cycle, although states are now looking at using treated wastewater to recharge aquifers [10,11].

Conclusion

Integrated water resources management is the most feasible option to overcome the serious crisis of water shortage. Water conservation should be an integral part of this option, with a clear focus on public awareness and participation without which the chances of success will be lessened.

Access to information related to water conservation and public awareness activities is lacking in Gulf countries both because of the limited number of these activities, and because of poor information management and exchange. Such access can be improved by adopting a strategy to identify sustainable water management practices, documenting the experiences in water conservation and public awareness. This Strategy should be made available to water specialists as well as to the public by networking at the regional and national levels.

Gulf States can meet the challenge of global warming with a three-pronged approach: reducing emissions of carbon pollution, minimizing human stresses on ecosystems, and adapting to the challenges to come. Reducing the dependence on fossil fuels by developing clean energy sources would reduce global warming gas emissions and create jobs and new economic opportunities for region. By implementing the best practices in land and water resource use, policy-and decision-makers can minimize ecologically harmful side effects of climate change. And finally, elected officials and government leaders can plan ahead by increasing their flexibility and adaptive capacity in managing the state's precious water resources, agriculture, forests, ecosystems, and coasts.

References

1. Al Zawad (2008) Impacts of Climate Change on Water Resources in Saudi Arabia.

2. Water and Energy Linkages in the Middle East (2009) Stockholm International Water Institute, Sweden.

3. Afifi, Madiha Moustafa (1996) Egyptian National Community Water Conservation Programme. Environmental Communication Strategy and Planning for NGOs ,Jordan Environment Society, Jordan.

4. Ayesh, Mohammed (1996) Awareness Project in Water. Environmental Communication Strategy and Planning for NGOs, Ma'ain, Jordan Environment Society, Jordan.

5. Akkad AA (1990) Water Conservation in Arabian Gulf Countries. Journal of the American Water Works Association 82: 40–50.

6. Al-Tamimi, Izz El Din (1991) Religion as a Power for Protection of the Environment (in Arabic). Environmental Research and Studies, Jordan Environment Society, Jordan.

7. Farshad A, Zinck JA (1995) The fate of agriculture in the Semi-arid Regions of Western Iran - A Case Study of the Hamadan Region. Annals of Arid Zone 34: 235-242.

8. Abdulrazzak MJ, Khan MZA (1990) Domestic Water Conservation Potential in Saudi Arabia. Journal of Environmental Management 14: 167–178.

9. Al-Sodi, Abdul Mahdi (1993) Attitudes of Jordanian Citizens towards Environmental Protection in the Sweileh and Naser Mountains Areas (in Arabic). Environmental Research and Studies, Jordan Environment Society, Jordan.

10. Madani, Ismail (1989) Islam and Environment. For Environmental Awareness in the Gulf Countries, Ministry of Information, Bahrain.

11. Hamdan M, Toukan Ali, Shaniek M, Abu Zaki M, Abu Sharar T, et al. (1997) Environment and Islamic Education. International Conference on Role of Islam in Environmental Conservation and Protection, Al-Najah University, Palestine.

Permissions

The contributors of this book come from diverse backgrounds, making this book a truly international effort. This book will bring forth new frontiers with its revolutionizing research information and detailed analysis of the nascent developments around the world.

We would like to thank all the contributing authors for lending their expertise to make the book truly unique. They have played a crucial role in the development of this book. Without their invaluable contributions this book wouldn't have been possible. They have made vital efforts to compile up to date information on the varied aspects of this subject to make this book a valuable addition to the collection of many professionals and students.

This book was conceptualized with the vision of imparting up-to-date information and advanced data in this field. To ensure the same, a matchless editorial board was set up. Every individual on the board went through rigorous rounds of assessment to prove their worth. After which they invested a large part of their time researching and compiling the most relevant data for our readers.

The editorial board has been involved in producing this book since its inception. They have spent rigorous hours researching and exploring the diverse topics which have resulted in the successful publishing of this book. They have passed on their knowledge of decades through this book. To expedite this challenging task, the publisher supported the team at every step. A small team of assistant editors was also appointed to further simplify the editing procedure and attain best results for the readers.

Apart from the editorial board, the designing team has also invested a significant amount of their time in understanding the subject and creating the most relevant covers. They scrutinized every image to scout for the most suitable representation of the subject and create an appropriate cover for the book.

The publishing team has been an ardent support to the editorial, designing and production team. Their endless efforts to recruit the best for this project, has resulted in the accomplishment of this book. They are a veteran in the field of academics and their pool of knowledge is as vast as their experience in printing. Their expertise and guidance has proved useful at every step. Their uncompromising quality standards have made this book an exceptional effort. Their encouragement from time to time has been an inspiration for everyone.

The publisher and the editorial board hope that this book will prove to be a valuable piece of knowledge for researchers, students, practitioners and scholars across the globe.

List of Contributors

MH Golabi and Kirk Johnson
College of Natural and Applied Sciences, University of Guam, Mangilao, Guam 96923, USA

Takeshi Fujiwara and Eri Ito
Solid Waste Management Research Center, Okayama University, 3-1-1 Tsushima Naka, Okayama, 7008530, Japan

Michelle MV Snyder
Energy and Environment Directorate, Pacific Northwest National Laboratory, Richland, WA 99354, USA

Wooyong Um
Energy and Environment Directorate, Pacific Northwest National Laboratory, Richland, WA 99354, USA
Division of Advanced Nuclear Engineering, Pohang University of Science and Technology (POSTECH), Pohang, South Korea (ROK)

Hristina Stevanovic Carapina, Jasna Stepanov, Dunja C Prokic, Ljiljana Lj Curcic and Natasa V Zugic
Faculty of Environmental Governance and Corporative Responsibility, Educons University, SremskaKamenica, Serbia

Andjelka N Mihajlov
Faculty of Technical Sciences, University of Novi Sad, Serbia

Queena K Qian
Endeavour Australia Cheung Kong Fellow, Center for Sustainable Design and Behaviour (sd+b), University of South Australia, Australia

Abd Ghani Bin Khalid
Professor, Faculty of Built Environment, University Technology Malaysia, Malaysia

Edwin HW Chan
Professor, Building and Real Estate Department, The Hong Kong Polytechnic University, Hong Kong S.A.R., China

Mesfin Tilaye
Ethiopian Civil Service University, Addis Ababa, Ethiopia

Meine Pieter van Dijk
Professor of urban management at ISS of Erasmus University, Netherlands

Randy Schroeder
Department of English, Mount Royal University, Canada

LI Si yuan, Hao Chun bo, Feng Chuan ping, Wang Li hua and LIU Ying
Key Laboratory of Groundwater Circulation and Evolution of Ministry of Education, China
School of Water Resources and Environment, Beijing 100083, China

Ramzi Taha, Okan Sirin and Husam Sadek
Department of Civil and Architectural Engineering College of Engineering Qatar University, Qatar

John J Harwood
Department of Chemistry, Tennessee Technological University, Cookeville, TN, USA

Clifton Curtis
Director, The Varda Group; and Policy Director, Cigarette Butt Pollution Project, USA

Susan Collins
President, Container Recycling Institute, USA

Shea Cunningham
Sustainability Policy, Research & Planning Consultant, Container Recycling Institute, USA

Paula Stigler
Assistant Professor, University of Texas Health Sciences, San Antonio Regional Campus, USA

Thomas E Novotny
Chief Executive Officer, Cigarette Butt Pollution Project and Professor of Epidemiology, Graduate School of Public Health, San Diego State University, USA

Daniel Bergerson, Magdy Abdelrahman and Mohyeldin Ragab
Department of Civil and Environmental Engineering, North Dakota State University, North Dakota State University, CIE 201, Fargo, USA

Dolores Queiruga
Department of Economics and Business Administration, University of La Rioja, Edificio Quintiliano. C/ La Cigüeña 60. 26006 Logroño, Spain

Araceli Queiruga-Dios
Department of Applied Mathematics, ETSII. University of Salamanca, Avda. Fernandez Ballesteros 2, 37700-Bejar, Salamanca, Spain

Henning Wilts and Alexandra Palzkill
Wuppertal Institute for Climate, Environment and Energy, Germany

Paschalidis X, Mouroutoglou X and Koriki A
Technological Educational Institute of Kalamata, Antikalamos, 24100, Kalamata, Greece

Ioannou Z and Kavvadias V
Hellenic Agricultural Organization- DEMETER, Soil Science Institute of Athens, Greece

Baruchas P
Technological Educational Institute of Western Greece, Koukouli, Patra, Greece

Chouliaras I
Mediterranean Agronomic Institute of Chania, AlsylioAgrokepio, Chania, Crete, Greece

Sotiropoulos S
Greek Agricultural Insurance Organization, Nafpliou & Al. Soutsou, 22100, Tripoli, Greece

Djelloul Bendaho, Tabet Ainad Driss and djillali Bassou
Laboratory of organic chemical-physical and macromolecular Faculty of exact sciences, University DjilaliLiabès, Algeria

Supono
Department of Aquaculture, Faculty of Agriculture, Lampung University, Indonesia

Johannes Hutabarat, Slamet Budi Prayitno and YS Darmanto
Department of Fisheries, Faculty of Fisheries and Marine Science, Diponegoro University, Indonesia

Max J Krause and Timothy G Townsend
Environmental Engineering Sciences, University of Florida, Gainesville, FL 32611, USA

Savariar Vincent
Loyola Institute of Frontier Energy, Chennai, India

Saffarzadeh A and Takayuki Shimaoka
Department of Urban & Environmental Engineering, Kyushu University, Fukuoka, 819-0395, Japan

Akihisa Aoyama, Kazuhiro Yamada, Yoshinobu Suzuki, Yuta Kato, Kazuo Nagai and Ryuichiro Kurane
Department of Biological Chemistry, College of Bioscience and Biotechnology, Chubu University, Japan

Farouk KM Wali
Chemical technology Department, The Prince Sultan Industrial College, Kingdom of Saudi Arabia

Nassereldeen Kabbashi, Optakun Suraj, Md Z Alam and Elwathig MSM
Bio-Environmental Research Centre, Department of Biotechnology Engineering; International Islamic University Malaysia, Malaysia

Sandra Tostar and Mark R StJ Foreman
Department of Industrial Materials Recycling, Chalmers University of Technology, 412 96 Gothenburg, Sweden

Erik Stenvall and Antal Boldizar
Department of Material and Manufacturing Technology, Chalmers University of Technology, 412 96 Gothenburg, Sweden

Effie Papargyropoulou, Rory Padfield, Parveen Fatemeh Rupani and Zuriati Zakaria
University Technology Malaysia (UTM), Kuala Lumpur, Malaysia

El-Sayed G. Khater
Agricultural Engineering Department, Faculty of Agriculture, Benha University, Egypt

Susanne Fischer, Meghan O'Brien, Henning Wilts, Sören Steger, Philipp Schepelmann and Bettina Rademacher
Research Group on Material Flows and Resource Management, Wuppertal Institute for Climate, Environment and Energy, Germany

Nino David Jordan
Institute for Sustainable Resources, University College London , London, UK

Ninghu Su
School of Earth and Environmental Sciences, James Cook University, Cairns, QLD 4870, Australia

Kartik Venkatraman
East Gippsland Shire Council, 273 Main Street, Bairnsdale, Victoria 3875, Australia

Nanjappa Ashwath
School of Medical and Applied Science, CQUniversity, Rockhampton, QLD 4702, Australia

Elnazir Ramadan
Department of Geography, Sultan Qaboos University, Muscat, Oman

Printed in the USA
CPSIA information can be obtained
at www.ICGtesting.com
J3HW031440221024
72173JS00006B/1595